Microwave and Optical Transmission

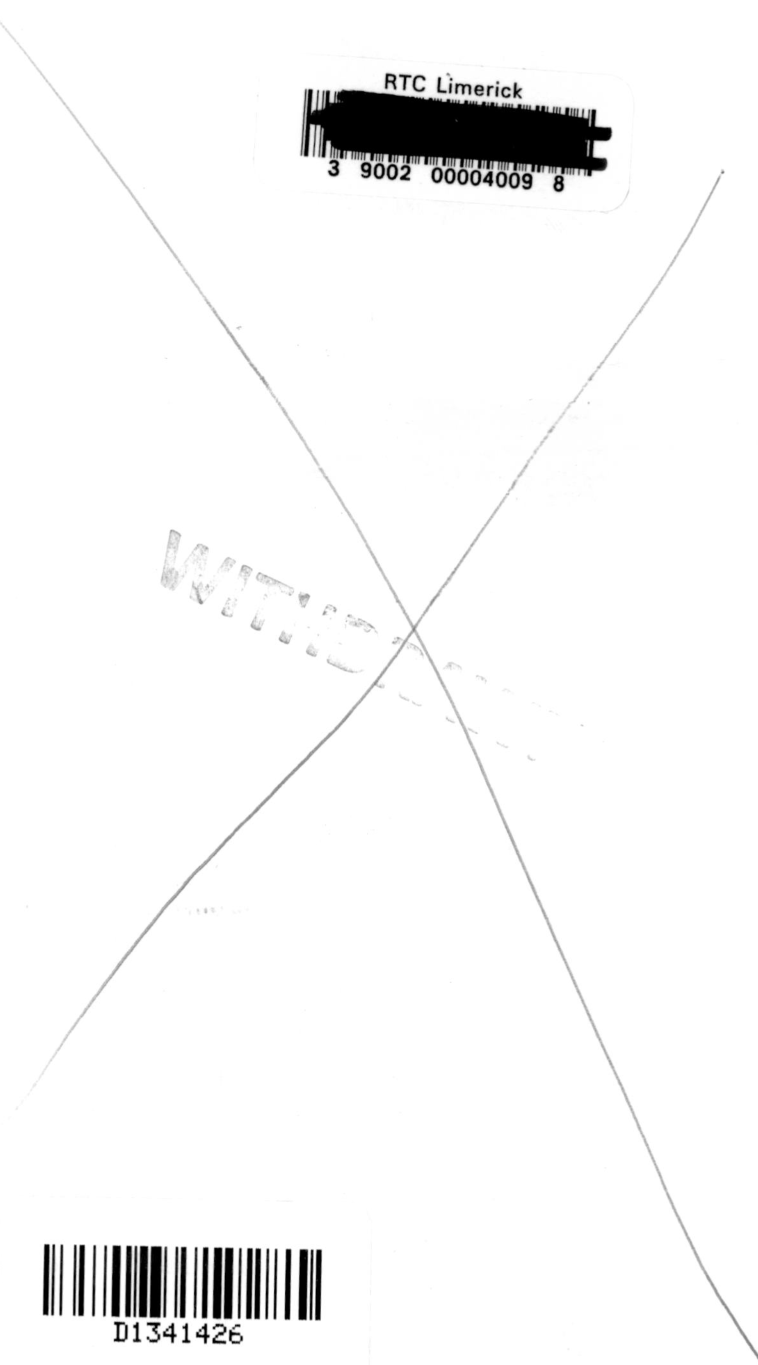

Microwave and Optical Transmission

A. DAVID OLVER

Queen Mary and Westfield College
University of London, UK

JOHN WILEY & SONS
Chichester · New York · Brisbane · Toronto · Singapore

Other Wiley Editorial Offices

John Wiley & Sons, Inc., 605 Third Avenue,
New York, NY 10158-0012, USA

Jacaranda Wiley Ltd, G.P.O. Box 859, Brisbane,
Queensland 4001, Australia

John Wiley & Sons (Canada) Ltd, 22 Worcester Road,
Rexdale, Ontario M9W 1L1, Canada

John Wiley & Sons (SEA) Pte Ltd, 37 Jalan Pemimpin #05-04,
Block B, Union Industrial Building, Singapore 2057

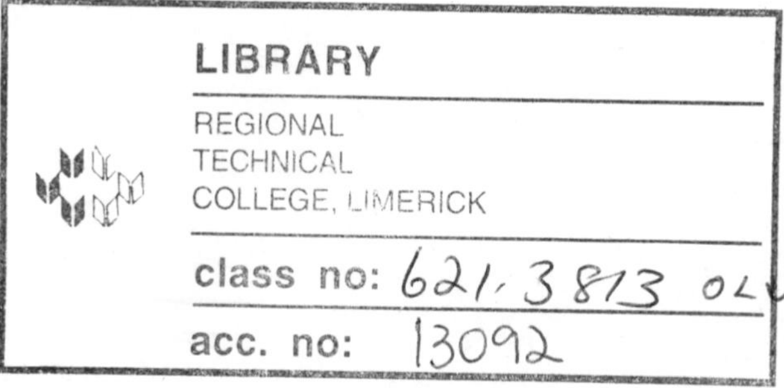

Library of Congress Cataloging-in-Publication Data

Olver, A.D. (A. David)
Microwave and optical transmission / A. David Olver.
p. cm.
Includes bibliographical references and index.
ISBN 0 471 93416 X (pbk); 0 471 93478 X
1. Optical communications. 2. Microwave communication systems.
I. Title.
TK5103.59.048 1992
621.382'7—dc20 91-47671
CIP

British Library Cataloguing in Publication Data

A catalogue record for this book is available from the British Library

ISBN 0 471 93416 X (pbk); 0 471 93478 X

Typeset in 10/12pt Times by Thomson Press (India) Ltd, New Delhi, India
Printed and bound in Great Britain by Courier International, East Kilbride

Contents

Preface

Communications systems, radar systems, radio navigation systems, remote sensing systems and many heating, control and recording systems use microwave or optical transmission as a fundamental part of their operation. The number of systems in use has grown rapidly and collectively they form an important section of electronic engineering and physics. This alone is justification for studying the theory and practice of microwave and optical transmission, but in addition there is intellectual stimulation to be obtained from a subject which can both be explained with mathematical rigour and leads to many useful applications as diverse as satellite communications, air traffic radar, microwave cooking, infrared remote control, laser surgery and audio compact discs. This book aims to provide a thorough introduction to the subject by combining theory with descriptions of applications.

The subjects of microwaves and optics are often treated in separate texts. The justification for treating them in one book is that the common foundation for both subjects is electromagnetic wave theory and one theory can be used to describe both microwave transmission and optical transmission. The book develops the theoretical aspects of the subject along classical lines using static electric and magnetic fields to evolve Maxwell's equations and then using electromagnetic waves to explain the principles of both free space and guided microwave and optical transmission. This approach provides a logical development of the subject which builds on previous knowledge to yield a strong foundation for describing practical microwave and optical systems.

The book is the outgrowth of a course of lectures given to second year electronic engineering, physics and avionics students. The book will support an intermediate module in an electronic engineering or physics degree course. It is expected that students using the book will have taken an introductory course in electricity and magnetism. A mathematical knowledge of differential equations and vector calculus is expected, though not essential since the mathematics is self contained and can be learnt from the material in the early chapters. Some students will go on after studying the subject at this level to work in the fields of microwaves or optics. In this case they will pursue more advanced study by taking modules which confine themselves to one aspect of the subject, such as optical communications, photonics, satellite communications, radar, antennas and electromagnetic wave theory.

The first chapter in the book is deliberately descriptive and not mathematical. This sets the scene for the succeeding chapters by describing the history and current applications of microwave and optical transmission. It should be read as introduction to the subject which tries to convey the considerable usefulness to modern society of microwave and optical systems.

Chapter 2 describes static electric and magnetic fields. The basic equations are formulated so that they can be combined to form Maxwell's equations. The treatment is not comprehensive but is in sufficient detail to lay the foundation for electromagnetic wave theory. The chapter could be omitted by those who have already studied in the subject. The foundation for electromagnetic wave theory is dealt with in Chapter 3 which describes Maxwell's equations and the electromagnetic wave equation. The next chapter introduces the boundary conditions which are needed to solve electromagnetic wave problems and then devotes the rest of the chapter to the solution of quasi-static fields. These might be considered to be somewhat peripheral to microwave and optical transmission, but they are worth studying because of their simplicity and ease with which physical insight can be obtained into the behaviour of electric fields.

The next five chapters treat microwaves and optical waves which radiate and travel in free-space or in a dielectric medium. Chapter 5 develops and explains the basic characteristics of all plane waves including the parameters which are used to describe the properties of radio waves, microwaves and optical waves. Chapter 6 studies the reflection and refraction of plane waves incident on dielectric and conducting boundaries, including standing waves and total internal reflection. The results are particularly applicable to optics but also apply at microwave frequencies. Chapters 7, 8 and 9 use the knowledge gained in the previous chapters to explain practical systems. Chapter 7 deals with microwave systems in free space—satellite and terrestrial communication systems, broadcasting and mobile radio, radar, navigation and remote sensing. Chapter 8 considers situations where microwaves energy is lost as heat. This covers the influence of water on microwave singals and the use of high power microwaves for heating and cooking. The basis of microwave radiation safety levels is also explained. Chapter 9 studies ray optical systems. This is a very large subject in its own right and the objective is not to be comprehensive but to show how the basic theory developed in earlier chapters explains the operation of optical components such as prisms, mirrors and lenses. Optical sources and detectors are surveyed, including the human eye which forms a high quality optical receiving systems for visible waves. Various examples of modern optical systems which implement the transmission theory are described.

Attention is then turned to guided waves and Chapters 10 to 13 study waveguides and methods of analysing them. The purpose of Chapter 10 is to explain the properties of guided waves in the same way as Chapter 5 explained the properties of plane waves. The principles apply to both microwave and optical waveguides. Chapter 11 studies an alternative method of studying waveguides using transmission line theory. This is widely used at microwave frequencies and is particularly useful for matching components to waveguides. The theory and practice of microwave cables, microstrip and waveguides are described in Chapter 12. Optical waveguides are studied in Chapter 13, including the factors which determine the performance of optical fibre cables. The final chapter in the book links together guided waves and free space radiated waves by describing the basic principles of antennas.

I have worked throughout my career on microwave systems and antennas and acknowledge a considerable debt of gratitude to the many engineers and scientists throughout the world who have helped to make this a rewarding and stimulating time. Particular thanks go to the authors of the classic electromagnetic and microwave textbooks whose effort in earlier times has taught and inspired myself and many others.

A. David Olver
London
October 1991

1 Microwave and optical systems

1.1 INTRODUCTION

Microwave and optical transmission is the sending and receiving of information using microwaves or optical waves. The information could be data, television, telephony, radar signals or radiation emitted by natural objects. The main interest in this book is the systems which are used to carry the information and the characteristics of the waves which carry the information. Knowledge of these characteristics will show why microwaves and optical waves can be used to carry information and show how the properties of the waves depend on the medium through which they travel. Microwave and optical transmission is important because of the large, and growing, use of microwave and optical systems. Thus to set the scene, this chapter describes some of the microwave and optical systems which are used.

Microwave transmission and optical transmission are usually treated in separate texts, but the theoretical basis for microwave systems and optical systems is the same electromagnetic wave theory. In a text at this level it is possible and desirable to study both microwaves and optical waves in one book. By starting from the basic electromagnetic wave theory and developing wave and ray theory, the applications can be explained in terms of the same theoretical foundation. In fact the commonality is even greater than that conveyed by the words *microwaves* and *optics*. Waves transmitted in any part of the electromagnetic spectrum have the same basic properties.

1.2 CHARACTERISTICS OF WAVES

Information can be transmitted either by discrete bundles of energy or by regular waves with the information superimposed on top of the waves. The former is a more general description but it is also more complicated to analyse. The latter is easier to study and covers most of the uses encountered in practical microwave and optical transmission and will be the focus for the theory developed in this book. The waves will be assumed to be sinusoidal in character so that if a microwave or optical wave is sketched against time, t, and distance, z, it has the forms shown in Figure 1.1. The wave is described by

$$V = A\sin[2\pi(ft + z/\lambda)]$$

or

$$V = A\cos[2\pi(ft + z/\lambda)]$$

or

$$V = A\exp[\mathrm{j}2\pi(ft + z/\lambda)]$$

and is characterised by three parameters; the amplitude of the wave, A; the *frequency*,

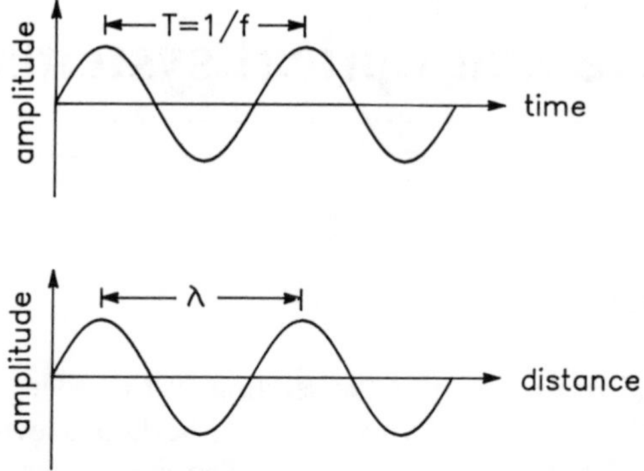

Figure 1.1. Waves in microwave and optical transmission

f, (or the *radian frequency*, $\omega = 2\pi f$); and the *wavelength*, λ. The frequency is the rate at which the wave oscillates in cycles per second and is the inverse of the time period, T, shown in Figure 1.1. The units are Hertz, abbreviated to Hz, and named after the discoverer of radio waves. The wavelength is the distance between two similar points on the wave and is measured in metres.

The product of the frequency, in Hertz, and the wavelength, in metres, is equal to the velocity of electromagnetic waves, in metres per second (m/s). All electromagnetic waves travel at the same velocity in space and this is universally known as the *velocity of light*,

Frequency × Wavelength = Velocity of light

or

$$f\lambda = c$$

The velocity of light has an internationally accepted value of $c = 2.997\,924\,58 \times 10^8$ m/s. The value is so close to 3×10^8 m/s, that for the vast majority of engineering purposes the velocity of light can be taken as 3×10^8 m/s. The relationship between frequency and wavelength shows that as the frequency increases the wavelength decreases and vice versa.

The range of both frequency and wavelength is many orders of magnitude so that prefixes are normally used in front of the basic units. The prefixes for frequency are as follows.

Name	Abbreviation	Value
kilohertz	kHz	10^3
megahertz	MHz	10^6
gigahertz	GHz	10^9
terahertz	THz	10^{12}
petahertz	PHz	10^{15}

The prefixes for wavelength are as follows.

Name	Abbreviation	Value
kilometre	km	10^3
millimetre	mm	10^{-3}
micrometre or micron	μm	10^{-6}
nanometre	nm	10^{-9}

Traditionally engineers used frequency as the method of describing waves and scientists used wavelength but this distinction has now become blurred and at present it tends to be the case that waves with frequencies up to about 100 GHz (wavelengths above 3 mm) are referred to primarily by their frequency and waves of higher frequency are referred to primarily by their wavelength. There is semantic inconsistency in the use of the word *microwave*, because the name clearly comes from the wavelength but the period of the waves is described by frequency! In addition, a *microwave* might be expected to have a shorter wavelength than a *millimetrewave*, whereas it is actually larger.

1.3 THE ELECTROMAGNETIC SPECTRUM

In theory it is possible to have electromagnetic waves with frequencies from zero through to infinity. In practice the range is more limited but still extends over many tens of decades. The range of usable frequencies is called the *electromagnetic spectrum*. It is shown diagrammatically in Figure 1.2. From the figure it is seen that *radio waves* extend from a few hundreds of kHz to about 1 GHz (300 mm). At this frequency *microwaves* take over and extend up to about 30 GHz (10 mm). Then come *millimetrewaves*,

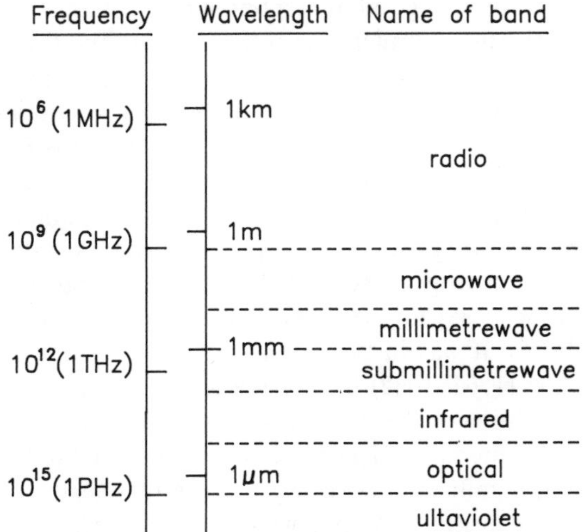

Figure 1.2. The electromagnetic spectrum

submillimetrewaves and *infrared waves. Optical waves* have frequencies of between about 100 THz to 1000 THz (3 μm to 300 nm). After that come *ultraviolet waves, X-rays* and *gamma rays*. The border between any two classification is very rough and in practice is determined more by the types of use than an absolute frequency. The corresponding wavelengths range from hundreds of metres for radio waves to a fraction of a micron for optical waves. Within the optical band there is a narrow range of frequencies over which the human eye is sensitive. This visible band of light extends from 380 THz to 770 THz. The bands of colour and their corresponding frequencies and wavelengths are shown below.

Colour	Frequency (THz)	Wavelength (nm) (in vacuum)
Red	380–480	790–625
Orange	480–500	625–600
Yellow	500–520	600–577
Green	520–610	577–492
Blue	610–660	492–455
Violet	660–770	455–390

The term *optical* refers to a wider band of frequencies than visible light and many of the modern uses of lasers and optical systems actually use wavelengths above the visible band. The three bands in between microwaves and optical waves—millimetre-waves, submillimetrewaves and infrared—have characteristics which are a mixture of microwave and optical waves.

1.4 HISTORY AND USES OF MICROWAVES AND OPTICS

Engineers and scientists are very good at responding to demands for new services. This is particularly true for the uses of microwaves and optics. The theory has been around for a long time, but the practical uses keep growing as the demand keeps expanding. This can be illustrated by a brief survey of the history of optics and microwaves. Optics—meaning the use of light—is of ancient origin. There is reference in the Book of Exodus (ca. 1200 BC) to the use of looking glasses and the Greeks evolved elementary theories of light. The Romans used focusing lenses for burning and magnification. There was a gap in development during the Dark Ages and before 1600 only a few developments took place. The Franciscan Roger Bacon (1215–1294) seems to have been the first to have initiated the use of lenses for correcting vision and thus give rise to the use of spectacles and eye-glasses. It was in the seventeenth century that the science of optics took off. Galileo Galilei (1564–1642) built a refracting telescope and used it to study the stars and planets. This was the real start to the science of astronomy. Johannes Kepler (1571–1630) discovered the phenomena of total internal reflection. Isaac Newton (1642–1727) had a major influence on many aspects of science, including optics. He studied dispersion with a prism and split white light up into its constituent colours. When he could not find a way of removing the distortions from refracting telescopes, he was led to invent the reflecting telescope. Later developments by others improved the performance and these designs remain the basis for optical telescopes and microwave reflector antennas. Theoretical research on optics continued in the eighteenth and nineteenth centuries and merged with research on electromagnetic waves, but the uses for optics remained static. It is only in the latter part of the twentieth century that there has been a veritable explosion in the applications of optical waves. This is due to a combination of factors. The demand for improved communications and computers, the development of digital technology requiring high performance displays, the invention of the laser and the invention of solid state components. All these have led to optical fibre communications, infrared imaging and remote control devices, laser printers, photocopiers, holography, laser surgery, electro-optical displays and other devices.

The history of microwaves is of more recent origin but from the beginning has been driven by demand and so has shown a steady growth. The forerunner of microwaves was the development of the theories for static electric and magnetic fields. Particularly important for the background to electromagnetic waves is the work of Charles Coulomb (1736–1806) who demonstrated the law of electrostatic force, Karl Friedrich Gauss (1777–1855) who discovered the divergence properties of electric fields and André-Marie Ampère (1775–1836) who discovered the relation between a steady current in a wire and the associated magnetic force. The foundations for time-dependent electromagnetics was laid by the brilliant experimentalist Michael Faraday (1791–1867) who, among many major discoveries, found that a changing magnetic field induced a current in a wire, and vice versa. This led the theoretician James Clerk Maxwell (1831–1879) to summarise and extend all the empirical knowledge on electromagnetics into a single set of mathematical equations, that have remained the basis for the theoretical study of microwaves and optical waves to the present day. Maxwell's theoretical predictions that light was *an electromagnetic disturbance in the form of waves* was confirmed experimentally by Heinrich Hertz (1875–1894) in a careful set of experiments conducted between 1885 and 1889. Working at a frequency of about 450 MHz, he not only verified Maxwell's equations but also *discovered* radio waves and showed that the optical laws of reflection and refraction applied to microwaves.

Microwave optics was researched in academic laboratories in the latter part of the nineteenth century but died out when Guglielmo Marconi (1874–1937) showed that lower frequency radio waves could be used for communications. Marconi was an entrepreneur as well as an engineer and was able to exploit the need for a method of communicating between ships and land which did not depend on optical sighting. In 1901 he demonstrated that transmission across the Atlantic was possible. This led to the rapid development of radio communications, first at low frequencies, then in the 1920s using high frequencies (HF or short waves). The 1930s saw the rediscovery of microwaves with a number of developments. Microwave point-to-point communications were born and microwave waveguides were developed. A microwave radio link across the English Channel was commissioned in 1931. In the same year, Karl G. Jansky constructed a large antenna in his garden and picked up *noise* from radio stars. This led to the birth of radio-astronomy and pioneering work on large antennas and sensitive receivers. Radar was developed in a number of countries and had an important impact on the Second World War. This was at very high frequencies (VHF) and ultra high frequencies (UHF). After the invention of the magnetron, microwave radar became the main type of radar because the higher frequencies offered greater angular resolution.

Since the 1940s radar technology has developed rapidly and has pioneered new microwave technologies such as electronically steered antennas. Digital signal processing of the radar signals has led to the sophisticated systems now available.

The concept of satellite communications was described by Arthur C. Clarke in 1945. The realisation of the ideas awaited the availability of suitable rockets, but since the 1960s, satellite communications has formed one of the most important uses of microwaves for communications. It has led to the development of high performance antennas and microwave solid state devices. In the early 1970s, the uses of microwaves was largely confined to radar, simple satellite systems and microwave radio links. Since then the potential and actual uses of microwaves has expanded rapidly. These include mobile communications, remote sensing and heating applications.

Microwave and optical systems can be divided into five basic categories:

- Communication systems
- Radar systems
- Radio navigation
- Remote sensing
- Heating and energy transfer
- Control and recording

A general description of the systems in each of these categories will now be presented in order to show the wide practical use of microwaves and optical transmission. The transmission aspects of the systems will be studied in later chapters.

1.5 COMMUNICATION SYSTEMS

Communication is the transmission of information by microwaves or optical waves between a source of information and a receiver for the information. The schematic of a general communication system is shown in Figure 1.3. The source of information could be sound or a picture or data. The transducer changes the source into an electrical signal, for example a microphone or camera. The converter converts the electrical signal at baseband into a form and frequency which are suitable for transmission over the system. The transmitter amplifies and then launches the converted signal into the transmission medium. There are two types of transmission-guided waves through cables or waveguides and radiation of electromagnetic waves in the air or space or a dielectric. There is a basic distinction between these two transmission paths. With an atmospheric path the signals will travel outwards from the transmitter and spread in all directions. The amount of spread depends on the antenna at the transmitter. Receivers placed anywhere within the beam radiated from the transmitter can pick up the signal. This is in contrast with a guided path where a waveguide or cable links the transmitter and receiver. In the guided case there is only one receiver.

The output side of the communication system mirrors the input side with a receiver which detects the signal and amplifies the low level signals. The second converter reconverts the radio, microwave or optical signal back into the electrical baseband by processing, decoding or demodulating the signal. Finally the transducer converts the

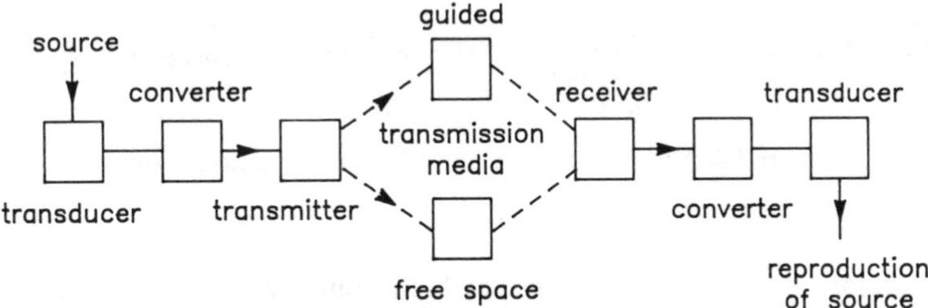

Figure 1.3. Schematic of a free space or guided communication system

electrical signal back into the original information. The transducers and converters are independent of the transmission medium and are the same for a microwave or optical system. The transmitter and receiver are designed to function for the specific operating frequency and application and are intimately tied to the characteristics of the transmission medium.

Five types of communication system will be described:

- Optical fibre communications
- Microwave radio links
- Satellite communications
- Mobile communications
- Broadcasting

OPTICAL FIBRE COMMUNICATIONS

Progress in optical communications has been very rapid over the last two decades. The concept of using optical fibres originated in the 1960s. At that time there were no solid state optical components and the attenuation of waves in good quality glass was orders of magnitude too high. Research sponsored by the telecommunication authorities soon led to many breakthroughs so that now optical systems are carrying an increasing proportion of the long haul telecommunications traffic. In the United Kingdom, optical fibre is the dominant transmission medium for the long line network. In 1988, 64% of the traffic went on optical fibres, with 13% on microwave radio and 23% on coaxial or pair cable. By 1995 it is projected that over 90% of the traffic will be over the optical network. The reasons for this growth are the fundamental benefits of optical fibre communications: low attenuation, low cost, high bandwidth. The costs of an optical system are partly controlled by the sources, detectors and couplers at the ends of the transmission lines.

A simple optical communications system is shown schematically in Figure 1.4. The block diagram includes a repeater, although the fibre loss is so low that they can be spaced many tens of kilometres apart. The repeater is actually a receiver and transmitter back-to-back. All present generation systems use power transmission and detection where a source generates a wide band of energy which is modulated by turning it on and off. Detection is by straight diode detection and there is no heterodyne, or coherent, detection using a local oscillator. This will come in the next generation of systems.

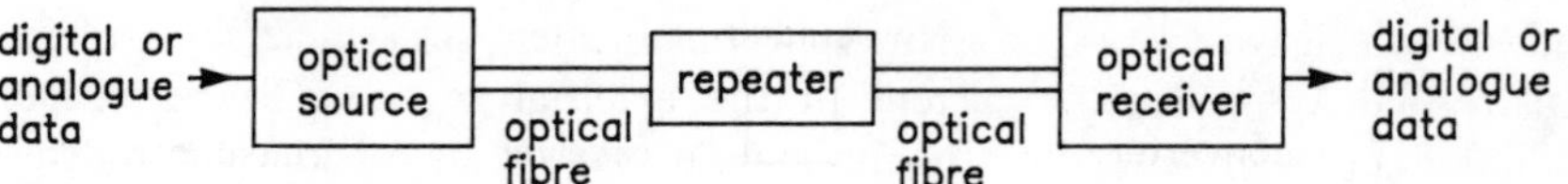

Figure 1.4. Schematic of an optical communication system

Semiconductor optical sources are either light emitting diodes (LEDs) or injection lasers. Both devices work by emitting photons when electrons fall from the conduction to the valence bands. For this reason the subject is called *photonics.* In an LED the action is spontaneous whereas in the laser it is by stimulated emission. The laser is consequently much brighter than the LED and has a faster response. The LED is cheap and the laser is moderately expensive. The other main difference is the line width of the output signal. An LED is a relatively broadband source with a bandwidth of typically 20 THz (centre frequency about 450 THz) and the injection laser is much narrower with a bandwidth of between a few GHz and 500 GHz. Note that in terms of microwave systems this still covers all bands up to submillimetrewaves! Future lasers using feedback and cavities to stabilises the signal may have bandwidths of a few MHz.

The connector between the source of receiver and the optical fibre is a significant part of an optical system because of the need to align the core to a high precision in the very small distances involved. Misalignment gives rise to coupling loss and can only be overcome by very accurate geometry. An optical fibre connector must terminate and protect the end of the fibre, act as an alignment guide and protect the cable from the environment.

The characteristics of the waves in the optical fibre will be studied in Chapter 13. A typical optical cable containing many separate fibres is shown in Figure 1.5. Early telecommunication systems used a low loss *window* at a wavelength of around 850 nm. Now systems use windows around either 1300 or 1550 nm, which are well into the infrared part of the electromagnetic spectrum. An optical receiver is usually a silicon avalanche

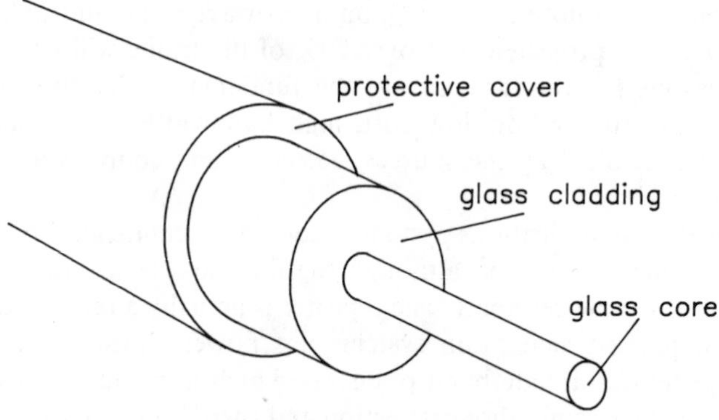

Figure 1.5. Optical fibre

photodiode which works by converting photons incident on a junction into electron–hole pairs. These produce a current in the external circuit. The performance of photodiodes is usually limited by the thermal noise in the diode which then determines the sensitivity.

There are many uses of optical fibre waveguides in addition to telecommunications. Optical fibres will be used in future aircraft to distribute command and control information because of their small size and weight. They are also used in industrial instrumentation which measures temperature, pressure, electric fields, and the presence of gases or smoke. An important benefit in these applications is the immunity from electrical interference. A fibre optic gyroscope performs very favourably compared to its mechanical counterpart.

MICROWAVE TERRESTRIAL COMMUNICATION SYSTEMS

Microwave terrestrial communications uses frequencies from about 1 GHz up to about 20 GHz. The upper frequency is currently set by technological limitations and by the fact that the frequency space below 20 GHz contains enough bandwidth to satisfy existing requirements. The growth in communication and broadcasting systems is, however, such that higher frequencies will gradually need to be used in order to satisfy demand. Closed metal waveguides (described in Chapters 10 and 12) or used for short lengths, particularly to and from an antenna because it has low loss and wide bandwidth, but microstrip or coaxial cable is preferred because it is generally smaller, has a lower cost and can be integrated with other components using MICs (microwave integrated circuits).

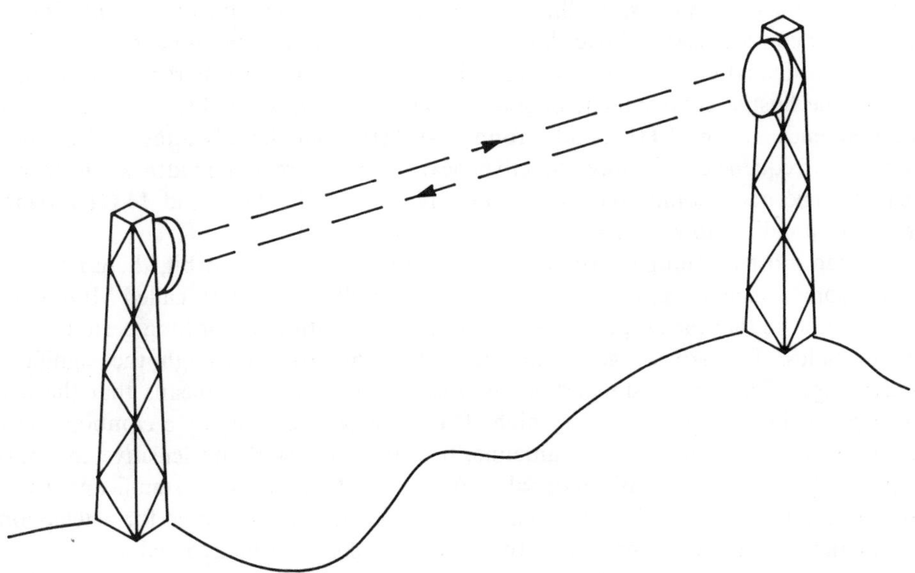

Figure 1.6. Microwave radio link

Terrestrial microwave communications is taken to mean point-to-point microwave radio links, Figure 1.6. These consist of a reflector antenna mounted on a high tower or building and fed from a high power microwave source which is modulated with the baseband information. The transmitting antenna points at a receiving antenna also on a tower or building so that a point-to-point link is established. Long distance communication can be established by cascading the individual links. The length of a link depends on the frequency of operation and the size of antenna. Higher frequency systems use smaller antennas but the distance apart is also shorter. A number of frequency bands are allocated to microwave radio links from 1.3 to 40 GHz. In each band the radiation pattern of the antenna must satisfy stringent specifications in order to avoid interference with other systems. The reflector antenna is a parabolic reflector which is often covered by a radome to protect it from the weather. The design of the system has to take account of the wind loading which increases as the fourth power of the reflector diameter.

Microwave radio links used to be the main type of long distance communication for trunk voice, video and data communication but they have been supplanted in dominance, first by satellite communications and then by optical cable communications. They still, however, have a place in telecommunication systems because a microwave radio link can be set up quickly and does not involve the long term capital infrastructure of cable systems.

SATELLITE COMMUNICATIONS

The first communication satellite was TELSTAR launched in 1962. This had an elliptical orbit which meant that communication across the Atlantic was limited to a short periods. The problem was overcome in the following year by using the geostationary orbit at an altitude of about 36 000 km where the satellite appears stationary in the sky. Continuous transmissions are possible and only three satellites are needed to cover the whole Earth. Since those early days, satellite communications has expanded to provide flexible, high quality communications over long distances. A modern communication satellite is a very sophisticated piece of engineering which is designed to operate remotely for many years in the hostile environment of space, Figure 1.7. Early satellite communications used frequencies in the VHF band, around 140 MHz, but the advantages of going to microwave frequencies of more directive beams and more bandwidth soon became apparent. The main communication bands are up-links in the 6 and 14 GHz bands, with corresponding down-links in the 4 and 11 GHz bands.

Although most communication satellites are in geostationary orbit, there are some Russian communication satellites which are in highly elliptical orbits, called MOLYNIA orbits, chosen so that the satellites appear to be nearly stationary for long periods above high latitudes. The advantage of the geostationary orbit outweigh the significant disadvantages. The long distance that the signals have to travel means that the path loss (explained in Chapter 7) is very high. This must be overcome by a combination of high transmitting power, high gain antennas (i.e. large diameter) and sensitive, low noise receivers. The long path distance also means that there is a significant round trip time delay of 0.25 s. Experience has shown that this is acceptable for a telephone conversation but a circuit containing two satellite links must be avoided.

The component parts of a satellite communication system are very similar to those of a terrestrial system except that the antennas need to be matched to the service to be

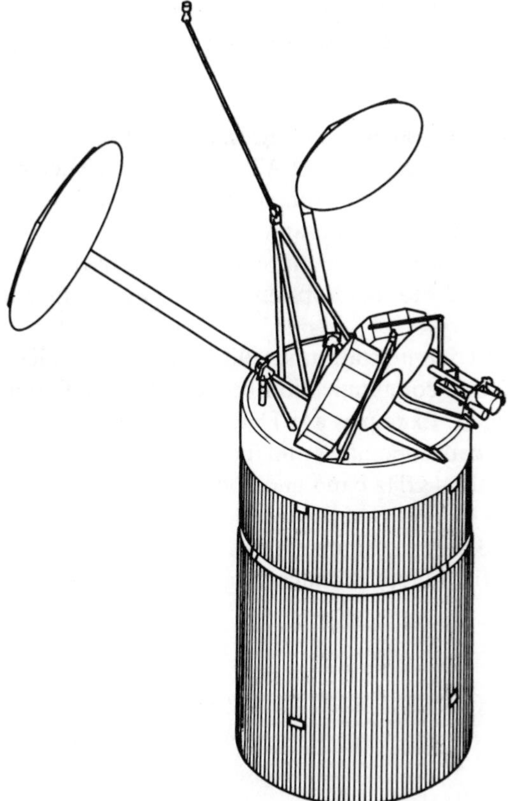

Figure 1.7. Communication satellite

provided and are generally larger in diameter. This is particularly true of the antennas on the spacecraft which are individually designed for the application and often very sophisticated with multiple beams or shaped beams so as to illuminate specified regions on the Earth. The design of spacecraft antennas is one of the most important area of microwave research and development. It must take account of many factors besides the electrical performance. These include the launch constraints, the space environment, the size and mass of the antenna and its associated microwave components.

Satellite communication systems can be divided into three classes.

(a) The intercontinental systems operated by INTELSAT (INTernational TELecommunication SATellite organisation) and INMARSAT (INternational MARitime SATellite organisation). The INTELSAT system is a high capacity system which originally used large, 30 m diameter reflector antennas as the Earth station antennas and operated in the 4/6 GHz bands. This band, however, is now fully occupied so that the 11/14 GHz band is increasingly being used. Antennas of 11 m diameter can be used in the 11/14 GHz band. There has also been improvements in microwave receiver technology and in the power which can be placed on the spacecraft in the microwave output amplifiers, which are travelling wave tubes. This has opened up the possibility of narrower band systems using smaller antennas.

(b) The second class of system is the regional and business satellite communications, run by consortia such as EUTELSAT (EUropean TELecommunication SATellite organisation) and by private companies. These mainly use the 11/14 GHz band and operate with Earth station antennas of about 3 m diameter, Figure 1.8. The antennas on the spacecraft are more directive than the intercontinental satellites and employ shaped beams to optimise power and reduce interference.

(c) A third class of communication satellites are those which are intended to have essentially one-way communication between the satellite and a large number of receiving terminals. Examples are TVRO (TeleVision Receive Only) for distributing television to cable operators and individuals. These use Earth station antennas of about 3 m in the 4 GHz band and about 1 to 2 m in the 11 GHz band. Another example is VSAT (Very Small Antenna Terminals) with diameters of under 1 m for narrow band data communications. This class of satellite communications is very similar to the Direct Broadcast Satellites.

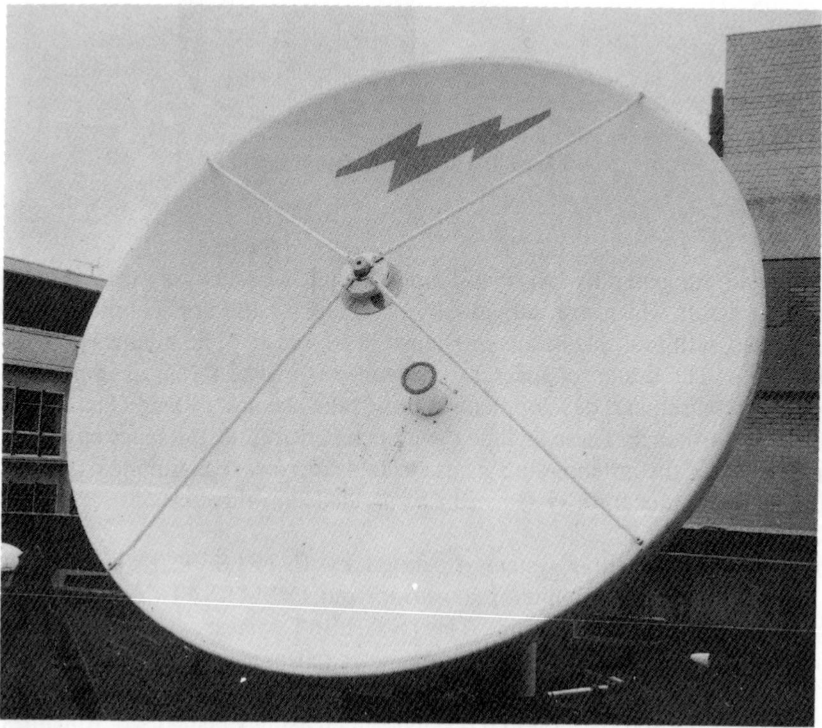

Figure 1.8. Satellite Earth station antenna

MOBILE COMMUNICATIONS

Mobile communication means communication between a fixed base station (on the ground or on a satellite) and a large number of mobiles which can both receive and transmit. The mobiles can be carried by a person, or be on a car, truck, plane or ship. Up to date most mobile communication has been at frequencies in the VHF and UHF ranges, because of the ease of propagation at these frequencies, the low cost of portable equipment and the small size of omni-directional antennas. The growth in mobile communications is, however, so rapid that new frequency allocations will be needed. There are a number of band between 1 and 3 GHz allocated for mobile communications as well as at millimetrewave frequencies. Most microwave antennas are small and directional, so that new types of antennas are being developed. The mobile unit also needs lightweight microwave components, sources and detectors.

BROADCASTING

Broadcasting systems have one transmitter and a large number of receivers. The Transmitting antenna must send adequate signals in the directions of the receivers so that only a small antenna is required to detect the television or radio signal. All terrestrial broadcasting currently takes place at frequencies below the microwave bands, but satellite broadcasting uses down-link frequencies around 12 GHz. This is called DBS or Direct Broadcast Satellites. Technically, DBS differs from the small satellite communications only in having a separate frequency band, higher power on the spacecraft and correspondingly smaller antennas for reception.

The microwave broadcast signals are received by a small antenna which is usually mounted on a wall or roof, the orientation of which is chosen to give a clear view of the geostationary satellite. An efficient antenna is required to maximise the amount of received signal. The principles of antennas are explained in Chapter 14. A reflector antenna is the prime choice because it is simple, cheap and relatively efficient. Flat plate antennas, using arrays of microstrip antennas are also available. These can be less environmentally obtrusive but are generally more expensive than reflector antennas. They offer the possibility of being able to electrically steer the receiving beam direction so that the antenna can be placed in a convenient physical location.

The antenna delivers the signal to an LNC or Low Noise Converter which is a combined amplifier and down-converter. This has a front-end microwave transistor with very low noise characteristics. Progress in the design of low noise gallium arsenide FETs has been rapid so that it is possible to use much smaller antennas than was envisaged in the early 1980s. These high performance devices are being produced in large numbers, so that the microwave parts of a DBS receiving system can be produced at low cost.

1.6 RADAR SYSTEMS

Radar is a prime user of microwave signals and most modern radars operate in the microwave part of the spectrum and depend on good quality microwave transmission of the signals. A radar system, Figure 1.9, consists of a transmitting antenna which emits a narrow beam of microwaves. A portion of the transmitted wave is intercepted by a reflecting object and reradiated in all directions. A portion of this reradiated signal travels back towards the transmitter where a receiving antenna (usually the same as the transmitting antenna) collects the weak signal and passes it to a receiver which is able to detect the presence of a target. The distance to the target is called the *range* and is determined partly by the time taken for the signal to travel to the target and back and partly by the transmitter and receiver. The relationship linking these factors is developed in Chapter 7. The angular position is determined by the direction of the narrow beam from the antenna. If the radar or the target are moving relative to each other then a shift in the carrier frequency occurs and the speed can be measured from the Doppler shift.

The most common radar waveform is a periodic pulse waveform where the pulses are very narrow and high power. The pulse travels out to the target and back again so the distance, or range, to the target can be determined. After a pulse has been transmitted, a second pulse cannot be sent until the first pulse has been received back from the target. Since this depends on distance there is an inverse relationship between the maximum pulse repetition frequency and the maximum range which can be detected.

The schematic of a typical pulse radar is shown in Figure 1.9. The transmitter will generate a train of pulses either by a solid state device for a low power radar or a microwave tube in a high power radar. Magnetrons and klystrons are used for high power microwave tubes. A radar for detecting aircraft over a range of a few hundred kilometres might use a microwave tube with a peak power of about a megawatt and an average power of a few kilowatts. It will probably have a pulse width of several microseconds and a repetition rate of a few hundred pulses per second. It is usual to have a single antenna for both transmitting and receiving so that a device known as a duplexer is necessary to stop the high power transmitted pulses from entering the sensitive receiver and to direct the received pulses to the receiver. This is either a gas-discharge device or a ferrite circulator.

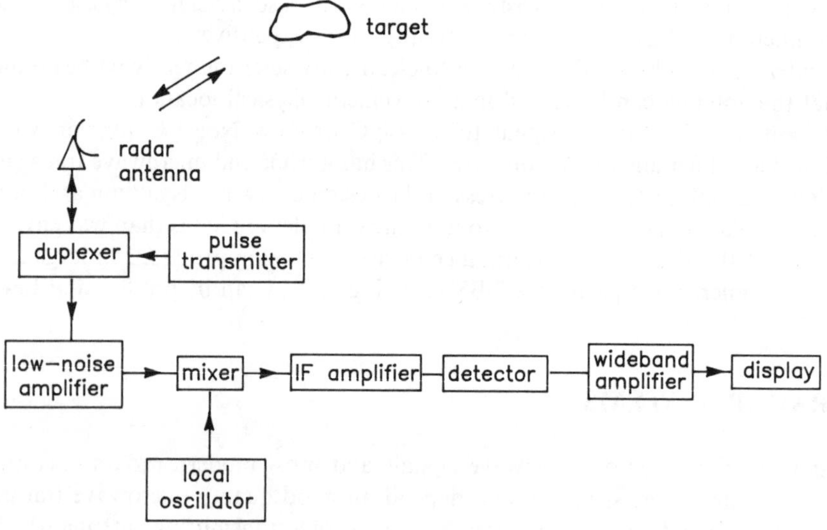

Figure 1.9. Schematic of a pulse radar system

The receiver usually operates in a superheterodyne mode, where a local oscillator signal is mixed with the incoming signal to produce a fixed intermediate frequency (IF). The mixer may be preceded by a microwave amplifier using a low noise transistor or cooled parametric amplifier. The IF amplifier is also a matched filter so that the maximum peak signal to mean noise power ratio is recovered. The output of the IF amplifier is passed to a second detector to extract the pulses from the pulse modulation. These are amplified by a wideband amplifier and passed to a suitable display such as a cathode ray tube. The most common form is called a PPI, or Plan Position Indicator, which has a circular face where radius represents the range and the intensity modulated beam rotates in angle as the antenna rotates.

The relative motion between a radar and a target can be measured with the Doppler principle. In this case only a single carrier wave needs to be transmitted and received, although Doppler information can be abstracted from any type of radar waveform. Frequency modulated carrier wave (FMCW) radar is used to measure distances accurately. It is used both for short distance measurements of a few metres and for aircraft altimeters to measure height above the ground. The principles of both Doppler radar and FMCW radar are described in Chapter 7.

APPLICATIONS OF RADAR

There are a wide variety of applications for radar. The majority operate between frequencies of 1 and 30 GHz with the higher power, longer distance radars at the lower end of this band and the lower power, precision, radars in the upper portions of the band. A list of some of the uses is as follows:

- Ships and sea position control
- Air traffic control
- Altimetry
- Position fixing
- Mapping
- Reconnaissance and surveillance
- Tracking
- Guidance and homing for weapons
- Obstacle avoidance
- Weather radar
- Distance measurement
- Speed detection
- Burglar detectors

Many of the applications are used by both the civil and military sectors, but it has usually been the military sector which has funded the large amount of research which has been done to develop improved radar systems. A brief description of aircraft radar will serve to illustrate a typical application.

Air traffic control radar is essential for air safety, Figure 1.10. It is used for control

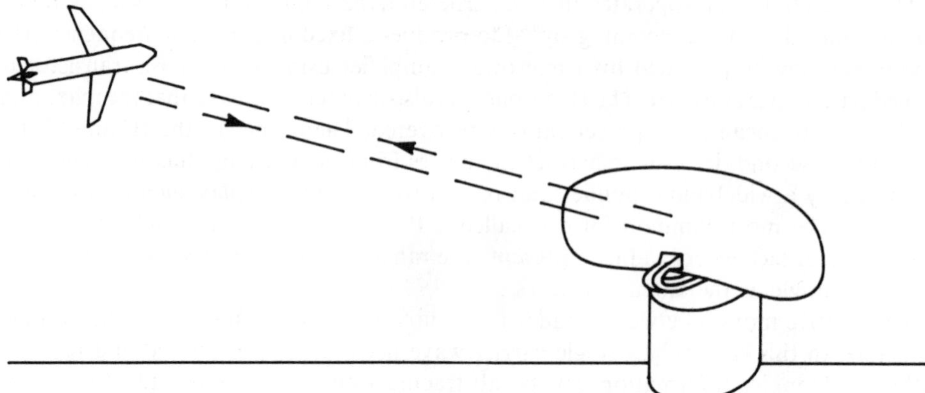

Figure 1.10. Air traffic control radar

of aircraft movements both in the air and on the ground by air traffic controllers, and as a landing aid by aircraft. High power primary radar using large antennas is used to locate aircraft. The signal received by a radar from an aircraft is, however, very weak and is embedded amongst a large amount of *clutter* due to clouds and rain. To overcome these problems, secondary surveillance radar was developed. Aircraft carry small transponders which detect the incoming radar signal and reradiate a coded signal on a different frequency. This can be picked up by the radar and provides not only a much more reliable detection and position location signal but also tells the air traffic controllers the identity of the aircraft. Most aircraft are fitted with weather radar so that the pilot can avoid storms. Radar is also used to assist in landing an aircraft in bad weather and the microwave landing system is a precision, semi-automatic landing system. The movements of the aircraft on the ground are monitored and controlled by a high precision airfield radar. The antenna for the radar is usually the largest, and most visible, part of the system. Positional information is obtained either by mechanically rotating the antenna, or by electronically steering the beam of a fixed antenna. The latter mode uses the principle of phased array antennas which enables fast and versatile scanning. It is, however, more expensive to build than a mechanically rotating antenna.

1.7 RADIO NAVIGATION

Radio navigation is essential for the determination of the position of ships and aircraft. Radar is used to provide accurate short distance position fixing, for example in and near harbours. The main radio navigation systems are, however, done by ships or aircrafts receiving radio signals from either terrestrial or satellite transmitters. The main systems in use are: the Decca hyperbolic system, the Loran C hyperbolic system, the US Navy TRANSIT satellite system and the US Global Positioning Satellite (GPS) system.

The Decca and Loran systems use a series of fixed base stations to radiate signals which are coordinated between pairs of stations. They are both based on the hyperbolic navigation principle (explained in Chapter 7) and use frequencies around 100 kHz. The

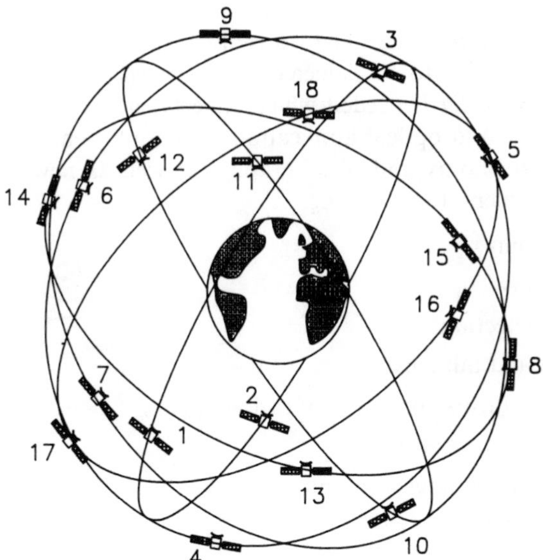

Figure 1.11. GPS navigation satellite system

Decca system is mainly used in European coastal waters whilst Loran-C stations exist throughout the northern hemisphere. The accuracy depends partly on the position of a ship relative to the base stations, and partly on the ionospheric propagation conditions.

The TRANSIT satellite system has been operational since the mid 1960s and uses a series of polar orbiting satellites which radiate fixed signals at 150 and 400 MHz. The user measures the Doppler shift and this together with a knowledge of the position and velocity of the satellite enables the users location to be determined. The TRANSIT system is being replaced by the much more accurate Global Positioning Satellite system. The full GPS system has a constellation of 18 satellites radiating at 1.228 and 1.575 GHz and orbiting at an altitude of 20 200 km, Figure 1.11. Each satellite carries a highly stable caesium or hydrogen laser clock. The user receives the signals from a number of satellites (normally four) and measures the time difference to determine position. The satellites radiate a low level signal using spread spectrum modulation. This has the advantage that very small antennas are needed to receive the signal, but the disadvantage that considerable processing is required to recover the coded time data. GPS will give position in three dimensions so that it can be used on land and in the air as well as at sea. Absolute accuracies of a few metres are possible.

1.8 REMOTE SENSING

Remote sensing is the detection of energy radiated by natural sources or energy reflected from the Sun or stars. The energy is a form of noise in which the signals are un-correlated and have components throughout the electromagnetic spectrum, but most objects peak in a particular frequency range.

The main parts of a remote sensing system are an antenna and a sensitive receiver. There is normally no transmitter. The transmission medium is either the atmosphere or space. Often the energy to be detected is very weak so that a special integrating receiver, called a *radiometer* is needed which collects energy over a period of time.

A list of microwave and optical applications of remote sensing includes classic uses of optics such as astronomy and photography, as well as new uses such as medical imaging and environmental monitoring.

- Optical astronomy
- Radio astronomy
- Infrared heat detection
- Imaging (land and humans)
- Military sensing (arms control verification etc.)
- Meteorological
- Geographical and geological monitoring
- Monitoring agriculture
- Pollution control
- Security surveillance

Optical remote sensing means the visual detection of optical or infrared energy either from distant sources (astronomy) or by indirect detection of the Sun's energy (weather forecasting, ground mapping). The essential element of all optical sensing is a lens or mirror which collects the energy. Optical sensing is very old and in the simplest system the receiver is the human eye, which is a sensitive optical antenna with a built in receiver and highly developed processing system. The optical characteristics of the eye are explained in Chapter 9.

Microwave remote sensing has been developed in the latter part of the twentieth century. The best most well known example is radio astronomy which was first developed in the VHF and UHF bands and later extended up in frequency into the microwave and millimetrewave band. Radio astronomers have been pioneers in the development of large, accurate reflector antennas and sensitive detectors. The millimetrewave and submillimetrewave parts of the spectrum are the last parts to become accessible for astronomical observations. There is considerable interest in developing telescopes to operate in this region because of the wealth of information about the universe available in this frequency range. The design and construction of radio telescopes to detect signals at millimetrewaves and submillimetrewaves is technically very challenging. Large reflector antennas are needed which must have surface accuracies of better than at least a tenth of a wavelength which translates into a few tens of microns. Low noise receivers must be developed to detect at these wavelengths.

Microwave remote sensing is also used to observe the ground, the sea and ice from satellites and to provide information which is complementary to that obtained from optical or infrared sensing. Microwave and millimetrewave meteorology is concerned with measurement of the constituents of the lower atmosphere using the resonant lines of oxygen, water vapour and other gases. Measurements using ground based, aircraft and balloon radiometers are already widely used and satellite monitoring have been shown to be capable of proving useful additional information.

1.9 HEATING AND ENERGY TRANSFER

Electromagnetic waves are efficient carriers of power. High power microwave and optical energy can be transmitted through the air with low loss and then used to heat materials. The principles are described in Chapters 8 and 9. Consequently there are a number of applications of specifically designed for high power systems.

- Microwave heating and drying
- Microwave cooking
- Plasma and gas diagnostics
- Medical hyperthermia
- Laser cutters
- Laser surgery
- Solar power satellites

High power microwave beams have considerable advantage when used as a way of removing extraneous water in materials (see Chapter 8). The microwave power penetrates into the material and interacts with the water molecules to dissipate the energy as heat. It is used in industry as a means of drying paper, hardening plastics, vulcanising rubber, sterilising foods and heating many materials. It is particularly good for drying applications because a material such as wet paper can be moved through a microwave beam as part of the production line. The most well known domestic application in this sphere is microwave cooking where the energy from a magnetron at a frequency of 2.45 GHz is directed onto food. The energy is confined to the microwave oven by an enclosed cavity. In a conventional gas or electric cooker, heat is applied to the outside of food by convection and to the inside of the food by subsequent conduction. In microwave cooking the energy penetrates deep into the food so that heating takes place simultaneously throughout the food.

Microwave power can also be used to study the characteristics of plasmas and gases. Many gases resonate in the millimetrewave part of the spectrum and interacting a high power beam with the gases forms a good diagnostic tool. Although a specialist use of microwaves, considerable research has been done because of the interest in future efficient energy generation. There are also applications in medicine. Carefully directed beams of microwave power have been used to treat cancers by heating the cancer so that it is killed without affecting the surrounding tissues.

High power optical energy can be generated with gas lasers and then used in a variety of applications ranging from metal cutters to laser eye cataract surgery. The advantage of the optical laser beam is that a very precise, and well defined beam is generated which can be carefully controlled. As will be explained in detail in Chapters 5 and 9, a laser beam makes an ideal example of a simple optical wave.

The main source of optical energy is the Sun which irradiates the Earth with a constant stream of electromagnetic waves. The level of power is high and on a clear day at the equator, this amounts to about 1.4 kW/m^2. Some of the energy is lost through the atmosphere and above the atmosphere the level is higher. Tapping this source of electromagnetic energy is a long term goal which will provide a limitless source of energy. Proposals have been made to build a large solar power satellite in space to collect some of the Sun's optical energy. The most efficient method of conveying the energy collected from the Sun to the Earth's surface is to transform the optical energy into microwave energy and use large antennas on the satellite and Earth to form a high power microwave beam.

Paramount in considerations of the use of high power electromagnetic waves are human safety considerations. The microwave energy penetrates into the human body whereas the shorter wavelength optical energy is dissipated in the skin. This has been a particularly public concern with microwave cooking but also applies to laser beams. Exposure levels have been agreed by many countries but because the long term effects are uncertain there is continual research into the effects of microwave and optical signals on the human body, as explained in Chapter 8.

1.10 CONTROL AND RECORDING

There are a number of important applications for optical and infrared waves which are concerned with control and recording. The transmission principles are described in Chapter 9.

- Remote control
- Bar code readers
- Compact Discs
- CD-ROMs
- Photocopiers
- Laser printers
- Cameras
- Video recorders

The infrared remote control for TVs and other domestic electronic equipment is well known. In this case an infrared light emitting diode generates a coded beam which is directed towards the remote receiver. Optical bar code readers use a light beam reflected off the bar code and detected by a lens system which focuses the signal onto a photodiode so that the bar code can be decoded.

A very sophisticated optical transmission system is contained in the growing band of optical recording systems. The most common is the compact disc (CD) digital audio system, which uses a series of optical transmission and processing components and is an excellent practical example of the theory of optical waves developed in this book. It is explained in Chapter 9. The CD was only introduced in the mid 1980s but immediately became popular in the domestic audio market. It shows how a sophisticated optical system can be made at relatively low cost in large numbers. The same principles of optical recording as used in the compact disc are also found in digital computer storage devises such as the CD-ROM (compact disc read-only-memory), and in the erasable optical disc. Both these devices have extremely high storage capacities due to the high packing density available at optical wavelengths.

Photocopiers use a system of lenses to transform an image from one piece of paper to another piece of paper. laser printers extend this principle to use a coded laser beam and optical transmission system to convert the digital data in a computer to hard copy.

Cameras and video recorders are widely used to record visual images. In a camera, the recording medium is chemical and in a video recorder it is electronic using an array of charge coupled photodiodes. A series of lenses are used to focus the reflected light from the sun or an artificial light source onto the recording medium.

1.11 SUMMARY

This chapter has set the scene for the study of microwave and optical transmission by reviewing the range of applications for microwaves and optical waves. The previous sections have briefly described an impressively wide range of uses which impinge on almost every aspect of modern life. The lists are not exhaustive and further applications will appear in the future. Some of the systems will be described in more detail in later chapters when they will be used as examples of the theory which will be developed. This electromagnetic wave theory needs to start at the fundamental level of electric and magnetic fields and build up so that the operation of any microwave or optical system can be understood.

2 Electric and magnetic fields

2.1 INTRODUCTION

Chapter 2 studies static electric and magnetic fields at a level which is sufficient for an understanding of the background to electromagnetic wave theory and Maxwell's equations which are the theoretical basis for microwave and optical transmission. It also introduces the notation and mathematics used throughout the book. The chapter can be omitted if electric and magnetic fields have been covered in a previous course or the reader is willing to accept Maxwell's equations as the starting point for the study of microwaves and optics.

Static electric fields (usually called electrostatics) and static magnetic fields are fields which are stationary in space with no time varying components. Electromagnetic wave theory links together the static electric and static magnetic fields through time, as shown schematically in Figure 2.1. As soon as a time varying situation occurs, there must be both electric and magnetic fields. This will become evident as the theory is developed. The treatment of static fields in this chapter will not be comprehensive. Many books have been written on the subject and some with extensive treatments are listed in the references at the end of the book.

Most practical applications of static electric and magnetic fields occur either for direct current uses or for low frequency systems. There are, however, situations even at microwave frequencies where the fields across an object are essentially constant because the object is small in wavelengths. This can be seen by considering a wave approaching an object in three situations, Figure 2.2. If the object is small by comparison with the wavelength, Figure 2.2(a) then the fields across the object will be nearly uniform. This is the quasi-static case where static electric and magnetic theory can be applied. If the object is comparable in size to a wavelength, Figure 2.2(b), then no approximations can be used in the theory and an exact analysis must be undertaken. This is the usual situation at microwave frequencies. Lastly, the wavelength may be very small by comparison with the object, Figure 2.2(c). Now the behaviour of the individual components of the wave do not matter and only the amplitude of the wave is important. This is what happens at optical frequencies, where typical wavelengths are a fraction of a micron. In this case the waves can be treated as a bundle of energy which travel as a ray of light. The exact theory can in principle be applied at all frequencies and will enable the physical properties of all types of waves to be explained. The solution of problems at low and high frequencies is, however, more conveniently done by applying simplifications and approximations to the exact theory. Later chapters will develop the exact theory and show how these approximations can be applied.

The order of the sections in this chapter follows the chronological development of electric and magnetic fields. Some knowledge of static electricity and magnetism had been known since ancient times but the modern era began in the eighteenth century when the pace of discovery of both electric and magnetic phenomena quickened. As

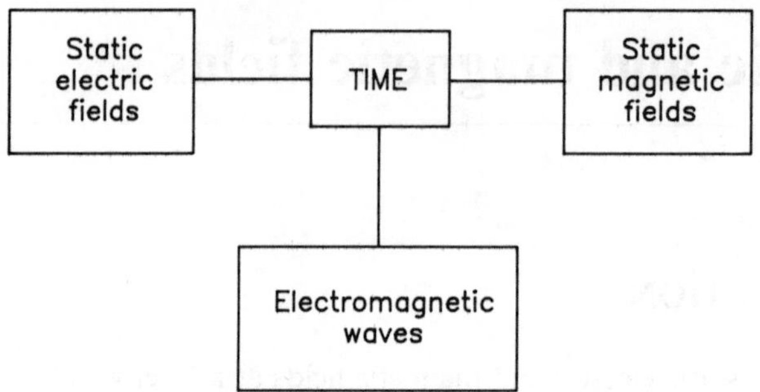

Figure 2.1. Relationship of static fields to electromagnetic waves

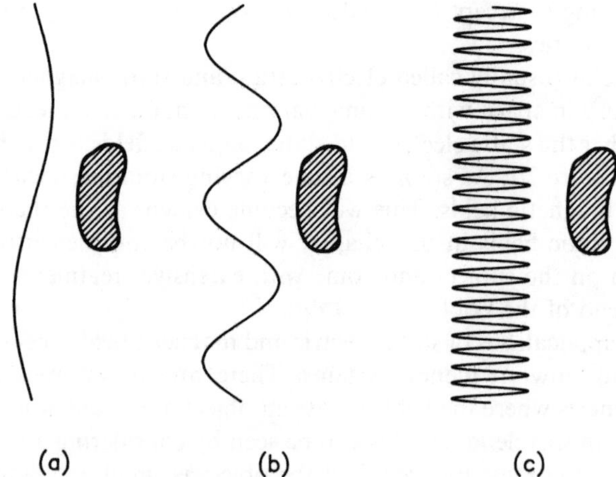

Figure 2.2. Relationship of wavelength to size of object: (a) quasi-static, (b) microwave, (c) optical

each discovery was reported, more scientists were stimulated to explore the new subjects. This led first to the classic laws of electrostatics and then to the laws of static magnetic fields. All the basic laws were discovered by experimental observations. Only after a law had been verified experimentally was a mathematical expression derived.

2.2 THE ELECTRIC FIELD

The basis of electrostatics is the experimentally observed law which Charles Coulomb (1736–1806) discovered in 1785. He used two charged metal spheres to determine the forces that act on electric charges. His experiments showed that two positive or two negative charges repel each other whereas opposite charges attract each other. He also demonstrated that the force between the charges is proportional to the product of their magnitudes and inversely proportional to the square of the distance between charges. Figure 2.3 shows two charges q_1 and q_2 which are separated by a distance r. The force

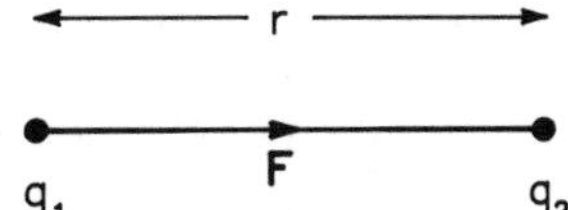

Figure 2.3. Force between two charges

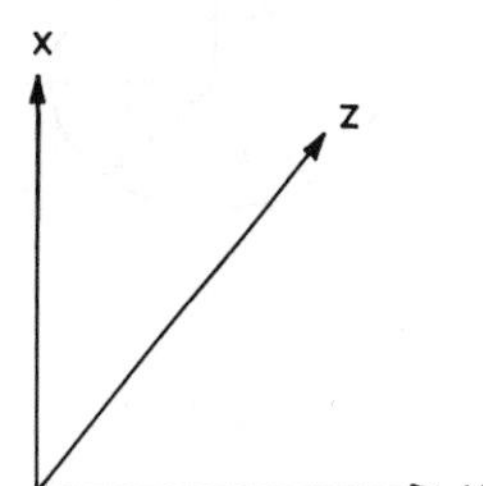

Figure 2.4. Rectangular coordinate system

between the charges acts along a line between the charges. The relationship which Coulomb discovered is known as either Coulomb's Law or the Inverse Square Law. It can be written down mathematically as

$$\boldsymbol{F} = \frac{k q_1 q_2}{r^2} \hat{\boldsymbol{r}} \tag{2.1}$$

where $\boldsymbol{F}$ is the force, $\hat{\boldsymbol{r}}$ a unit vector pointing from one charge directly to the other charge and k a constant of proportionality. Each of these terms will now be examined.

The charges q_1 and q_2 are scalar quantities which mean that they have only a magnitude. As with many of the basic parameters in electrical engineering, the units are named after the discoverer of the basic law, so the units of charge are measured in coulombs (C). The force $\boldsymbol{F}$ is a vector because force must always be acting in a particular direction. A vector is described by both its magnitude and its direction. This is reflected in the units which are newtons (N) and equal the force required to accelerate 1 kg by 1 m/s^2.

A vector represents both magnitude and direction, thus two numbers are needed to describe a vector. This is not easy notationally, so a vector is written in two parts. A symbol represents the magnitude and a second symbol represents the direction. The direction is actually specified as a vector with unit magnitude, called a *unit vector*. The force vector $\boldsymbol{F}$ is written as $F\hat{\boldsymbol{r}}$ where F is the magnitude and $\hat{\boldsymbol{r}}$ is the unit vector in direction r.

In a rectangular coordinate system, Figure 2.4, the unit vectors normally point in the x, y and z directions. These are designated by $\hat{\boldsymbol{x}}, \hat{\boldsymbol{y}}, \hat{\boldsymbol{z}}$, respectively. The vector $\boldsymbol{F}$ can then be broken up into three component parts with magnitudes F_x, F_y and F_z:

$$\boldsymbol{F} = F_x\hat{\boldsymbol{x}} + F_y\hat{\boldsymbol{y}} + F_z\hat{\boldsymbol{z}}$$

The same principle can be applied in the other two main coordinate systems, Figure 2.5. In the cylindrical coordinate system with coordinates r, ϕ, z the unit vectors are $\hat{\boldsymbol{r}}, \hat{\boldsymbol{\phi}}, \hat{\boldsymbol{z}}$ respectively. In the spherical coordinate system with coordinates r, θ, ϕ the unit vectors are $\hat{\boldsymbol{r}}, \hat{\boldsymbol{\theta}}, \hat{\boldsymbol{\phi}}$ respectively.

Forces, as with any vectors can be added and subtracted in the same way as normal

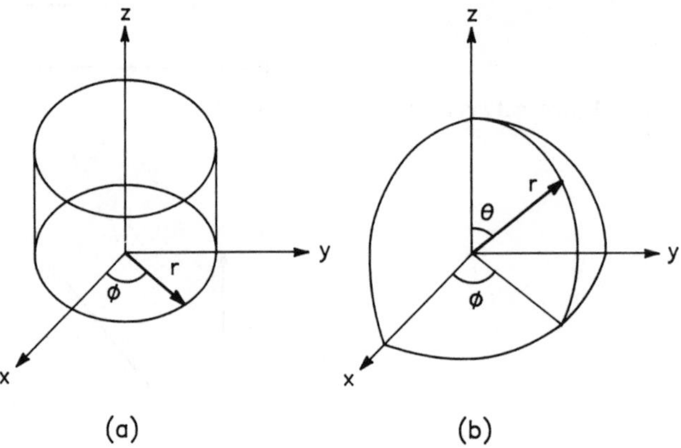

Figure 2.5. (a) Cylindrical coordinate system. (b) Spherical coordinate system

algebra by adding and subtracting the component parts. Thus if there are two forces $\boldsymbol{F}_1$ and $\boldsymbol{F}_2$

$$\boldsymbol{F}_1 + \boldsymbol{F}_2 = (F_{1x} + F_{2x})\hat{\boldsymbol{x}} + (F_{1y} + F_{2y})\hat{\boldsymbol{y}} + (F_{1z} + F_{2z})\hat{\boldsymbol{z}}$$

The constant of proportionality, k, in equation (2.1) is given by

$$k = \frac{1}{4\pi\varepsilon} = \frac{1}{4\pi\varepsilon_0\varepsilon_r}$$

The 4π is due to the International System of units (called SI after the French names Système Internationale). ε is the electrical permittivity of the medium and expresses the way in which the medium holding the charges alters the force between the charges. If the medium is a vacuum, the permittivity is called the *absolute permittivity*, ε_0, and has a value

$$\varepsilon_0 = 8.854 \times 10^{-12}\,\mathrm{F/m}$$

If the medium is not a vacuum, the permittivity is expressed as two parts—the absolute permittivity ε_0, and the *relative permittivity*, ε_r. The value of ε_r depends on the medium containing the charges. The relative permittivity of air at normal atmospheric pressure is 1.0006. This is so close to unity that it is the accepted practice to say that $\varepsilon_r = 1$ for air. A common situation is where the medium is a dielectric. The relative permittivity is sometimes called the *dielectric constant*. In optics, a related term called the *refractive index* is used. The refractive index of a pure dielectric is the square root of the relative permittivity.

Coulomb's law, equation (2.1), can now be written in its full form

$$\boldsymbol{F} = \frac{q_1 q_2}{4\pi\varepsilon_0\varepsilon_r r^2}\hat{\boldsymbol{r}} \tag{2.2}$$

Example 2.1 Force due to charges

An example of the use of Coulomb's law is provided by calculating the force on a positive charge of $10\,\mu$C placed at the origin due to a negative charge of $1\,\mu$C placed

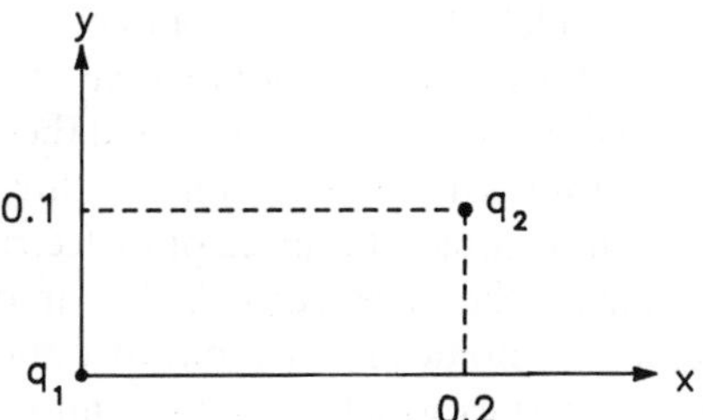

Figure 2.6. Charges for Example 2.1

at a point 100 mm from the x axis and 200 mm from the y axis. The charges are in air.

The negative charge, q_2, is at $x = 0.2$ m, $y = 0.1$ m, Figure 2.6. (Note that the units of distance must be in metres.) The distance from the origin is thus $r = (x^2 + y^2)^{1/2} = 0.224$ m. Substituting the values into equation (2.2):

$$\boldsymbol{F} = \frac{10^{-5} \times (-10^{-6})}{4\pi \times 8.854 \times 10^{-12} \times (0.224)^2} \hat{\boldsymbol{r}}$$

$$= -1.79\, \hat{\boldsymbol{r}} \text{ N}$$

The force acts in the negative direction at an angle to the x axis given by $\theta = \tan^{-1}(y/x) = 26.6°$.

Coulomb's law specifies the force for two charges but the deduction can be made that it applies for any number of charges and therefore applies when one charge is near to a system of charges. This leads to the concept of the *electric field* which is defined as the force per unit charge for each point in a region of charges,

$$\boldsymbol{E} = \frac{\boldsymbol{F}}{q} = \frac{q}{4\pi\varepsilon_0\varepsilon_r r^2} \hat{\boldsymbol{r}} \tag{2.3}$$

The units of the electric field are volts per metre (V/m). It is more correctly called the electric field intensity, but this is rarely used nowadays.

Example 2.2 Electric field due to a charge

Calculate the electric field in air at a point which is at a distance of 10 m along the negative y axis due to a positive charge of 10^{-8} C which is situated at the origin of a rectangular coordinate system. Substituting the values into equation (2.3):

$$\boldsymbol{E} = \frac{10^{-8}}{4\pi \times 8.854 \times 10^{-12} \times 10^2} \hat{\boldsymbol{y}}$$

$$= 0.9 \text{ V/m in the positive } y \text{ direction.}$$

The electric field is a measurable quantity which is widely used in practical microwave and optical engineering. Equation (2.3) shows that the electric field is inversely proportional to the relative permittivity. All dielectrics have relative permittivities which are greater than unity, so that, for a given charge, the electric field is less in a dielectric.

A dielectric is a medium in which the atoms and molecules form a system of bound charges that interacts with the electric field. If the relative permittivity at a point in the dielectric is independent of the direction of the electric field then the dielectric is described as isotropic. If the relative permittivity depends on the direction of the electric field, the dielectric is described as anisotropic. If a piece of dielectric has constant properties throughout, then it is described as homogeneous. At low frequencies and microwaves, most dielectrics are isotropic and homogeneous, but at optical frequencies, a number of materials have dielectric properties which are either inhomogeneous or anisotropic. The relative permittivity of some dielectrics commonly encountered in microwaves and optics are given in the Appendix.

Some materials, for example water, change properties considerably as the frequency changes. In other materials, for example glass, the relative permittivity depends on the amount of impurity and the exact composition. A survey of the table indicates that plastics and other common materials such as wood and glass have relative permittivities in the range 2 to 4. Foam plastics are mixtures of plastic and air and the relative permittivity is thus proportional to the density.

2.3 CONSERVATION PROPERTY OF ELECTRIC FIELDS

If a charge in the vicinity of other charges is moved there will be an energy exchange as the force experienced by the moving charge must come from the other charges. Considering the situation shown in Figure 2.7 where a charge is moved from A to B by an arbitrary path in an electric field generated by a system of charges, the component of force at a point P is $F\cos\theta$, where θ is the angle between the electric field and the direction of movement. The work done is the force times the distance moved, so the differential force at P for a small distance moved δl is

$$\delta W = -F\cos\theta\,\delta l \tag{2.4}$$

The negative sign indicates that work has been done. Equation (2.4) is the definition of a *scalar product* between two vectors and is written as $\boldsymbol{F}\cdot\boldsymbol{dl}$ where $\boldsymbol{dl} = \delta l\hat{l}$. The result of a scalar product is a scalar. The component form can be obtained by noting that the scalar product for unit vectors in a rectangular coordinate system are given by

$$\hat{x}\cdot\hat{y} = \hat{x}\cdot\hat{z} = \hat{y}\cdot\hat{z} = 0$$
$$\hat{x}\cdot\hat{x} = \hat{y}\cdot\hat{y} = \hat{z}\cdot\hat{z} = 1$$

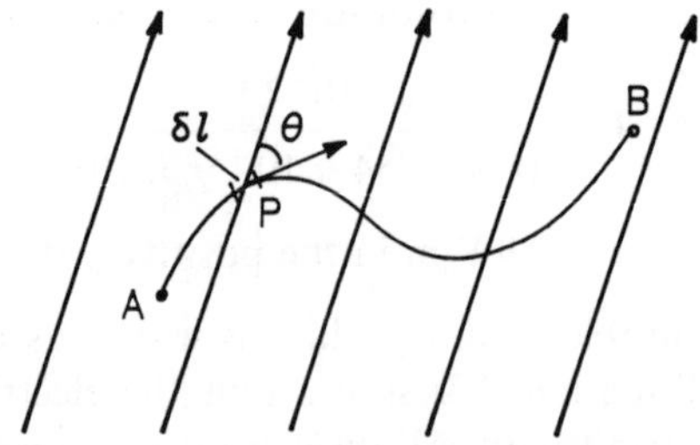

Figure 2.7. Electric charges moving from A to B

These are derived by putting $\theta = 90°$ into equation (2.4). The scalar product of two vectors then becomes the product of their respective components

$$\boldsymbol{F} \cdot \boldsymbol{dl} = F_x \, \mathrm{d}l_x + F_y \, \mathrm{d}l_y + F_z \, \mathrm{d}l_z$$

The total work done, W, in moving along the path from A to B is the summation of the differential work done along all sections of the line from A to B,

$$W = -\sum_{\mathrm{A}}^{\mathrm{B}} F \cos\theta \, \delta l = -\int_{\mathrm{A}}^{\mathrm{B}} \boldsymbol{F} \cdot \boldsymbol{dl} \tag{2.5}$$

The integral in equation (2.5) is called a *line integral* and if the end points are unknown is written as $\int_l$. Notice that the work done depends only on the end points and not on the intermediate path. This can be seen by the following example.

Example 2.3 Work done due to a moving charge

Calculate the work done in moving a charge from the origin to the point (2, 1) if the electric field is orientated along the x axis using the two paths shown in Figure 2.8.

For the straight line paths, equation (2.5) becomes $W = \sum (Fl \cos\theta)$ where l is the length of each section of the path. Only the parts of the path in the x direction contribute, since $\cos\theta = 0$ for the parts of the path in the y direction. The work done for the two paths are

$$W_1 = F \times 3 + F \times 0 + F \times (-1) = 2F$$
$$W_2 = F \times (0) + F \times (2) + F \times (0) = 2F$$

The work done is clearly independent of the paths.

Another conclusion from the example is that if the charge is returned to its initial position, the work done will be zero and there will be no net transfer of energy. This is called the conservation property of the electric field.

The force quantity in equation (2.5) can be replaced by the electric field (equation (2.3)) to give

$$W = -q \int_{\mathrm{A}}^{\mathrm{B}} \boldsymbol{E} \cdot \boldsymbol{dl} \tag{2.6}$$

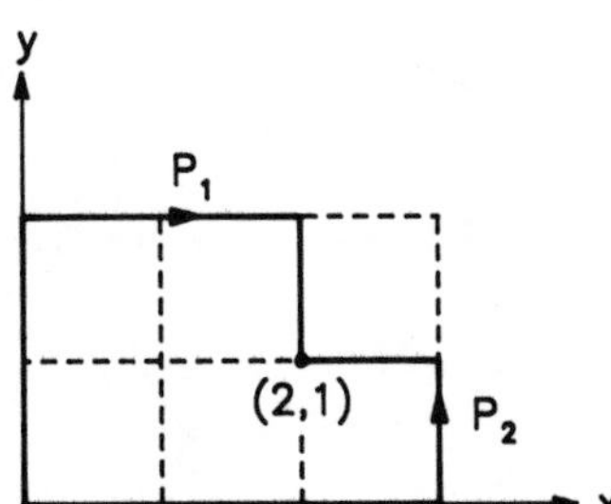

Figure 2.8. Movement of charge for Example 2.3

2.4 ELECTRIC POTENTIAL DIFFERENCE

Energy considerations lead directly to the very useful concept of potential difference, V. This is defined as the work done on a unit charge in moving from a point A to a point B. Thus $V = W/q$ and from equation (2.6)

$$V_{\mathrm{B}} - V_{\mathrm{A}} = -\int_{\mathrm{A}}^{\mathrm{B}} \boldsymbol{E}\cdot d\boldsymbol{l} \tag{2.7}$$

Only a *difference* of potential has been defined. The absolute value at a point has no significance and the end points are chosen for the situation being studied. The negative sign is thus present because of the definition of the start and end points. The end points could equally well have been chosen the other way around, in which case the potential difference would be positive. The definition is, however, written with the negative sign to show that the work is done in moving against a positive electric field. It is useful to have one of the end points as a reference point. Since the reference point is arbitrary it is defined as being at infinity where the potential must be zero. Equation (2.7) becomes

$$V = V_{\mathrm{B}} = -\int_{\infty}^{\mathrm{B}} \boldsymbol{E}\cdot d\boldsymbol{l} \tag{2.8}$$

All points are now taken with reference to infinity. This is what happens in circuit theory where the potential difference is always expressed by a single parameter V.

Example 2.4 Potential due to a point charge

Find the potential between two points a distance R_{A} and R_{B} from a point charge q, Figure 2.9.

The electric field is given by equation (2.3), which substituting into equation (2.7) gives,

$$V_{\mathrm{B}} - V_{\mathrm{A}} = -\int_{\mathrm{A}}^{\mathrm{B}} \frac{q}{4\pi\varepsilon_r\varepsilon_0 r^2}\hat{\boldsymbol{r}}\cdot(\hat{\boldsymbol{r}}\,\mathrm{d}r)$$

$$= \frac{-q}{4\pi\varepsilon_0\varepsilon_r}\left(\frac{1}{r_{\mathrm{B}}} - \frac{1}{r_{\mathrm{A}}}\right) \tag{2.9}$$

Notice that the difference in potential depends only on the radius of the points A and B and not on their absolute position. This means that point B does not have to be

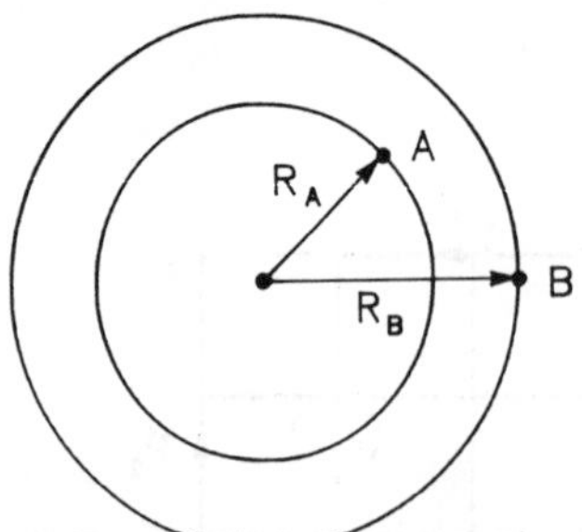

Figure 2.9. Potential between points for Example 2.4

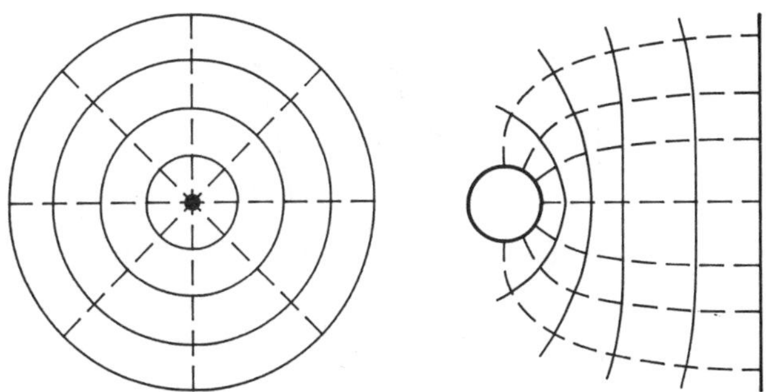

Figure 2.10. Equipotential (——) and electric field (- - -) for (a) a point charge, and (b) flat conducting sheet and cylinder

on the same radial line connecting the point charge and point A. The reason is that the potential is constant over all of a spherical surface surrounding q, so it is only the difference in the sizes of the two spherical surfaces which is relevant.

If the point A is moved to infinity where $V_A = 0$, the result is

$$V = \frac{-q}{4\pi\varepsilon_0\varepsilon_r r_B} \tag{2.10}$$

It is helpful to be able to draw a plot of the variations of electric field and the potential difference in systems of charges and conductors. The electric field lines follow the force lines, as shown by equation (2.3). In order to draw the potential difference it is most convenient to plot contours along which the potential is constant. These are called *equipotential* lines or surfaces. For a small change in potential, equation (2.7) becomes

$$\delta V = -E\cos\theta\,\delta l \tag{2.11}$$

If $\theta = 0°$ then $\partial V/\partial l = -E$. This shows that the gradient of the potential difference is equal to the electric field. The electric field and potential difference are thus everywhere orthogonal, that is they are at right angles to each other. This makes the drawing of the equipotential lines and electric field lines relatively simple. Figure 2.10(a) shows them for a point charge where the electric field lines are radial and the equipotentials are circles. Figure 2.10(b) shows them for the case of a flat conductor and a circular conductor. The electric field lines terminate at right angles to the conductor. The reason for this will be explained in a later chapter. The equipotential lines can be easily drawn by ensuring that they cross the electric field lines everywhere at right angles. The potential difference is a measurable quantity and can be recorded by probing the electric field of a system of charges. In practice the equipotential lines can thus be tracked and recorded and the electric field lines deduced from the results.

2.5 DIFFERENTIAL FORM OF RELATION BETWEEN ELECTRIC FIELD AND POTENTIAL

For the purposes of manipulating the fields in microwave and optical situations, it is more convenient to express the relation between the electric field and the potential difference in vector differential form. A small portion of the electric field is shown in

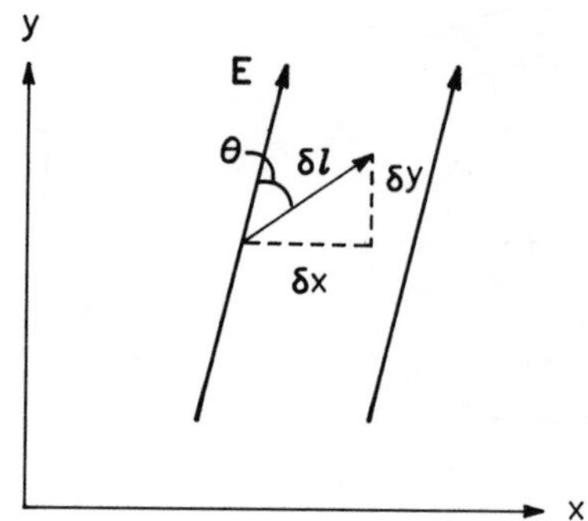

Figure 2.11. Differential form of relation between electric field and potential

Figure 2.11. Equation (2.11) can then be resolved into components. In the rectangular coordinate system the component parallel to the x axis is $\delta V = -E_x \delta x$ or in differential form

$$E_x = -\frac{\partial V}{\partial x} \tag{2.12}$$

There will be similar relation in the other two directions, so that the total electric field is

$$\boldsymbol{E} = -\left(\frac{\partial V}{\partial x}\hat{x} + \frac{\partial V}{\partial y}\hat{y} + \frac{\partial V}{\partial z}\hat{z}\right) \tag{2.13}$$

This is the differential form of the relation between $\boldsymbol{E}$ and V, but it can be expressed in a more compact notation by using the vector differential operator, ∇, pronounced *del*

$$\nabla = \frac{\partial}{\partial x}\hat{x} + \frac{\partial}{\partial y}\hat{y} + \frac{\partial}{\partial z}\hat{z}$$

Equation (2.13) is written as

$$\boldsymbol{E} = -\nabla V \tag{2.14}$$

The right-hand side is called the *gradient* of V. Note that although the definition of ∇ makes it looks like a vector, it is only an operator on scalars or vectors to the right of ∇. It does however obey the rules of vector mathematics. The advantage of the compact form of equation (2.14) is its simplicity so that when it is combined with other equations, cumbersome mathematics are avoided.

2.6 GAUSS'S LAW AND ELECTRIC FLUX DENSITY

One of the most useful laws in electrostatics is that established by Karl Gauss (1777–1855) in 1813. This links the charge in a region or on a surface to the electric field and enables many practical problems to be easily solved. Before discussing Gauss's Law, the *electric flux density*, $\boldsymbol{D}$, will be introduced. From equation (2.3) it can be seen that the electric

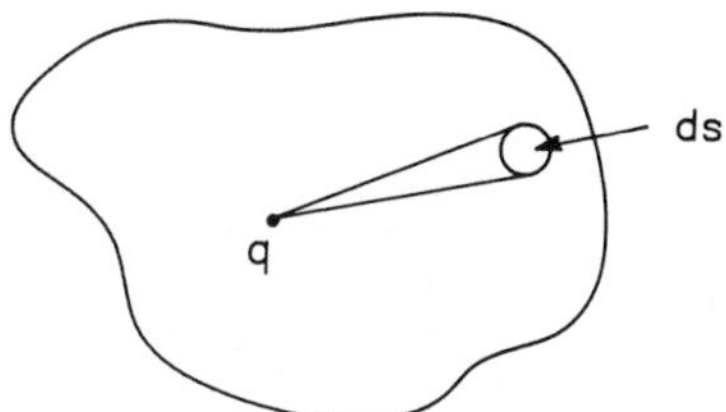

Figure 2.12. Volume surrounding a charge

field times the permittivity is equal to the charge divided by $4\pi r^2$. The quantity $4\pi r^2$ is the surface area of a sphere surrounding the charge. $\boldsymbol{E}\varepsilon_0\varepsilon_r$ is thus a density and it is convenient to give this combination the name, *electric flux density*. This is a measure of the electric field flowing out of the sphere. In summary

$$\boldsymbol{D} = \varepsilon_0\varepsilon_r\boldsymbol{E} \tag{2.15}$$

The sphere of radius r surrounds the charge and encloses the point charge q. The electric flux flowing out of the sphere is thus equal to the charge enclosed. By extension, if this is true for one charge then it will be true for any number of charges inside the sphere. It will also be true if the sphere is replaced by a closed surface of arbitrary shape. This general situation is shown in Figure 2.12. A general volume region, v, surrounds a closed surface containing a total charge Q. This leads to the general statement of Gauss's law:

The electric flux flowing out of a closed surface is equal to the charge enclosed.

The law can be expressed mathematically as

$$\int_s \boldsymbol{D}\cdot\boldsymbol{ds} = Q \tag{2.16}$$

In many cases the charge is uniformly distributed within the volume region. Then the charge enclosed by a surface is not a convenient quantity to use in calculations and it is of more use to replace it by a charge density per unit volume, ρ, integrated over the volume, v. Then Gauss's law becomes

$$\int_s \boldsymbol{D}\cdot\boldsymbol{ds} = \int_v \rho\,\mathrm{d}v \tag{2.17}$$

Gauss's law can be used to solve many electrostatic problems which are simplified examples of structures encountered in microwaves. The problems divide into those with analytical solutions where the integrals in equations (2.16) or (2.17) can be easily evaluated, and those which need to be solved using numerical integration. Only the former will be discussed here. The procedure for the solution to a problem is to draw the surface enclosing the charge or charges. This is called a *Gaussian surface*. Then Gauss's law is used to evaluate to electric flux density. The electric field and potential can at the Gaussian surface then be found. This will be illustrated with a number of examples.

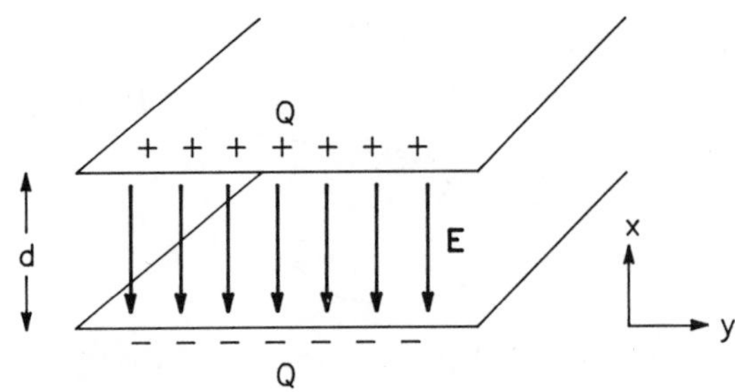

Figure 2.13. Potential between parallel plates for Example 2.5

Example 2.5 Potential between parallel plates

Calculate the potential between two parallel plates which are distance d apart and have a surface area S. There are charges on the two plates of $+Q$ and $-Q$, as shown in Figure 2.13.

The Gaussian surface is the area of each of the plates. For this simple geometry, equation (2.16) becomes

$$\boldsymbol{D} \times \text{area of plates} = \text{charge on plates}$$

or

$$\boldsymbol{D}S = Q$$

From equation (2.15)

$$E = \frac{Q}{\varepsilon_0 \varepsilon_r S}$$

Inserting this into the definition of potential equation (2.7) gives

$$V = \frac{-Q}{\varepsilon_0 \varepsilon_r S} \int_0^d (-1)\,\mathrm{d}x = \frac{Qd}{\varepsilon_0 \varepsilon_r S} \tag{2.18}$$

The result shows that the potential between two parallel plates is proportional to the distance apart of the plates and inversely proportional to the area of the plates. Parallel plates form the basis of microwave microstrip transmission lines and will be described later in the book.

Example 2.6 Electric field and potential due to a charged sphere

Find the distribution of electric field and potential (a) inside a sphere of radius a, and (b) outside the sphere which contains a uniform distribution of charge of value ρ C/m^3, Figure 2.14.

In this example, the Gaussian surface is an imaginary sphere either inside the spherical surface containing the charge, or outside the sphere. The charge enclosed is ρ times the volume of the sphere, i.e.

$$Q = \rho \frac{4\pi}{3} r^3 \tag{2.19}$$

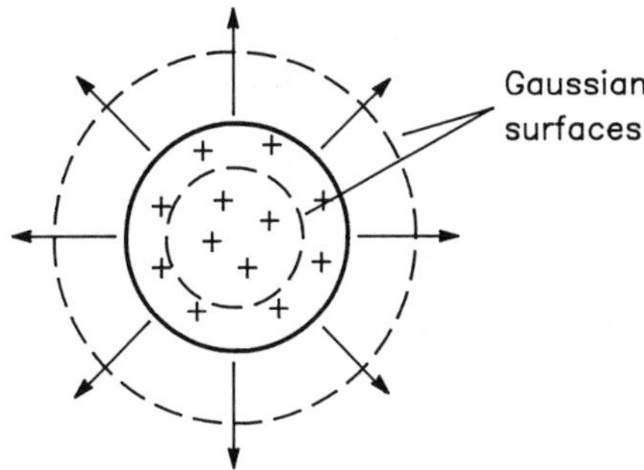

Figure 2.14. Sphere containing a uniform distribution of charge

The flux flowing out of the Gaussian surface must be in the radial direction and is the flux density times the surface area of the sphere,

$$\int \boldsymbol{D}\cdot d\boldsymbol{s} = D_r 4\pi r^2 \tag{2.20}$$

Gauss's law equates equations (2.19) and (2.20). There are two cases to consider. First when the Gaussian surface is inside the sphere and secondly when it is outside the sphere. For $r < a$ only a portion of the charge enclosed is covered by the Gaussian surface so the electric field follows from equations (2.16), (2.19) and (2.20) as

$$E_r = \frac{\rho r}{3\varepsilon_0\varepsilon_r}$$

For the second case, $r \geqslant a$, the Gaussian surface is outside the sphere so the charge enclosed is constant and the electric field is then given by

$$E_r = \frac{\rho a^3}{3\varepsilon_0\varepsilon_r r^2}$$

These two cases show that inside the sphere the electric field increases linearly with radius but outside the sphere it decreases as the square of the radius.

For points outside the sphere the potential is given by equation (2.8)

$$V = -\int_{\infty}^{r} E_r \, \mathrm{d}r = \frac{a^2\rho}{3\varepsilon_0\varepsilon_r r}$$

Similarly for points inside it is given by

$$V = -\int_{\infty}^{a} E_r \, \mathrm{d}r - \int_{a}^{r} E_r \, \mathrm{d}r = \frac{\rho}{2\varepsilon_0\varepsilon_r}\left(a^2 - \frac{r^2}{3}\right)$$

The variation of electric field and potential with radius is sketched in Figure 2.15.

Example 2.7 Electric field and potential due to a long rod

Find the electric field and potential for a long rod of radius a with a surface charge per metre of q, Figure 2.16.

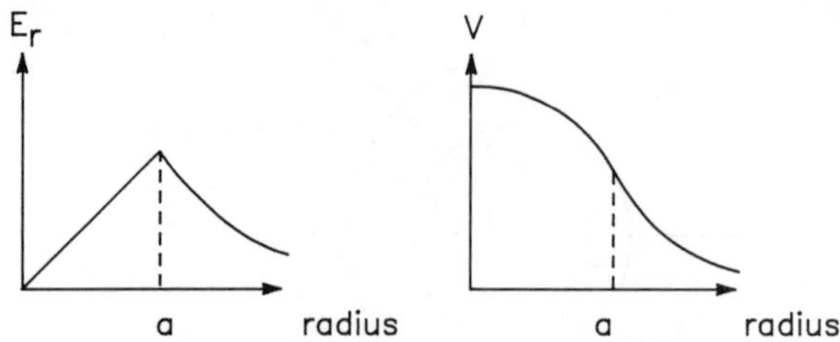

Figure 2.15. Electric field and potential of a charges sphere for Example 2.6

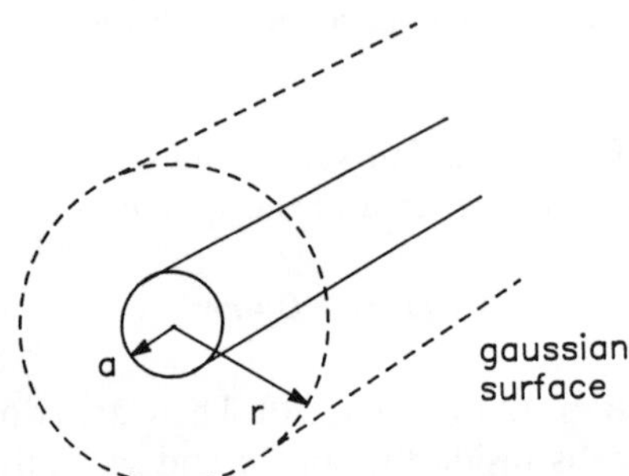

Figure 2.16. Long rod with a surface charge

The coordinate system necessary to solve this case is a cylindrical system and the Gaussian surface is an imaginary cylinder of radius r surrounding the rod. The surface area of the cylinder is the circumference times the length, l. Symmetry shows that the electric field must be directed radially outwards from the centre of the rod. The charge on the rod is totally enclosed by the imaginary cylinder so Gauss's law gives

$$D_r 2\pi r l = q l$$

The electric field is

$$E_r = \frac{q}{2\pi\varepsilon_0\varepsilon_r r} \tag{2.21}$$

Notice that the radius of the rod does not appear in the equation and the result would be the same if there was an infinitely thin line of charge or if there was a cylinder of charge. In all cases the charge appears to be along the axis of the coordinate system. The potential difference does depend on the radius of the rod because it is obtained by integrating between the radius of the rod and the observation radius r

$$V = -\int_a^r E_r \,\mathrm{d}r = -\int_a^r \frac{q}{2\pi\varepsilon_0\varepsilon_r r}\,\mathrm{d}r = \frac{-q}{2\pi\varepsilon_0\varepsilon_r}\ln\left(\frac{r}{a}\right) \tag{2.22}$$

Example 2.8 Potential due to coaxial cylinders

Find the potential difference between the conductors of a coaxial cylinder, Figure 2.17, with radii a and b.

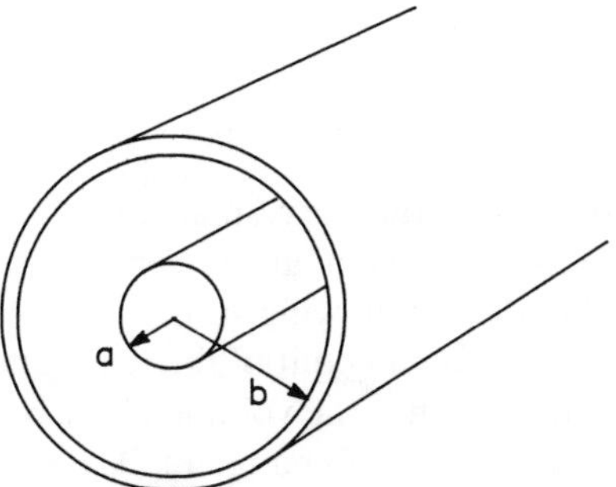

Figure 2.17. Coaxial cylinders

This is an important case because it is the basis of the coaxial cable which is widely used at RF and microwaves. If the charge on the inner cylinder is q C/m, the electric field is identical to the previous case of the long rod because as far as the imaginary Gaussian cylinder is concerned, the situation is identical. It does not matter what, if any, charge exists on the outer cylinder. The potential difference is given by equation (2.22) with radius b replacing radius r,

$$V = \frac{-q}{2\pi\varepsilon_0\varepsilon_r}\ln\left(\frac{b}{a}\right) \tag{2.23}$$

The potential depends on the logarithm of the *ratio* of a and b, not on their absolute values. If the coaxial cylinders are filled with a dielectric, then the potential decreases by comparison with an air-filled coaxial cylinder.

2.7 DIFFERENTIAL FORM OF GAUSS'S LAW

Gauss's law is one of the basic equations describing electromagnetic waves, but for the purposes of incorporating it with other equations it is more use to express it in differential form. This can be derived by considering a small portion of the electric field and using a small rectangular box as the Gaussian surface, Figure 2.18. The box has sides dx, dy

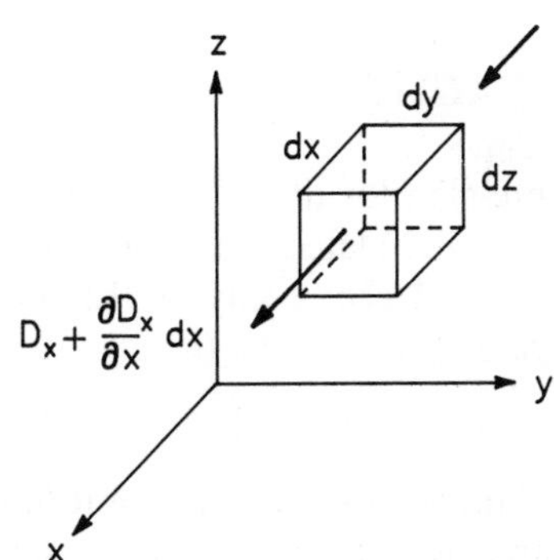

Figure 2.18. Differential form of Gauss's law

and dz so the volume of the box is $dv = dx\,dy\,dz$. From equation (2.17), Gauss's law becomes

$$\boldsymbol{D}\cdot\boldsymbol{ds} = \rho\,dv \tag{2.24}$$

The integral disappears because of the very small region. The aim is to express the left-hand side of equation (2.24) in rectangular differential components. Considering the component of the electric flux density flowing in the positive x direction (as shown in Figure 2.18), the total electric flux flowing out of the rear surface will be the electric flux density times the area, or $D_x\,dy\,dz$. Since the volume of the box is very small, the change in D_x through the box will be the rate of change of D_x (i.e. $\partial D_x/\partial x$) times the width of the box dx. This will be added to the flux flowing through the rear surface, so the flux flowing out of the front surface is

$$\left(D_x + \frac{\partial D_x}{\partial x}dx\right)dy\,dz$$

The total flux flowing in the x direction is the difference between the flux out of the front surface and out of the rear surface, or

$$\text{Flux difference} = \left(D_x + \frac{\partial D_x}{\partial x}dx\right)dy\,dz - D_x\,dy\,dz = \frac{\partial D_x}{\partial x}dx\,dy\,dz$$

By extrapolation, the total flux flowing out of all surfaces of the box is

$$\boldsymbol{D}\cdot\boldsymbol{ds} = \left(\frac{\partial D_x}{\partial x} + \frac{\partial D_y}{\partial y} + \frac{\partial D_z}{\partial z}\right)dx\,dy\,dz$$

Equating this to the charge, $\rho\,dx\,dy\,dz$ gives

$$\frac{\partial D_x}{\partial x} + \frac{\partial D_y}{\partial y} + \frac{\partial D_z}{\partial z} = \rho \tag{2.25}$$

This is Gauss's law in differential form. The left-hand side of equation (2.25) is called the *divergence of D*. It can be expressed more compactly by using the ∇ operator and written as $\nabla\cdot\boldsymbol{D}$. Hence

$$\nabla\cdot\boldsymbol{D} = \rho \tag{2.26}$$

This will be the form used when Maxwell's equations are written down in Chapter 3.

2.8 CAPACITANCE

The expression for the potential of a point charge, equation (2.3) or the potential of a pair of parallel plates, equation (2.18) indicates that the ratio of the potential and the charge, V/Q, depends only on the geometry and the dielectric permittivity. The inverse of this ratio is defined as the *capacitance*,

$$C = Q/V \tag{2.27}$$

It has units of Coulombs per volt (C/V). In SI units this is given the name Farad (F). One Farad is a large capacitance and practical values are usually in μF or pF. The geometric volume is called a *capacitor*. A capacitor is an energy storage device as can be seen be reference back to Section 2.3. Two common examples relevant to microwaves are a parallel plate capacitor and a coaxial cable.

Example 2.9 Capacitance of parallel plates

Find the capacitance of a parallel plate capacitor where the plates have sides 10 mm by 5 mm and the gap between the plates is filled with dielectric of relative permittivity 2.5 and thickness 1 mm.

The capacitance is given by inserting equation (2.18) into equation (2.27) as

$$C = \frac{\varepsilon_0 \varepsilon_r S}{d} \tag{2.28}$$

where S is the area of the plates.

Inserting the values, gives $C = (8.854 \times 10^{-12} \times 2.5 \times 10^{-2} \times 5 \times 10^{-3})/10^{-3} = 1.11 \times 10^{-12} = 1.11\,\text{pF}$. This shows that a high value of capacitance needs either a very narrow gap between the plates or a large area of the plates, or a dielectric with a high relative permittivity. Equation (2.28) neglects any field outside the edges of the plates, the fringing fields. If the plates are large and close together this will be a good approximation, otherwise some allowance will need to be made for fringing capacitance.

Example 2.10 Capacitance of a coaxial cable

Calculate the capacitance of a coaxial cable with inner diameter 1 mm and outer diameter 5 mm which is filled with polythene with relative permittivity 2.25.

The capacitance per unit length is given by inserting equation (2.23) into equation (2.27)

$$C = \frac{2\pi\varepsilon_0\varepsilon_r}{\ln(b/a)} \tag{2.29}$$

Inserting the values gives, $C = 7.78 \times 10^{-11} = 77.8\,\text{pF/m}$. This is the typical capacitance of an coaxial cable used at microwaves or RF.

2.9 CONDUCTORS AND ELECTRIC FIELDS

Materials can be divided into *insulators*, or *dielectrics* in which discrete current will not flow and *conductors* in which direct current flows relatively freely. There are two special cases of conductors, namely *semiconductors* and *superconductors*. Semiconductors have partial conductance which depends on the doping of the material. Superconductors have very low conductance. Most superconductors operate at temperatures near to absolute zero, but recently superconducters nearer to room temperature have been discovered.

Materials which are conductors are characterised by the *conductivity* which is a quantity depending only on the material properties. It is given the symbol σ and has SI units of Siemens per metre (S/m). The inverse of the conductivity is the *resistivity*

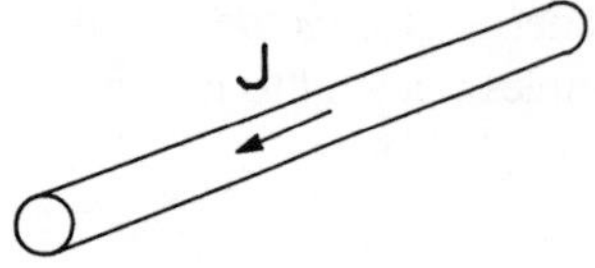

Figure 2.19. Current density flowing long a conducting rod

with units of Ohm metres (Ω m). The *resistance*, R, of a block of conductor of length l, and cross-section area S, is given by

$$R = \frac{l}{\sigma S} \Omega \tag{2.30}$$

A good conductor has a low resistance and a high conductivity. The conductivity is a function of temperature and varies nearly linearly with the absolute temperature. Thus all conductors have infinite conductivity at a temperature of absolute zero. Some typical conductivities are listed in the Appendix.

In electromagnetic wave theory and microwaves the concept of a perfect conductor is often used, which has an infinite conductivity. This is a good practical approximation because, as can be seen from the table, most metals have very high conductivities.

The electric field, $\boldsymbol{E}$, inside a conductor is related to the current density, $\boldsymbol{J}$, by

$$\boldsymbol{J} = \sigma \boldsymbol{E} \tag{2.31}$$

This is the vector form of Ohm's law. Notice that if the conductivity is infinite the electric field is zero, so there are no electric fields inside a perfect conductor.

Example 2.11 Ohm's law from field theory

Show that Ohm's law is a special case of equation (2.31) by considering a long conducting rod of length L and cross-section area S, carrying a current density $\boldsymbol{J}$, Figure 2.19.

The voltage difference between the ends is

$$V = \int \boldsymbol{E} \cdot d\boldsymbol{l} = \int_0^L \frac{J}{\sigma} \cdot d\boldsymbol{l} = \frac{JL}{\sigma}$$

For a uniform rod with a uniform current density, $J = I/S$. Also the resistance of the rod is $R = L/\sigma S$, so

$$V = IR$$

Starting with a relationship in field theory a relationship has been obtained that applies to circuit theory. It is generally true that circuit theory is a special case of field theory.

2.10 STATIC MAGNETIC FIELDS

The relations between the components of the static magnetic fields are analogous to those between static electric fields, but they are more complicated. This is because a

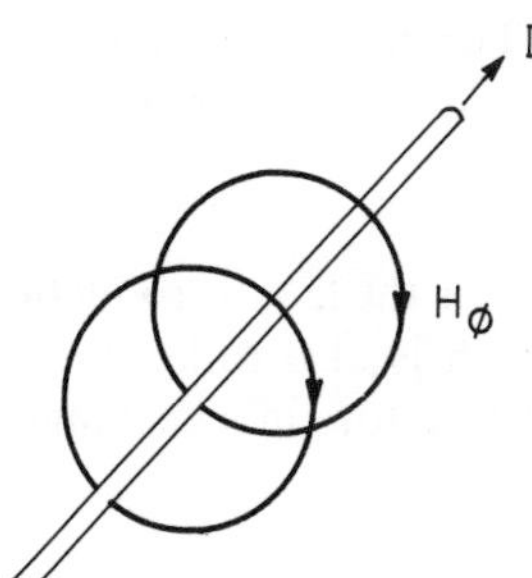

Figure 2.20. Magnetic field lines formed by a current in a wire

static magnetic field is generated by a current flowing *along* a wire, Figure 2.20, whereas a static electric field is generated by a charge at a *point*. The extra dimension implied by the flow of current along a wire makes the relationships more complex. Fortunately electromagnetic wave theory only needs an elementary picture of static magnetic fields.

The fundamental discovery of the relationship between the current in a wire, the magnetic field surrounding the wire and the force on the wire was made by Hans Christian Oersted (1777–1851) around 1820. He used a compass to observe that the compass needle always pointed perpendicular to the wire and perpendicular to a radial line from the wire. The needle aligns itself with the magnetic field so that the observation means that the magnetic field must form closed loops around the wire, as shown in Figure 2.20. The direction of the magnetic field is given by the *right-hand rule*, which is that if the current flows away from an observer, the direction of the magnetic field is clockwise, as viewed by the observer.

The elemental force $\mathrm{d}f$ produced by a small current element of length $\mathrm{d}l$ carrying a current I is such that, Figure 2.21,

$$\mathrm{d}f = I\,\mathrm{d}l\,B\sin\phi \tag{2.32}$$

B is the quantity describing the magnetic field. Since the magnetic field has direction as well as magnitude, the full description is a vector quantity called the *magnetic flux density*, $\boldsymbol{B}$. It has units of webers per square metre ($\mathrm{Wb/m^2}$) or teslas (T). ϕ is the angle between $I\,\mathrm{d}l$ and $\boldsymbol{B}$.

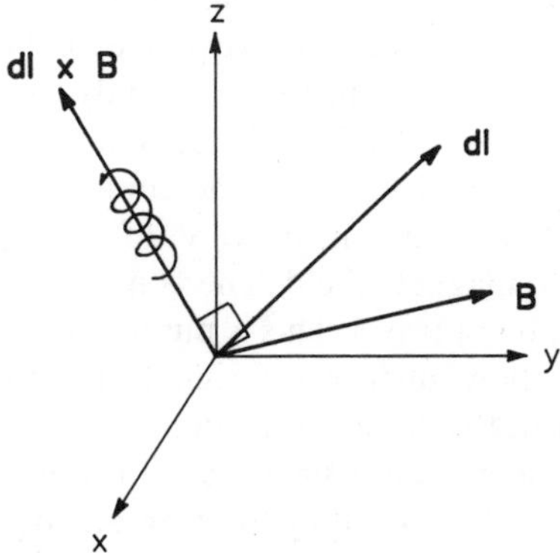

Figure 2.21. Rotational motion of an electron in a magnetic field

The right-hand side of equation (2.32) is the definition of a *vector product*, and the equation can be rewritten as

$$d\boldsymbol{f} = I d\boldsymbol{l} \times \boldsymbol{B} \tag{2.33}$$

This is the first introduction of the vector product so it is worth reviewing the operation, which is also called a cross product. If the definition in equation (2.32) is applied to unit vectors in a rectangular coordinate system, the results are

$$\hat{x} \times \hat{y} = \hat{z} \quad \hat{y} \times \hat{z} = \hat{x} \quad \hat{z} \times \hat{x} = \hat{y}$$
$$\hat{x} \times \hat{x} = \hat{y} \times \hat{y} = \hat{z} \times \hat{z} = 0$$

Notice that the vector product of two orthogonal unit vectors is a unit vector in the orthogonal direction to the two unit vectors. The actual direction is given by the right-hand rule. If two vectors $\boldsymbol{E}$ and $\boldsymbol{H}$ are vector multiplied together then the result, using the above definitions is

$$\boldsymbol{E} \times \boldsymbol{H} = (E_y H_z - E_z H_y)\hat{x} + (E_z H_x - E_x H_z)\hat{y} + (E_x H_y - E_y H_x)\hat{z}$$

The order of the terms can be remembered by noting that the $\hat{x}$ component contains only y and z terms of E and H. Alternatively the vector product can be written as a 3×3 determinant

$$\boldsymbol{E} \times \boldsymbol{H} = \begin{vmatrix} \hat{x} & \hat{y} & \hat{z} \\ E_x & E_y & E_z \\ H_x & H_y & H_z \end{vmatrix}$$

The magnetic flux density, $\boldsymbol{B}$, is the fundamental quantity in static magnetic fields, but it is useful to define a *magnetic field intensity*, $\boldsymbol{H}$ (usually called just the *magnetic field*) by

$$\boldsymbol{B} = \mu_0 \mu_r \boldsymbol{H} \tag{2.34}$$

This is analogous to the relation between electric flux density and electric field. $\boldsymbol{H}$ has the SI units of amps per metre. $\mu_0 \mu_r$ is the *permeability* and is a measure of the magnetisation of materials. μ_0 is the *absolute permeability* in a vacuum and has a value

$$\mu_0 = 4\pi \times 10^{-7}\ \text{H/m}.$$

μ_r is the *relative permeability*. The values for some materials are given in the Appendix. Non-ferrous metals and dielectrics have a relative permeability which is very nearly one.

As far as microwave engineers are concerned, the only significant material which has a non-unity relative permeability is ferrite. The high conductivity of the metals means that there is usually insignificant field inside the conductor. Ferrite is an anisotropic material which means that μ_r is a tensor with different permeabilities in different directions relative to a static magnetic field. The consequence is that the properties of waves depends on the direction of travel. This is put to good use to build non-reciprocal microwave devices which allow microwave signals to travel un-attenuated in one direction but absorbs them in the reverse direction.

In electromagnetic wave theory, it is usual to take the electric field $\boldsymbol{E}$ and the magnetic field $\boldsymbol{H}$ as the basic quantities because they form an analogous pair. As shown above, however, the fundamental quantities are actually $\boldsymbol{E}$ and $\boldsymbol{B}$. For this reason some textbooks take these two as the basic quantities.

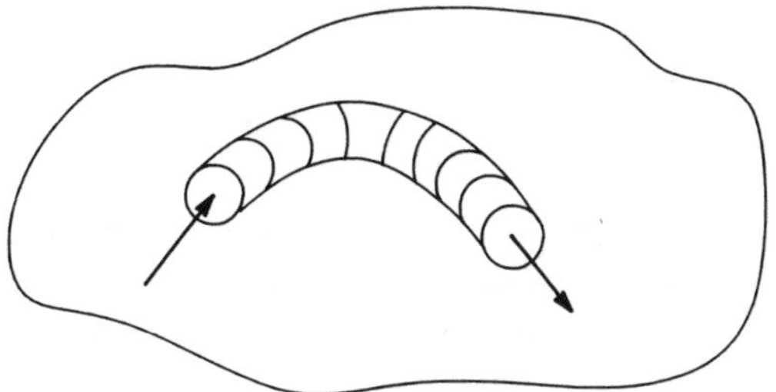

Figure 2.22. Tube of flux flowing through a volume

The magnetic flux is the magnetic flux density $\boldsymbol{B}$ times the area of a surface cut by the magnetic field. For a general surface this is

$$\psi = \int_S \boldsymbol{B} \cdot d\boldsymbol{s} \tag{2.35}$$

It was shown earlier in the chapter that lines of electric field start and end on charges. In contrast magnetic field lines always form closed loops. If a tube of magnetic flux is considered, as shown in Figure 2.22, a tube which enters a closed volume must come out again and therefore cut the closed surface twice. Hence the magnetic flux out of a volume must be zero, or

$$\int \boldsymbol{B} \cdot d\boldsymbol{s} = 0 \tag{2.36}$$

This is analogous to Gauss's law and by deduction the differential form can be written down by comparison as

$$\nabla \cdot \boldsymbol{B} = 0 \tag{2.37}$$

Both equations (2.36) and (2.37) are expressions describing the continuous nature of $\boldsymbol{B}$ and they show a fundamental difference between magnetic fields and electric fields. The electric field is generated by electric charges or sources, but there are no equivalent magnetic sources of charges. The magnetic field must have zero divergence everywhere and is generated by a current which flows in a conducting medium.

2.11 AMPÈRE'S LAW

In 1820, André-Marie Ampère (1775–1836) followed up the discovery of Oersted and used a series of experiments to demonstrate a relationship between the magnetic field intensity and a system of direct currents. Shortly afterwards Jean-Baptiste Biot (1774–1862) and Félix Savart (1791–1841) derived a law of force which quantified Ampère's experimental deductions. The law was originally expressed in terms of a system of current elements, but a more useful form for electromagnetic wave theory is here called Ampère's law. It is also called the *magnetic circuit law* and states that the line integral of the static magnetic field taken along a closed loop equals the current enclosed by the loop,

$$\int \boldsymbol{H}\cdot \boldsymbol{dl} = \int_{s} \boldsymbol{J}\cdot \boldsymbol{ds} \tag{2.38}$$

The right-hand side is the current enclosed. It has been written for the general case of a current density integrated over a surface. The law can be applied to simple problems to find the magnetic field created by the current in a system of wires.

Example 2.12 Magnetic field due to a wire

Find the magnetic field at a radius r, due to a wire carrying a current I.

This is the situation shown in Figure 2.20 and straightforward application of equation (2.38) gives

$$\int \boldsymbol{H}\cdot \mathrm{d}l = \int_{0}^{2\pi} H_{\phi}\, r\, \mathrm{d}\phi = H_{\phi} 2\pi r = I$$

Thus $H_{\phi} = I/2\pi r$ and the magnitude of the magnetic field is proportional to the current in the wire and is inversely proportional to distance from the wire. If the magnetic field is required at a radius which is inside the wire, the equation still applies, except that only a proportion of the current is inside the closed loop. I is replaced by $I(r^2/a^2)$ where a is the radius of the wire.

2.12 DIFFERENTIAL FORM OF AMPÈRE'S LAW

The process of obtaining the differential form of Ampère's law is analogous to the method used to obtain the differential form of Gauss's law. That involved the operation of divergence which was defined as the surface integral taken about an infinitesimal surface divided by the volume enclosed by the surface. For Ampère's law another vector operation, called the *curl* of a vector, is needed. This is defined as the line integral divided by the area enclosed by the closed loop. The curl is a vector operation because the surface area enclosed by a loop can have any orientation with respect to the loop.

To obtain the differential form of equation (2.38) in rectangular coordinates the equation is applied to a small rectangular loop in the yz plane through which a current density flows, Figure 2.23. This will give the x component and by analogy the other components can be found.

For small lengths, the right-hand side of equation (2.38) is

$$\boldsymbol{J}\cdot \boldsymbol{ds} = J_x\, \mathrm{d}y\, \mathrm{d}z \tag{2.39}$$

The left-hand side of equation (2.38) is obtained by summing around the four sides of the rectangle. Along the left side of the rectangle, $\boldsymbol{H}\cdot \boldsymbol{dl} = H_z\, \mathrm{d}z$, and along the right side of the rectangle the field will have changed by $\partial H_z/\partial y$ times the length $\mathrm{d}y$, that is

$$\boldsymbol{H}\cdot \boldsymbol{dl} = H_z\, \mathrm{d}z + \frac{\partial H_z}{\partial y}\, \mathrm{d}y\, dz$$

Adding up all the contributions from the four sides, starting at the bottom left-hand

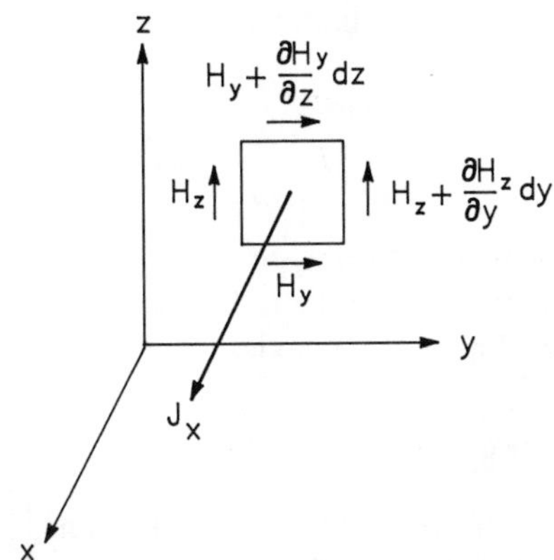

Figure 2.23. Differential form of Ampère's law

corner and moving clockwise gives

$$\boldsymbol{H}\cdot\boldsymbol{dl} = H_z\,\mathrm{d}z + \left(H_y + \frac{\partial H_y}{\partial z}\mathrm{d}z\right)\mathrm{d}y - \left(H_z + \frac{\partial H_z}{\partial y}\mathrm{d}y\right)\mathrm{d}z - H_y\,\mathrm{d}y$$

$$= \frac{\partial H_z}{\partial y} - \frac{\partial H_y}{\partial z} \tag{2.40}$$

The current flow in the y and z directions will produce similar equations, so Ampère's law in differential form is obtained by equating equations (2.39) and (2.40) for all three orthogonal components:

$$\left(\frac{\partial H_z}{\partial y} - \frac{\partial H_y}{\partial z}\right)\hat{\boldsymbol{x}} + \left(\frac{\partial H_x}{\partial z} - \frac{\partial H_z}{\partial x}\right)\hat{\boldsymbol{y}} + \left(\frac{\partial H_y}{\partial x} - \frac{\partial H_x}{\partial y}\right)\hat{\boldsymbol{z}} = J_x\hat{\boldsymbol{x}} + J_y\hat{\boldsymbol{y}} + J_z\hat{\boldsymbol{z}} \tag{2.41}$$

The left-hand side of equation (2.41) is defined as the curl of $\boldsymbol{H}$ or $\nabla \times \boldsymbol{H}$ so Ampère's law in vector differential form is

$$\nabla \times \boldsymbol{H} = \boldsymbol{J} \tag{2.42}$$

In component form, this can either be written out as (2.41) or as a determinant

$$\nabla \times \boldsymbol{H} = \begin{vmatrix} \hat{\boldsymbol{x}} & \hat{\boldsymbol{y}} & \hat{\boldsymbol{z}} \\ \dfrac{\partial}{\partial x} & \dfrac{\partial}{\partial y} & \dfrac{\partial}{\partial z} \\ H_x & H_y & H_z \end{vmatrix} \tag{2.43}$$

2.13 INDUCTANCE

Inductance is the magnetic analogy of capacitance and an inductor is a storage device for magnetic energy in the same way as a capacitor is a storage device for electrical energy. Inductance, L, is defined in a similar manner to capacitance as the ratio of the magnetic flux, ψ, to the current, I,

$$L = \frac{\psi}{I} \tag{2.44}$$

The SI units are Weber per metre (Wb/m) or Henrys (H). The magnetic flux is found from equation (2.35). The application of the inductance equation will be illustrated with a parallel plate transmission line and a coaxial transmission line.

Example 2.13 Inductance of parallel plates

Find the inductance of a parallel plate transmission line, Figure 2.24, with plates of width w and spacing apart d. Assume that the field is uniform within the conductors so that fringing fields are neglected.

The magnetic field between the plates is given by Ampère's law, equation (2.38) with current I flowing in a conductor and the line integral taken around one of the conductors.

$$I = \int \boldsymbol{H} \cdot \boldsymbol{dl} = H_x w$$

The surface cut by the flux lines is an imaginary rectangle along the transmission line of height d and of unit length, as shown in Figure 2.24. The flux is given by equation (2.35) as $\psi = \boldsymbol{B}(d \times 1) = \mu_0 H_x d$. Then the inductance, from equation (2.44), is

$$L = \mu_0 \frac{d}{w} \text{ H/m} \tag{2.45}$$

This assumes that the material filling the plates has $\mu_r = 1$. The inductance can be increased by increasing the spacing of the plates or by reducing the width of the plates.

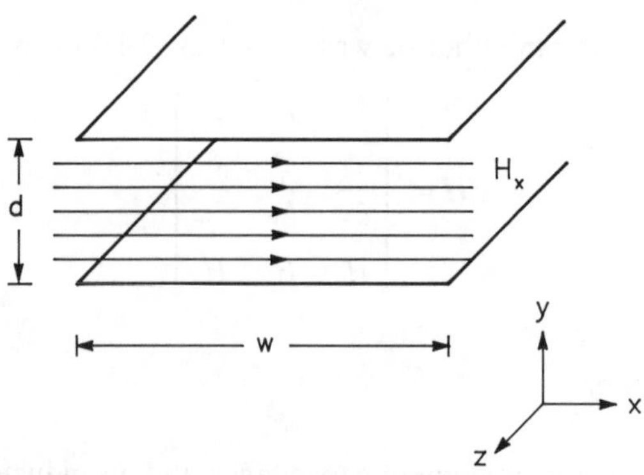

Figure 2.24. Parallel plate transmission line

Example 2.14 Inductance of a coaxial cable

Find the inductance of the coaxial cable used for Example 2.10 with a current I flowing along the inner conductor and back along the outer cylinder.

The magnetic field between the cylinders is the same as that due to a long wire, so at a radius r $(a < r < b)$

$$H_\phi = \frac{I}{2\pi r}$$

The magnetic flux is given in the same manner as the previous example. The surface cut by the flux is of unit length and goes from $r = a$ to $r = b$.

$$\psi = \int_a^b \mu_0 H_\phi \, \mathrm{d}r = \frac{\mu_0 I}{2\pi} \ln\left(\frac{b}{a}\right)$$

The inductance per unit length is thus

$$L = \frac{\mu_0}{2\pi} \ln\left(\frac{b}{a}\right) = 2 \times 10^{-7} \ln\left(\frac{b}{a}\right) \tag{2.46}$$

The inductance value for $a = 0.5\,\text{mm}$ and $b = 2.5\,\text{mm}$ is $0.32\,\mu\text{H/m}$.

PROBLEMS

1. Two small positively charged conducting spheres with charges of 5×10^{-7}C and -1×10^{-7}C, respectively are 30 cm apart. What is the force between them? What will the force be if they are brought into contact and then separated by the same distance as before.

2. A thunderstorm contains electric charges of $+30$ C at a height of 10 km and -30 C at a height of 5 km. If the charges can be considered to be at a point, find the force of attraction between the two charges.

3. A helium atom has two protons with a charge of 1.6×10^{-19} C, separated by a distance of 2×10^{-15} m. Find the force of repulsion between the two protons.

4. Two charges of value $+q$ are placed at opposite corners of rectangle of side 2 m by 4 m and two more charges of value $-2q$ are placed at the other two corners of the rectangle. Find the force on one of the $+q$ charges.

5. Find the electric field at the surface of a uranium nucleus with a radius of 7.5×10^{-15} m and with a charge of -1.5×10^{-17} C, situated at the centre of the nucleus.

6. An aircraft at an altitude of 7 km flies through the thunderstorm described in Problem 2. Find the electric field at the aircraft when it is 10 km, 5 km and 0 km from the thunderstorm.

7. A charge of 1 C moves in an electric field of 10 V/m directed from the point (0, 0) to the point (10, 5). The path is constrained to either the x or y directions. Find the work done and show that it depends only on the end points.

8. A sphere of radius R carries a total charge Q uniformly distributed over the surface of the sphere. Find the electric potential inside and outside the sphere.

9. An electrostatic dipole is formed by two charges of equal magnitude, q, but opposite sign. If the two charges are at $+x$ and $-x$, show that the potential at a point r, ϕ in the xy plane $(r \gg x)$ of a cylindrical coordinate system is $V(r, \phi) = qx\cos\phi/(2\pi\varepsilon_0 r^2)$, where ϕ is the angle with respect to the x axis.

10. Sketch the equipotential lines for the electrostatic dipole of the previous problem.

11. Find the electric field, $\boldsymbol{E}$, for a potential of (a) $V = 5y^2 + 10x$ V and (b) $V = 2\sqrt{y}$ V.

12. Use Gauss's law to deduce the electric field due to a point charge.

13. Find the electric field using Gauss's law for a large flat sheet with a uniform surface charge density of ρ C/m^2.

14. A rod of radius 10 mm and $\varepsilon_r = 1$ is charged to 10 μC. Find the electric field against radius for two cases: (a) if the charge is distributed uniformly throughout the volume and (b) if the charge is distributed uniformly over the surface of the cylinder.

15. A thin glass rod of length l, is placed along the x axis with one end at the origin. It is electrified uniformly along its length with a total charge Q. Find the potential and electric field at any point on the x axis beyond the end of the rod.

16. A charge q_1 is placed on the inner conductor of a coaxial line and a charge $-q_2$ is placed on the outer conducting cylinder. Find and plot the electric field for $0 < r < \infty$.

17. A metal sphere is to be charged to 10^6 V. It is well removed from other conductors and from Earth. Air breaks down at approximately 3×10^6 V/m. What size of sphere would be suitable?

18. An electric field is given by $\boldsymbol{E} = x\hat{\boldsymbol{x}} + (x + y^2)\hat{\boldsymbol{y}}$. Find the charge density which generated the field in a dielectric of $\varepsilon_r = 2$.

19. Find the capacitance in a coaxial cable with inner diameter 3 mm and outer diameter 10 mm filled with (a) air and (b) polyethelene with $\varepsilon_r = 2.25$. If air breaks down at 3×10^6 V/m and polyethelene at 20×10^6 V/m, find the maximum voltages to which the cables can be charged.

20. Design a parallel plate capacitor to give a capacitance of 100 pF using a sheet of paper with $\varepsilon_r = 3$ and thickness 100 μm between the plates.

21. Show that the capacitance of a spherical capacitor consisting of two concentric spheres of radii a and b $(a < b)$, and filled with a dielectric of relative permittivity ε_r is $C = 4\pi\varepsilon_0\varepsilon_r/(1/a - 1/b)$.

22. Calculate the capacitance of the Earth, assuming that it can be treated as a conducting sphere of radius 6400 km.

23. A parallel plate capacitor contains a sandwich of two dielectrics between the plates with relative permittivities ε_1 and ε_2 and thicknesses d_1 and d_2. Find the total capacitance and hence deduce the relationship for capacitances in series.

24. Calculate the length of a rod of diameter 5 mm needed to give a resistance of 1 Ω if the rod is made of (a) copper, (b) stainless steel, (c) carbon and (d) pure silicon. (Conductivity values are listed in the Appendix.)

25. Explain physically the difference between Gauss's law for electrostatics and the equivalent law for static magnetic fields.

26. Sketch the radial distribution of magnetic field both inside and outside a long wire carrying a current I.

27. Find the magnetic field due to a metal sheet in the xz plane carrying a current $\boldsymbol{J}_s\hat{\boldsymbol{z}}$.

28. A toroid consists of a ring with a coil on N turns of wire wound around it and carrying a current I. Show that the magnetic field is $H_\phi = NI/(2\pi r)$ where r is the radius of the ring and ϕ is the azimuthal direction with the origin at the centre of the ring.

29. A magnetic field has $\boldsymbol{H} = (3 + y^2)\hat{\boldsymbol{x}} + (x + y^2)\hat{\boldsymbol{y}} - 4\hat{\boldsymbol{z}}$. Use the differential form of Ampère's law to find the current density producing the magnetic field.

30. Find the inductance of the coaxial cables described in Problem 19.

3 Maxwell's equations and electromagnetic waves

3.1 INTRODUCTION

This chapter studies the fundamentals of electromagnetic wave theory. The physical basis of electromagnetic wave theory are time varying electric and magnetic fields. The variations with time cause the electric and magnetic fields to be intimately coupled together. In the static theory outlined in Chapter 2, electric fields could exist independently of magnetic fields and vice versa. When an electric field varies with time, the magnetic field is automatically excited, and vice versa. The variation with time can take any form, but the most common one, and the only case considered for microwaves and optical waves in this book, is when the variations are sinusoidal waves. The theoretical basis of electromagnetic wave theory is a set of equations known as *Maxwell's* equations. These are based on the laws of static electric and magnetic fields plus two new factors. These are the important contribution of Faraday's law of induction and an additional time dependent term. The theory in this chapter is quite general so there are relatively few practical examples. These come in later chapters when the general equations are simplified for specific situations in microwaves and optical transmission.

The chapter starts by describing Faraday's law which is the first electromagnetic law involving time dependent terms. The historical development of electromagnetic wave theory is fascinating and will be briefly described. Maxwell's equations are stated in various forms which are useful for understanding the contribution that the equations make to electromagnetic wave theory. The physical meaning of the displacement current introduced by Maxwell is examined. The electromagnetic wave equations which will be used in later chapters are developed. Finally the power flow of electromagnetic waves is described and analysed in terms of its component parts.

3.2 FRADAY'S LAW OF ELECTROMAGNETIC INDUCTION

The *father* of time varying electromagnetic fields can be considered to be Michael Faraday (1791–1867) because he was the first person to discover that a time varying magnetic field will produce an electric current in a circuit which is immersed in a magnetic field. Faraday was a remarkable scientist whose formal education stopped at the age of thirteen, but he had a continual thirst for knowledge and taught himself by studying as many books as possible. At the age of twenty he was taken on by Sir Humphrey Davy as an assistant at the Royal Institution in London. There he established himself both as a premier chemist and as a pioneer in electromagnetism. He was an experimentalist who kept meticulous accounts of all his experiments and his notebooks are model examples of laboratory record keeping. Oersted, Biot and Savart and Ampère had shown that a steady electric current in a wire produced a magnetic field. Faraday wanted to demonstrate the converse effect. He tried a number of times between 1824 and 1830 without success, but in 1831 he made the breakthrough that a *changing* magnetic field was necessary to produce a current in a circuit. This he did by moving a bar magnet near the end of a cylindrical coil of wire with a galvanometer attached to the ends. The galvanometer needle moved as he moved the bar magnet. He recorded that the important part of the experiment was the moving of the magnet, or in other words that the magnetic field must be varying with time. The voltage induced at the ends of the coil is called an *electromotive force or emf.* Faraday's law of electromagnetic induction may be stated mathematically as

$$emf = -\frac{d\psi}{dt} \tag{3.1}$$

where ψ is the magnetic flux. This equation can be used directly to find the voltage induced in the secondary coil of a transformer by calculating the flux due to a single turn of the coil. The total flux is the number of turns times the flux due to one turn.

The negative sign in equation (3.1) is present in order to indicate that the direction of the electromotive force is such as to oppose the change in magnetic flux. This observation was made by Heinrich Lenz (1804–1865) and so that part of the equation is sometimes known as Lenz's law.

In a general situation the magnetic flux is the flux density integrated over a surface which has the path as its boundary,

$$\psi = \int_s \boldsymbol{B} \cdot \boldsymbol{ds} \tag{3.2}$$

but in most practical problems the geometry is simple and then $\psi = \boldsymbol{B} \times$ Area. An illustration of its use is provided by the following example.

Example 3.1 Emf due to a loop in a magnetic field

Find the emf induced in a circular coil of radius r which is normal to a sinusoidally varying magnetic field given by $B_z = B_0 \sin \omega t$ ($\omega = 2\pi \times$ frequency), Figure 3.1.

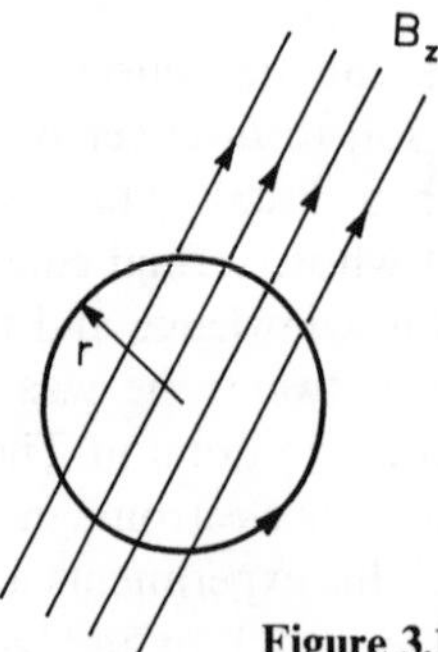

Figure 3.1. Circular coil in a magnetic field

In this case the simple geometry means that equation (3.1) can be used. The flux, ψ, is the flux density, B_z, times the area of the coil (πr^2). Therefore

$$emf = \frac{dB_z}{dt}\pi r^2 = \pi r^2 \omega B_0 \cos \omega t$$

The emf induced in the coil varies as $\cos \omega t$ which is $\sin(\omega t + 90°)$. This shows that the emf is 90° out of phase with respect to the magnetic field. The magnitude of the emf is proportional to the frequency of the magnetic field.

Faraday's law shows that there three practical cases which link an electric current and a magnetic field.

(a) A conducting circuit is moved through a steady magnetic field.

(b) A stationary circuit is immersed in a changing magnetic field.

(c) A combination of (i) and (ii).

Case (a) is represented by a d.c. motor, case (c) is represented by an a.c. motor. Case (b) is the most common and includes the transformer and the detection of electromagnetic waves.

The electromotive force is a voltage and any voltage has an associated electric field, as shown in Section 2.4. This means that equation (3.1) can generalized to any path in space by replacing the *emf* by the line integral of the electric field, using equation (2.7),

$$emf = \int \boldsymbol{E} \cdot d\boldsymbol{l} \tag{3.3}$$

For static electric fields the line integral is independent of the path and equal to the potential difference between the end points. This is not true for time varying fields as demonstrated by Faraday's law.

Equating equations (3.1) and (3.3) and inserting equation (3.2) gives Faraday's law in its general integral form

$$\int \boldsymbol{E} \cdot d\boldsymbol{l} = -\frac{d}{dt}\int_s \boldsymbol{B} \cdot d\boldsymbol{s} \tag{3.4}$$

This version of Faraday's law shows that an electric field can be created by a continuous changing magnetic field. In electrostatics, the electric field is created by unique electric charges which mean that electric field lines start and finish on charges. In electromagnetism, there are no magnetic charges, so the magnetic field lines must form closed loops. This is shown in Figure 3.2.

In order to incorporate Faraday's law into Maxwell's equations it is more use to express it in differential form. This may be done by noting that the form of equation (3.4) is the same as that of Ampères law, equation (2.38), so by direct analogy Faraday's law may be written down in differential form as

$$\nabla \times \boldsymbol{E} = -\frac{d\boldsymbol{B}}{dt} \tag{3.5}$$

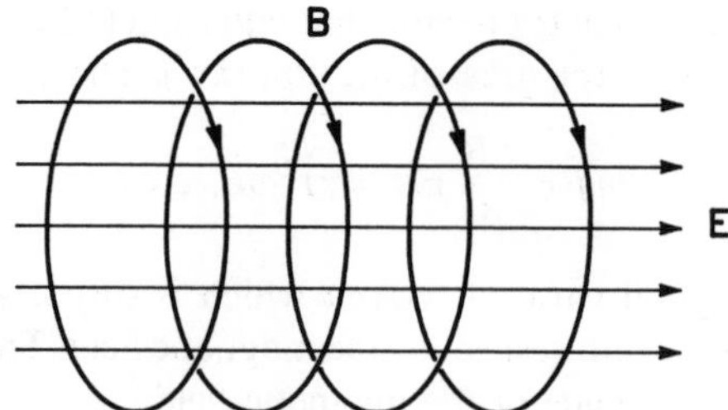

Figure 3.2. Magnetic field forms closed loops

3.3 CONTINUITY OF CHARGE AND CURRENT

A relationship is needed between time varying charges and currents. This is derived from the fact that the current flowing out of a closed volume must be matched by a decrease in the charge inside the volume, Figure 3.3. Conversely if current flows into a volume the charge will be increased. This can be expressed as current equals rate of decrease of charge, which is a definition of electric current

$$I = -\frac{dQ}{dt} \tag{3.6}$$

Generalizing the equation by replacing the left-hand side with the current density integrated over a surface and replacing the charge on the right-hand side by the charge per cubic metre, ρ, integrated over the volume containing the charge,

$$\int_s \boldsymbol{J}\cdot d\boldsymbol{s} = -\frac{d}{dt}\int_v \rho \, dv \tag{3.7}$$

This gives the continuity relationship between the charge in a volume region and the current density flowing out of the region. The differential form can be written down by

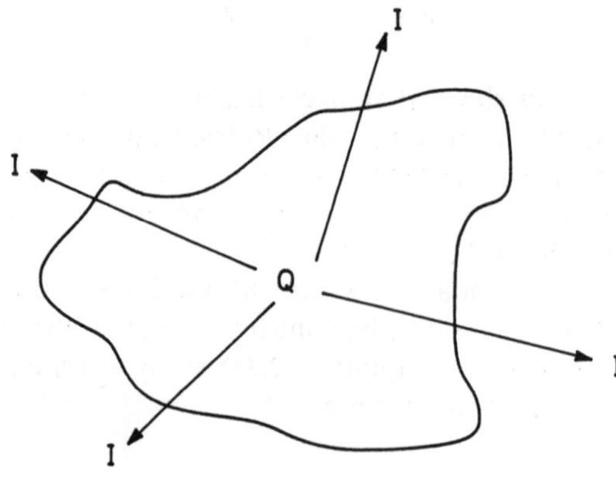

Figure 3.3 Current flowing out of a closed volume

noting that equation (3.7) is similar to Gauss's law equations (2.17) and (2.26)

$$\nabla \cdot \boldsymbol{J} = -\frac{\mathrm{d}\rho}{\mathrm{d}t} \tag{3.8}$$

This is called the *continuity equation* of current and charge. The equation clearly indicates that a current density is only created when there are time varying charges. Static charges do not produce currents.

3.4 MAXWELL'S EQUATIONS

The discoveries of Faraday were done experimentally. It was James Clerk Maxwell (1831–1879) who provided a firm theoretical basis for Faraday's law and also demonstrated the unity of the electric and magnetic field laws which had been discovered independently. Maxwell's career was in contrast to Faraday. He was born to a relatively well-off Scottish family and was educated at Edinburgh University and Cambridge University as a mathematical physicist. He was a theoretician not an experimentalist. At the age of twenty four he started work on expressing Faraday's ideas in mathematical form and this led him to study all the previous work on electric and magnetic fields.

Over the period from 1850 to 1861 he demonstrated the unity of electric and magnetic fields when time varied and postulated that light was just another form of electromagnetic wave. In doing this he challenged the established view of electricity and magnetism which depended on the concept of action at a distance. This stated that if a force was created at one point, the effects would be felt instantly at another point due to direct action of that force. Light was not associated with electric and magnetic fields, partly because the light could be seen and appreciated whereas electric and magnetic fields were unseen and had arisen from static conditions. Maxwell had to throw away the prevailing view of electricity and magnetism and consequently he had to be certain in his own mind that he had discovered an important fact. Eventually in 1861, he was confident enough to write:

> We can scarcely avoid the inference that *light consists in the transverse undulations of the same medium which is that cause of electric and magnetic phenomena*

Maxwell thus laid the foundations for electromagnetic wave theory and consequently for all radio, microwaves and optics. His work is printed in a number of papers and in a book *A Treatise of Electricity and Magnetism* published in 1864. This is not always easy to follow partly because of its detailed mathematical reasoning and partly because he did not have the benefit of hindsight which has enabled later workers to express his equations in a neat and simplified manner.

The *Maxwell* equations used in this book are in the now universally recognized form which was developed by Heinrich Hertz (1857–1894) and Oliver Heaviside (1850–1925). There are four basic equations from which all other relationships can be derived. Maxwell, however, actually developed extra equations which can be deduced from the four basic equations. This does not detract from the supreme importance of Maxwell's work as instanced by the fact that the four equations are still the starting point for all theoretical studies of electromagnetic wave theory and hence microwave and optical transmission.

When Maxwell started studying electromagnetic theory he had available the following four laws describing the behaviour of electric and magnetic fields:

(1) Faraday's law of electromagnetic induction, equation (3.5)

$$\nabla \times \boldsymbol{E} = -\frac{\mathrm{d}\boldsymbol{B}}{\mathrm{d}t} \tag{3.9}$$

(2) Ampère's law of static magnetic fields, equation (2.42),

$$\nabla \times \boldsymbol{H} = \boldsymbol{J} \tag{3.10}$$

(3) Gauss's law of static electric fields, equation (2.26),

$$\nabla \cdot \boldsymbol{D} = \rho \tag{3.11}$$

(4) The conservation property of magnetic fields, equation (2.37)

$$\nabla \cdot \boldsymbol{B} = 0 \tag{3.12}$$

Only equation (3.9) contains a time-dependent term. If these equations are correct for electromagnetic fields, they should be self-consistent. Maxwell found that they were not consistent and postulated an extra term which needs to be added to equation (3.10). The inconsistency can be shown as follows. An elimination between equations (3.10) and (3.11) should lead to the continuity equation, equation (3.8). Taking the divergence ($\nabla\cdot$) of equation (3.10) gives,

$$\nabla \cdot (\nabla \times \boldsymbol{H}) = \nabla \cdot \boldsymbol{J} \tag{3.13}$$

It is, however, a property of any vector, say $\boldsymbol{A}$, that $\nabla \cdot (\nabla \times \boldsymbol{A}) \equiv 0$, (this can be proved by expanding with components), therefore

$$\nabla \cdot (\nabla \times \boldsymbol{H}) = 0 \tag{3.14}$$

Equations (3.13) and (3.14) are inconsistent and it can be concluded that equations (3.10) is incorrect for some situations. Maxwell postulated that there ought to be an additional term in the equation which became zero for the static magnetic fields situation. This would be provided by a term which contained the derivative of time. The additional term is $\mathrm{d}\boldsymbol{D}/\mathrm{d}t$ so that equation (3.10) becomes

$$\nabla \times \boldsymbol{H} = \boldsymbol{J} + \frac{\mathrm{d}\boldsymbol{D}}{\mathrm{d}t} \tag{3.15}$$

Now taking the divergence of this equation

$$\nabla \cdot (\nabla \times \boldsymbol{H}) = \nabla \cdot \boldsymbol{J} + \frac{\mathrm{d}}{\mathrm{d}t}(\nabla \cdot \boldsymbol{D}) \tag{3.16}$$

Inserting equation (3.11) gives

$$\nabla \cdot (\nabla \times \boldsymbol{H}) = \nabla \cdot \boldsymbol{J} + \frac{\mathrm{d}\rho}{\mathrm{d}t} \tag{3.17}$$

The right-hand side is now the continuity equation and both sides of equation (3.17) are zero. Hence all the equations are consistent.

The modified set of equations are universally called *Maxwell's equations*:

$$\nabla \times \boldsymbol{E} = -\frac{\mathrm{d}\boldsymbol{B}}{\mathrm{d}t} \tag{3.18}$$

$$\nabla \times \boldsymbol{H} = \boldsymbol{J} + \frac{\mathrm{d}\boldsymbol{D}}{\mathrm{d}t} \tag{3.19}$$

$$\nabla \cdot \boldsymbol{D} = \rho \tag{3.20}$$

$$\nabla \cdot \boldsymbol{B} = 0 \tag{3.21}$$

They are the most important set of equations in microwave and optical wave theory. Physically they state that a time rate of change of an electric field produces a magnetic field and vice versa. Thus in any situation where the fields change with time, both electric and magnetic fields must exist together. There are three constituent relations which need to be used in conjunction with Maxwell's equations. These were derived in Chapter 2 and incorporate the electric and magnetic properties of the medium containing the fields. They are

$$\boldsymbol{D} = \varepsilon_0 \varepsilon_\mathrm{r} \boldsymbol{E} \tag{3.22}$$

$$\boldsymbol{B} = \mu_0 \mu_\mathrm{r} \boldsymbol{H} \tag{3.23}$$

$$\boldsymbol{J} = \sigma \boldsymbol{E} \tag{3.24}$$

All physical problems can be solved using equations (3.18) to (3.24) as a starting point, however, to solve them exactly is often unnecessary, because for particular problems simplifications and approximations can be made. This will be seen to be the case for the wave transmission described in later chapters.

3.5 MAXWELL'S EQUATIONS FOR TIME HARMONIC WAVES

Most situations in microwave and optical work are concerned with fields which are periodic with time. The time varying fields are described by sinusoidally varying fields or *waves*, for example

$$\boldsymbol{E} = \boldsymbol{E}_0\,\mathrm{e}^{\mathrm{j}\omega t} \tag{3.25}$$

This situation leads to a simplification of Maxwell's equations because the exponential factor is not changed by differentiation with time, namely

$$\frac{\mathrm{d}\boldsymbol{E}}{\mathrm{d}t} = \mathrm{j}\omega\,\boldsymbol{E}_0\,\mathrm{e}^{\mathrm{j}\omega t} = \mathrm{j}\omega\,\boldsymbol{E} \tag{3.26}$$

The $\mathrm{e}^{\mathrm{j}\omega t}$ can be cancelled from all equations and $\mathrm{d}/\mathrm{d}t$ replaced by $\mathrm{j}\omega$. Equations (3.18) to

(3.21) become

$$\nabla \times \boldsymbol{E} = -\mathrm{j}\omega \boldsymbol{B} \tag{3.27}$$

$$\nabla \times \boldsymbol{H} = \boldsymbol{J} + \mathrm{j}\omega \boldsymbol{D} \tag{3.28}$$

$$\nabla \cdot \boldsymbol{D} = \rho \tag{3.29}$$

$$\nabla \cdot \boldsymbol{B} = 0 \tag{3.30}$$

The symbols used for the electric and magnetic field in equations (3.27) to (3.30) represent the complex multiplier of $e^{j\omega t}$. This is different from the symbols used in equations (3.18) to (3.21) where they represent the instantaneous values of the vectors. In practice there is no need for confusion because any field can always be represented by time periodic fields. The real part of $\boldsymbol{E}e^{j\omega t}$ is $\boldsymbol{E}\cos(\omega t)$ and the imaginary part is $\boldsymbol{E}\sin(\omega t)$.

3.6 MAXWELL'S EQUATIONS IN INTEGRAL FORM

To complete the description of Maxwell's equation the integral form of the equations will be written down. These are simply the original form of the electric and magnetic field laws as stated in Chapter 2 with the addition of Maxwell's extra term. They can be verified by the same elimination procedure used in Section 3.4.

$$\int \boldsymbol{E}\cdot d\boldsymbol{l} = -\frac{\mathrm{d}}{\mathrm{d}t}\int \boldsymbol{B}\cdot d\boldsymbol{s} \tag{3.31}$$

$$\int \boldsymbol{H}\cdot d\boldsymbol{l} = \int \boldsymbol{J}\cdot d\boldsymbol{s} + \frac{\mathrm{d}}{\mathrm{d}t}\int \boldsymbol{D}\cdot d\boldsymbol{s} \tag{3.32}$$

$$\int \boldsymbol{D}\cdot d\boldsymbol{s} = \int \rho\, \mathrm{d}v \tag{3.33}$$

$$\int \boldsymbol{B}\cdot d\boldsymbol{s} = 0 \tag{3.34}$$

The integral form are not often used, except in cases where the specific field behaviour is required. An example is given in the next section.

3.7 DISPLACEMENT CURRENT DENSITY

The extra term added to the static field equations to form Maxwell's equstions, $(\mathrm{d}\boldsymbol{D}/\mathrm{d}t)$, is called the *displacement current density*, $\boldsymbol{J}_{\mathrm{d}}$. The other term on the right-hand side of equations (3.19) or (3.28) or (3.32) is called the *conduction current density*, $\boldsymbol{J}_{\mathrm{c}}$ thus

$$\nabla \times \boldsymbol{H} = \boldsymbol{J}_{\mathrm{c}} + \boldsymbol{J}_{\mathrm{d}} \tag{3.35}$$

An understanding of the physical significance of the displacement current density can be obtained by examining the physical meaning and typical values. The displacement current density is zero when there are no time dependent terms, which is the static

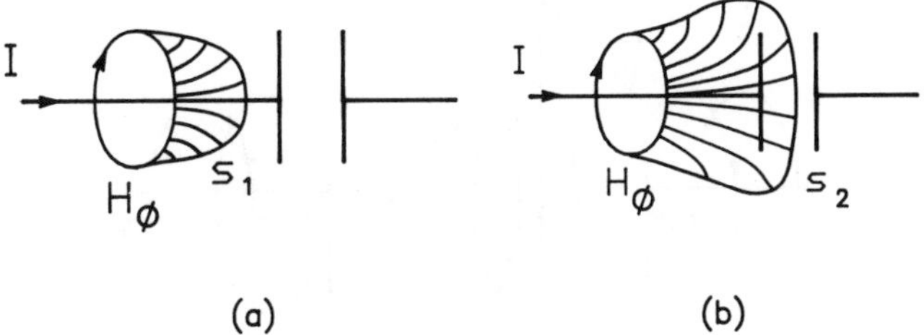

Figure 3.4. Capacitor in a circuit illustrating the need for displacement current. (a) Surface cuts wire. (b) Surface cuts gap between plates

situation. It must be present in time varying situations. This can be shown physically by considering the example of a capacitor in a circuit, Figure 3.4. A current I flowing in the circuit produces loops of magnetic field. Two surfaces, s_1 and s_2 are drawn with the loop as a boundary, surface s_1 cuts the wire and surface s_2 passes through the capacitor, then applying the integral form of Ampère's law (equation (3.32)) without the last term to surface s_1 gives

$$\int \boldsymbol{H}\cdot \boldsymbol{dl} = \int \boldsymbol{J}\cdot \boldsymbol{ds} = I$$

If applied to surface s_2, however, it gives

$$\int \boldsymbol{H}\cdot \boldsymbol{dl} = \int \boldsymbol{J}\cdot \boldsymbol{ds} = 0$$

This is because there is no conduction current in the air between the capacitor plates. Adding the displacement current term gives

$$\int \boldsymbol{H}\cdot \boldsymbol{dl} = \int \boldsymbol{J}\cdot \boldsymbol{ds} + \frac{\mathrm{d}}{\mathrm{d}t}\int \boldsymbol{D}\cdot \boldsymbol{ds} = \frac{\mathrm{d}}{\mathrm{d}t}(DS)$$

where S is the area of the plates of the capacitor. The right-hand side can be shown to be equal to the current by replacing $\boldsymbol{D}$ with $\varepsilon_0 \boldsymbol{E}$ and $\boldsymbol{E}$ with V/d, where d is the separation distance between the plates

$$\frac{\mathrm{d}}{\mathrm{d}t}(DS) = \frac{\mathrm{d}}{\mathrm{d}t}(\varepsilon_0 ES) = \varepsilon_0 \frac{S}{d}\frac{\mathrm{d}}{\mathrm{d}t}(dE) = C\frac{\mathrm{d}V}{\mathrm{d}t} = I$$

The displacement current thus accounts for the electromagnetic fields in the air gap of the capacitor whereas the conduction current accounts for the electromagnetic fields around the wire. The fields are shown in Figure 3.5.

The displacement current is the term in Maxwell's equations which gives rise to *radiation*. This can be seen by considering the case of two loops in free space with a source connected to one loop and a detector connected to the second loop, Figure 3.6. The signal induced in the second loop comes partly from the conduction current and partly from the displacement current. The conduction current part represent mutual inductance coupling between the loops and dies away very rapidly as the loops are moved apart. The displacement current part of the detected signal is due to radiation of electromagnetic waves from the first loop.

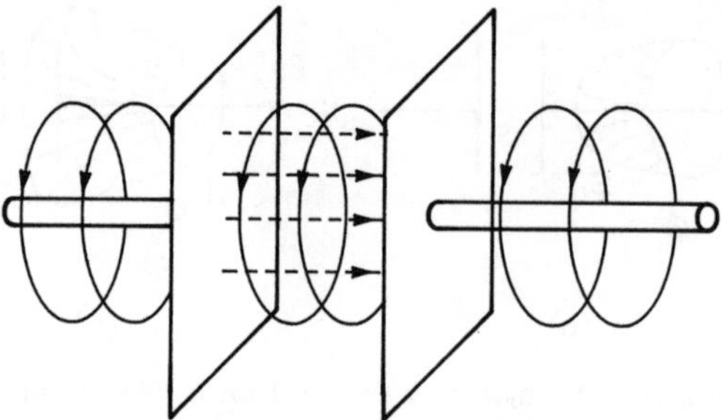

Figure 3.5. Electric and magnetic fields for a capacitor in a circuit. ——— magnetic field, - - - electric field

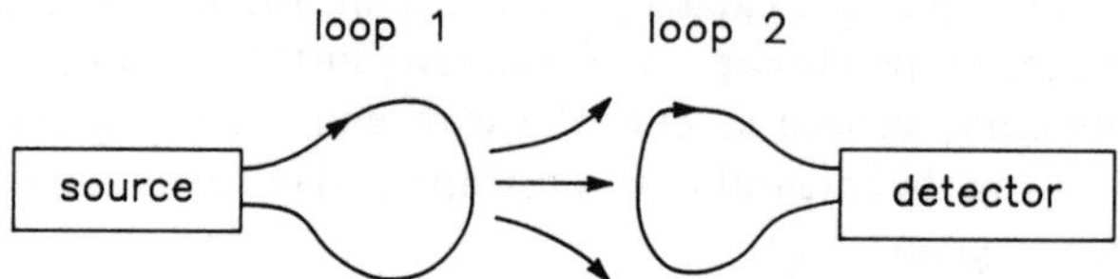

Figure 3.6. Radiation from loop 1 detected by loop 2

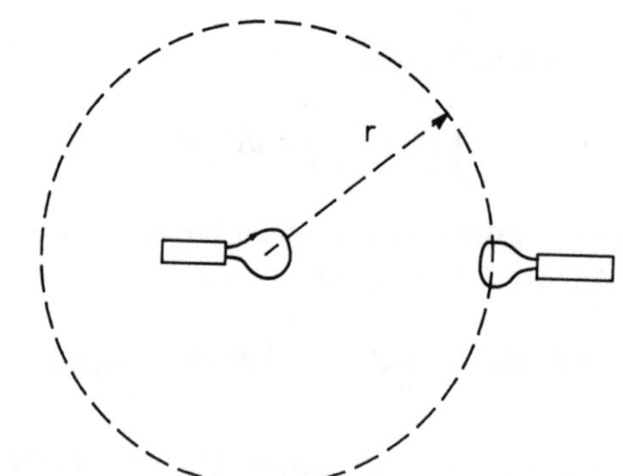

Figure 3.7. Loops separated by long distances

If the two loops are separated by a large distance, the radiation spreads outwards from the transmitting loop as if it were at the centre of a sphere, Figure 3.7. With the receiving loop on the imaginary sphere, the signal detected is proportional to the power on the surface of the sphere. The surface area of a sphere is $4\pi r^2$ where r is the radius of the sphere, so the power density at the receiver is

$$\text{Received power density} = \frac{\text{Power radiated}}{4\pi r^2}$$

The power decays thus as the square of the distance apart of the loops. The electric field is proportional to the square root of the power so the received electric field is inversely proportional to the distance apart of the loops. All free-space radio waves, microwaves and optical waves decay away from the source according to these relationships.

It is important to emphasise that the displacement current term is only present when the fields vary with time. There is no radiation unless the transmitting fields are varying with time. In practice this means time periodic fields which can be at any frequency in the electromagnetic spectrum from very low frequencies to optical frequencies.

The general name for the loops are *antennas*. The analysis and design of antennas varies according to the frequency and application desired. Antenna theory and design is a major area of study in its own right. The signal received by an antenna are dependent on the characteristics of the medium between the transmitter and receiver. These are called the *propagation* characteristics and depend on the frequency of use and the atmospheric and ionospheric properties. The study of the radio wave propagation is another major area of study. An introduction to antennas is given in Chapter 14.

It is instructive to quantify the relative magnitudes of the conduction and displacement current parts of Maxwell's equations for propagation in different media. Assuming time periodic fields, the displacement current density, from equation (3.28) is $\boldsymbol{J}_d = j\omega\boldsymbol{D}$. Incorporating equation (3.22) and the relationship between current density and electric field ($\boldsymbol{J}_c = \sigma\boldsymbol{E}$) this becomes

$$\boldsymbol{J}_d = j\omega\varepsilon_0\varepsilon_r\boldsymbol{E} = \frac{j\omega\varepsilon_0\varepsilon_r}{\sigma}\boldsymbol{J}_c \tag{3.36}$$

The ratio of displacement current to conduction current is independent of the fields. It is proportional to frequency so that it is likely that it will only be significant for high frequencies. This turns out to be true and at low frequencies the displacement current can often be ignored. Conversely at extremely high frequencies the conduction current can often be ignored. A few examples will illustrate this deduction.

Example 3.2 Displacement current for waves in free space

Find the displacement current density for radio waves with an electric field of 10 μV/m at 1 MHz and for a laser beam with an electric field of 30 kV/m at 10^{15} Hz.

In free-space there is no conduction current so the left part of equation (3.36) applies and $J_d = 5.56 \times 10^{-11} fE$ (A/m^2) where f is the frequency in Hz. Substituting the frequencies and electric fields specified gives that at 1 MHz, $J_d = 5.56 \times 10^{-10}$ A/m^2 and at 10^{15} Hz, $J_d = 1.67 \times 10^9$ A/m^2. The displacement current density is thus very small for the radio wave and very large for the laser beam.

Example 3.3 Displacement current in copper

Find the ratio of the displacement current to the conduction current in copper.

Copper has a conductivity of 5.8×10^7 S/m and a relative permittivity of unity so equation (3.36) gives $J_d/J_c = 9.6 \times 10^{-19} f$ where f is the frequency. This is a very small ratio, even for optical waves, so the displacement current is negligible for all frequencies below those in the short X-ray region.

Example 3.4 Displacement current in water

Find the ratio of the displacement current density to the conduction current density in sea water for waves at frequencies of 1 MHz, 1 GHz and 1 THz.

Water is a medium which changes its characteristics with frequency. In this example a grossly simplifying assumption will be made that the conductivity and relative permittivity are constant with values of 4 S/m and 80, respectively. This gives $J_d/J_c = 1.11 \times 10^{-9} f$. Substituting the frequencies above gives

Frequency	J_d/J_c
1 MHz	0.0011
1 GHz	1.1
1 THz	1100

This demonstrates that at 1 MHz, radio waves in sea water are dominated by the conduction current, at 1 GHz the conduction and displacement currents are nearly equal and at 1 THz (submillimetrewaves) the displacement current is dominant. It will be shown in Chapter 8 that the attenuation of the waves can be calculated from the conduction current.

3.8 THE ELECTROMAGNETIC WAVE EQUATION

Maxwell's equations describe the complete relationship between electric and magnetic fields but they are not convenient by themselves for solving electromagnetic wave problems because they involve both electric and magnetic field quantities. A subsidiary equation will now be derived that involves only the electric field or only the magnetic field. This is called the *electromagnetic wave equation* or often just the *wave equation.* The assumption will be made that there are no sources or charges in the region of interest. This means that the charge density, ρ, is zero. The additional assumption is also made that the waves are in free space or in a dielectric which is non-conducting and loss-less. This means that the conduction current density, $\boldsymbol{J}_c$, is zero. Maxwell's equations simplify to

$$\nabla \times \boldsymbol{E} = -j\omega \mu_0 \mu_r \boldsymbol{H} \tag{3.37}$$

$$\nabla \times \boldsymbol{H} = j\omega \varepsilon_0 \varepsilon_r \boldsymbol{E} \tag{3.38}$$

$$\nabla \cdot \boldsymbol{E} = 0 \tag{3.39}$$

$$\nabla \cdot \boldsymbol{H} = 0 \tag{3.40}$$

The flux densities $\boldsymbol{D}$ and $\boldsymbol{B}$ have been eliminated with equations (3.22) and (3.23). Under these assumptions of no currents and no charges, only the first two equations are necessary because equation (3.40) is implicit in equation (3.37) since $\nabla \cdot \nabla \times \boldsymbol{E} \equiv 0$. Similarly equation (3.39) is implicit in equation (3.38).

In order to derive the wave equation, the magnetic field is eliminated by combining equations. Taking the curl of the first Maxwell equation, equation (3.37), gives

$$\nabla \times \nabla \times \boldsymbol{E} = -j\omega\mu_r\mu_0 (\nabla \times \boldsymbol{H}) \tag{3.41}$$

The curl operates only on the spatial variables which is why the time factor $j\omega$ on the right-hand side can be taken outside the curl vector. The left-hand side may be expanded using the vector identity

$$\nabla \times \nabla \times \boldsymbol{E} \equiv \nabla(\nabla \cdot \boldsymbol{E}) - \nabla \cdot (\nabla \boldsymbol{E}) \tag{3.42}$$

The first term on the right-hand side of equation (3.42) is zero due to equation (3.39). The second term on the right-hand side is written as $\nabla^2 \boldsymbol{E}$ because in rectangular coordinates it is

$$\nabla \cdot (\nabla \boldsymbol{E}) = \nabla^2 \boldsymbol{E} = \frac{\partial^2 \boldsymbol{E}}{\partial x^2} + \frac{\partial^2 \boldsymbol{E}}{\partial y^2} + \frac{\partial^2 \boldsymbol{E}}{\partial z^2}$$

Equation (3.41) now becomes

$$\nabla^2 \boldsymbol{E} = \mathrm{j}\omega\, \mu_0 \mu_r (\nabla \times \boldsymbol{H}) \tag{3.43}$$

Substituting the second Maxwell equation, equation (3.38), will eliminate $\boldsymbol{H}$ and give an equation for $\boldsymbol{E}$:

$$\boxed{\nabla^2 \boldsymbol{E} = -\omega^2 \mu_0 \mu_r \varepsilon_0 \varepsilon_r \boldsymbol{E}} \tag{3.44}$$

A similar derivation can be done by starting with the second Maxwell equation and this will lead to an identical equation for $\boldsymbol{H}$, namely

$$\boxed{\nabla^2 \boldsymbol{H} = -\omega^2 \mu_0 \mu_r \varepsilon_0 \varepsilon_r \boldsymbol{H}} \tag{3.45}$$

Equations (3.45) and (3.46) are the *electromagnetic wave equations*, Since they have the same form it can be deduced that only one equation is needed and the other fields can be obtained from the appropriate Maxwell equation. The wave equations completely specify the characteristics of time periodic electromagnetic waves in free space or dielectric. The procedure for solution is to expand either of the wave equations into their component parts for the coordinate system appropriate to the problem being solved. In general this can be quite lengthy, but for the free-space waves and guided waves studied in later chapters, simplifications can be made. The wave equations will be solved in Chapter 5 in order to study the characteristics of plane waves and in Chapter 10 to study the characteristics of guided waves.

The wave equations have been derived in this section for time periodic waves, but it is possible to start without this restriction in which case the result is similar except that $-\omega^2$ is replaced by $\mathrm{d}^2/\mathrm{d}t^2$.

3.9 POWER IN ELECTROMAGNETIC WAVES AND THE POYNTING VECTOR

The power in an electromagnetic wave is analogous to the power in a circuit. If there is a voltage V across a component and a current I flowing in the component then the power flow is $P = VI$. Voltage and current are scalars, so the power flow is also a scalar. By contrast, the electric and magnetic fields are vectors, so the power or energy flow is the vector product of $\boldsymbol{E}$ and $\boldsymbol{H}$ and the result is a vector. For an electromagnetic wave,

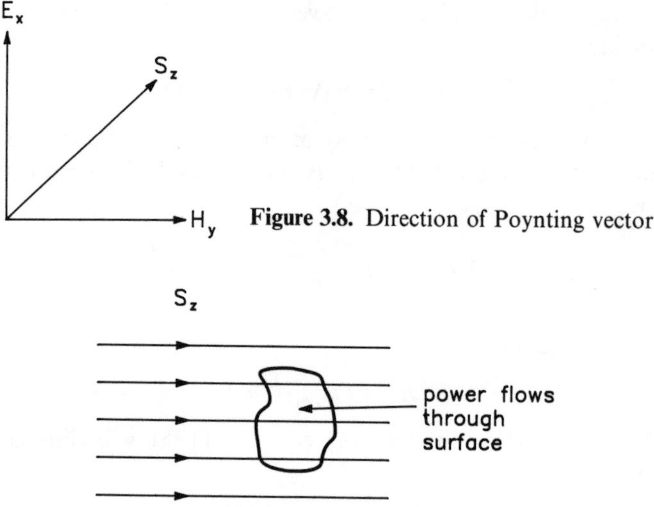

Figure 3.8. Direction of Poynting vector

Figure 3.9. Power flowing through a surface

the power density per unit area of a wave is defined as the *Poynting vector*, $\boldsymbol{S}$,

$$\boldsymbol{S} = \boldsymbol{E} \times \boldsymbol{H} \qquad \text{W/m}^2 \tag{3.46}$$

$\boldsymbol{E}$ and $\boldsymbol{H}$ are instantaneous values. The vector product indicates that the direction of vector $\boldsymbol{S}$ is orthogonal to the plane containing $\boldsymbol{E}$ and $\boldsymbol{H}$. For example, If $\boldsymbol{E} = E_x\hat{x}$ and $\boldsymbol{H} = H_y\hat{y}$, then $\boldsymbol{S} = S_z\hat{z}$, Figure 3.8. This will be seen to be important when plane waves are described in Chapter 5. Equation (3.46) also demonstrates that there is no power flow if either the electric field is zero or the magnetic field is zero. This is another confirmation of the deduction inherent in Maxwell's equations, that both electric and magnetic fields must be present.

The Poynting vector gives the power density. To obtain the power flow, P, in watts, the Poynting vector must be integrated over a region through which the power is flowing, Figure 3.9. In general

$$P = \int_s (\boldsymbol{E} \times \boldsymbol{H}) \cdot \boldsymbol{ds} \tag{3.47}$$

Example 3.5 Power flow in a conducting rod

Find the power flow in a rod of radius a and resistance R per unit length, carrying a current I_z, Figure 3.10.

This is a simple example which demonstrates the calculation of the power flow. The magnetic field of a conducting rod was derived in Section 2.11,

$$H_\phi = \frac{I_z}{2\pi a}$$

The electric field inside the rod is given by Ohm's law

$$E_z = I_z R$$

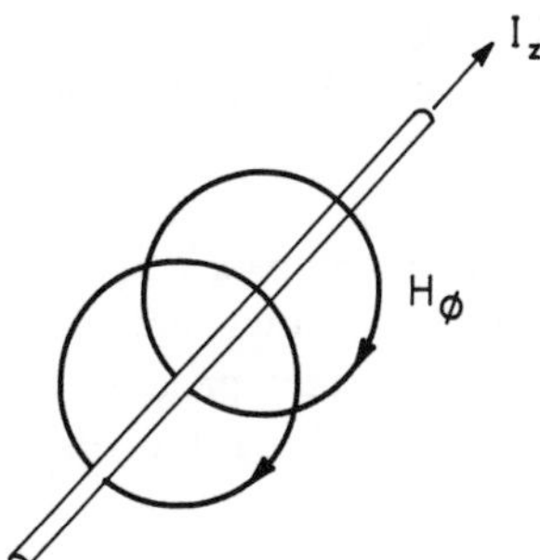

Figure 3.10. Power flow in a conducting rod

The Poynting vector is obtained from expanding $\boldsymbol{E} \times \boldsymbol{H}$ in cylindrical coordinates as

$$S_r = -E_z H_\phi = \frac{-RI_z^2}{2\pi a}$$

The surface through which the power flows is the cylindrical surface of the rod. The Poynting vector has only a radial component and is negative so the power flow per unit length is directed radially inwards. The power flow is the Poynting vector multiplied by the circumference. It leads to the well known result,

$$P_r = \frac{-RI_z^2}{2\pi a}(2\pi a) = -RI_z^2$$

The negative sign indicates that power is coming into the rod where it is dissipated as heat.

COMPLEX POYNTING VECTOR

The main interest in this book is the power flow of a wave characterized by a time dependence $e^{j\omega t}$. The electric and magnetic fields are, in general complex quantities with real and imaginary parts. When dealing with sinusoidally time-dependent fields and power, a complication arises which means that the Poynting vector must be modified. This is because the power is the product of two complex quantities which means that the product of two complex waves is not the product of the spatial parts of the waves. This can be shown for complex electric and magnetic fields as follows. The time dependent fields are defined as

$$\boldsymbol{E}(t) = \mathrm{Re}(\boldsymbol{E}\,e^{j\omega t}) = \mathrm{Re}(E\,e^{j\phi}\,e^{j\omega t}) \tag{3.48}$$

$$\boldsymbol{H}(t) = \mathrm{Re}(\boldsymbol{H}\,e^{j\omega t}) = \mathrm{Re}(H\,e^{j\Phi}\,e^{j\omega t}) \tag{3.49}$$

The magnitudes, E and H, are now the root mean square (RMS) value of the fields. Then

$$\boldsymbol{E}(t) \times \boldsymbol{H}(t) \neq \mathrm{Re}(\boldsymbol{E} \times \boldsymbol{H}\,e^{j\omega t})$$

The dilemma can be resolved by noting that, for sinusoidal waves, it is the time average power over one period ($\omega t = 2\pi$) of the wave which is required,

$$P = \frac{1}{2\pi}\int_0^{2\pi} \boldsymbol{E}(t) \times \boldsymbol{H}(t)\,\mathrm{d}(\omega t) \tag{3.50}$$

Substituting equations (3.48) and (3.49) into equation (3.50) and evaluating the integral shows that it can be expressed as

$$P = \tfrac{1}{2}\mathrm{Re}(\boldsymbol{E} \times \boldsymbol{H}^*) \tag{3.51}$$

where $\boldsymbol{H}^*$ is the complex conjugate of $\boldsymbol{H}$ (i.e. if $H = a + \mathrm{j}b, H^* = a - \mathrm{j}b$).

Equation (3.51) shows that the time average complex power flow is given by half the real part of the vector product of the electric field and the conjugate of the magnetic field. The complex Poynting vector is therefore

$$\boldsymbol{S} = \boldsymbol{E} \times \boldsymbol{H}^* \tag{3.52}$$

3.10 POWER FLOW IN A TIME HARMONIC ELECTROMAGNETIC WAVE

In this section Poynting's vector will be derived for a time harmonic electromagnetic wave and the power flow through a general surface obtained in order to show how it can be broken into component parts which balance out the source power with the losses and energy storage. The Poynting vector is derived from the first and second Maxwell equations, including the conduction current term. From equations (3.27) and (3.28) these are

$$\nabla \times \boldsymbol{E} = -\mathrm{j}\omega\mu_0\mu_r\boldsymbol{H} \tag{3.53}$$

$$\nabla \times \boldsymbol{H} = \mathrm{j}\omega\varepsilon_0\varepsilon_r\boldsymbol{E} + \boldsymbol{J} \tag{3.54}$$

Taking the scalar product of $\boldsymbol{H}^*$ with equation (3.53) and the scalar product of $\boldsymbol{E}$ with the complex conjugate of equation (3.54) and subtracting,

$$\boldsymbol{H}^* \cdot \nabla \times \boldsymbol{E} - \boldsymbol{E} \cdot \nabla \times \boldsymbol{H}^* = -\mathrm{j}\omega\mu_0\mu_r\boldsymbol{H} \cdot \boldsymbol{H}^* + \mathrm{j}\omega\varepsilon_0\varepsilon_1\boldsymbol{E} \cdot \boldsymbol{E}^* - \boldsymbol{E} \cdot \boldsymbol{J}^* \tag{3.55}$$

Making use of a vector identity, the left-hand side can be replaced by $\nabla \cdot (\boldsymbol{E} \times \boldsymbol{H}^*)$. Also $\boldsymbol{H} \cdot \boldsymbol{H}^*$ can be written as $|\boldsymbol{H}|^2$ and $\boldsymbol{E} \cdot \boldsymbol{E}^*$ as $|\boldsymbol{E}|^2$, therefore equation (3.55) becomes

$$\nabla \cdot (\boldsymbol{E} \times \boldsymbol{H}^*) = -\mathrm{j}\omega\mu_0\mu_r|\boldsymbol{H}|^2 + \mathrm{j}\omega\varepsilon_0\varepsilon_r|\boldsymbol{E}|^2 - \boldsymbol{E} \cdot \boldsymbol{J}^* \tag{3.56}$$

This represents the energy at a point in a medium and is seen to be the divergence of the Poynting vector. To find the power flow through a closed surface, equation (3.56) must be integrated over a volume v. The integral is over a volume because $\boldsymbol{E}$ has the units of volts per metre (V/m) and $\boldsymbol{J}$ has the units of amps per square metre (A/m^2). Thus $\boldsymbol{E} \cdot \boldsymbol{J}^*$ has the dimensions of watts per cubic metre. The volume v is shown in Figure 3.11 with a surface s and an outward unit vector $\hat{\boldsymbol{n}}$ which is normal to the surface. Integrating equation (3.56) over the volume gives

$$\frac{1}{2}\int_{\mathrm{s}} \hat{\boldsymbol{n}} \cdot (\boldsymbol{E} \times \boldsymbol{H}^*)\mathrm{d}s = -\mathrm{j}\omega \int_{\mathrm{v}} \frac{1}{2}(\mu_0\mu_r|\boldsymbol{H}|^2 - \varepsilon_0\varepsilon_r|\boldsymbol{E}|^2)\mathrm{d}v - \frac{1}{2}\int_{\mathrm{v}} (\boldsymbol{E} \cdot \boldsymbol{J}^*)\mathrm{d}v \tag{3.57}$$

where use has been made of the following integral transform:

$$\int_{\mathrm{v}} \nabla \cdot \boldsymbol{A}\,\mathrm{d}v = \int_{\mathrm{s}} \hat{\boldsymbol{n}} \cdot \boldsymbol{A}\,\mathrm{d}s \tag{3.58}$$

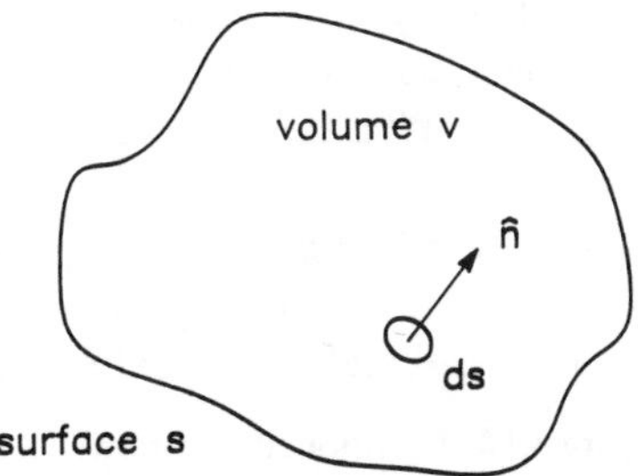

Figure 3.11. Power flowing out of a volume region

Equation (3.57) is called the Poynting theorem and it is a statement of power conservation. The physical meaning of all the terms can be deduced by separating the real and imaginary parts.

THE REAL PARTS OF THE POYNTING THEOREM

The real parts of equation (3.57) give

$$\frac{1}{2}\int_s \hat{\boldsymbol{n}}\cdot \mathrm{Re}(\boldsymbol{S})\,\mathrm{d}s = -\frac{1}{2}\int_v \mathrm{Re}(\boldsymbol{E}\cdot\boldsymbol{J}^*)\,\mathrm{d}v \tag{3.59}$$

The left-hand side of equation (3.59) is the Poynting vector integrated over the surface, s, and represents the power flowing out of the volume v. The right-hand side represents the time average power flows as can be seen by analogy with the product VI in circuit theory.

The right-hand side of equation (3.59) has two parts, one due to the source of the energy and one due to the losses in the conducting medium. The two parts of the current density $\boldsymbol{J}$ are $\boldsymbol{J}_s$ and $\boldsymbol{J}_c$. The source current density, $\boldsymbol{J}_s$, represents the current density due to any sources. An example is a dipole antenna where the current flow on the surface of the dipole gives rise to radiation. The conduction current density, $\boldsymbol{J}_c$, is the current which flows if the medium has finite conductivity. It is equal to the conductivity times the electric field ($\boldsymbol{J}_c = \sigma\boldsymbol{E}$). The right-hand side of equation (3.59) can now be written

$$-\frac{1}{2}\int_v \mathrm{Re}(\boldsymbol{E}\cdot\boldsymbol{J}^*)\,\mathrm{d}v = -\frac{1}{2}\int_v \mathrm{Re}(\boldsymbol{E}\cdot\boldsymbol{J}_s^*)\,\mathrm{d}v + \frac{1}{2}\int_v \sigma|\boldsymbol{E}|^2\,\mathrm{d}v \tag{3.60}$$

The first term on the right-hand side of equation (3.60) is the source power and the second term is any power dissipated by the finite conductivity as heat.

Equation (3.59) thus states that the power flowing out of a volume containing electromagnetic waves is equal to the difference between the power generated and the power dissipated.

THE IMAGINARY PARTS OF THE POYNTING THEOREM

These can similarly be split into its component parts. Equation (3.57) yields

$$\frac{1}{2}\int_s \hat{\boldsymbol{n}}\cdot \mathrm{Im}(\boldsymbol{S})\,\mathrm{d}s = -\mathrm{j}\omega\int_v \frac{1}{2}(\mu_0\mu_r|\boldsymbol{H}|^2 - \varepsilon_0\varepsilon_r|\boldsymbol{E}|^2)\,\mathrm{d}v - \frac{1}{2}\int_v \mathrm{Im}(\boldsymbol{E}\cdot\boldsymbol{J}_s^*)\,\mathrm{d}v \tag{3.61}$$

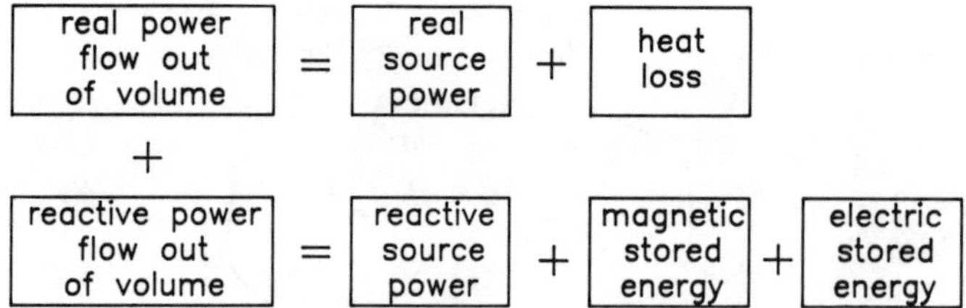

Figure 3.12. Components of power flow

The left-hand side of this equation is the imaginary, or reactive, part of the power flow out of the volume v. The first term on the right-hand side represents the reactive magnetic energy stored in volume v. Similarly the second term represents the electric energy stored in volume v. The last term is the reactive power generated by the source current density.

The complete power flow equation is shown diagrammatically in Figure 3.12. The real, or usable, or measurable power, is given by equation (3.59) whereas the stored, or reactive power, is given by equation (3.61).

PROBLEMS

1. A single loop of wire of area $1\,\text{m}^2$ is situated in air with a uniform magnetic field normal to the plane of the loop. If the flux density is changing at $2\,\text{Wb/m}^2\text{/s}$ what is the potential difference at the ends of the loop of wire?

2. A fixed square coil with five turns is placed with its lower left corner at the origin. It has sides a and b, both equal to 10 cm. If the magnetic flux density, $\boldsymbol{B}$, is normal to the plane of the coil and has a space variation $B = 3\sin(\pi x/a)\sin(\pi y/b)\ \text{Wb/m}^2$ and a sinusoidal time variation of frequency 1 kHz, find the potential difference between the ends of the coil.

3. A Boeing 747 aircraft with a wind span of 60 m is flying horizontally at a speed of 900 km/h. The Earth's magnetic field has a vertical component at the aircraft's position with a magnetic flux density of $6 \times 10^{-5}\,\text{Wb/m}^2$. Find the induced emf between the wing tips.

4. The rate of flow of liquid in a plastic pipe can be measured by placing the pipe in a magnetic field and measuring the emf induced by the motion of the liquid. If a pipe has a diameter of 50 mm containing liquid flowing at 2 m/s and is placed in a magnetic field with a value perpendicular to the pipe of $1.5 \times 10^{-5}\,\text{A/m}$, find the emf induced across the pipe.

5. The two Maxwell curl equations completely specify the electromagnetic field with the addition of the continuity equation. Prove this fact by deriving the two Maxwell divergence equations from the curl equations.

6. Why are Maxwell's equations not completely symmetrical?

7. Show that $\nabla \cdot \boldsymbol{D} = 0$ in a perfect conductor by using the second Maxwell equation and Ohm's law for electric fields.

8. Verify that the integral form of Maxwell's equations are self-consistent.

9. A subsidiary magnetic vector potential, $\boldsymbol{A}$, can be defined with $\boldsymbol{B} = \nabla \times \boldsymbol{A}$. Use the first Maxwell equation to obtain the following time-dependent form of the relation between $\boldsymbol{E}$ and V: $\boldsymbol{E} = -\nabla V - \mathrm{d}\boldsymbol{A}/\mathrm{d}t$.

10. Calculate the displacement current density for microwaves at 6 GHz with an electric field of 1 mV/m in (a) stainless steel ($\sigma = 1.1 \times 10^6$ S/m) and (b) silicon ($\sigma = 1.2 \times 10^3$, $\varepsilon_r = 12$).

11. A parallel plate capacitor has a capacitance of 10 pF and is in a circuit with a voltage of RMS amplitude 100 mV and frequency of 100 MHz. Find the displacement current density in between the plates of the capacitor.

12. Verify that the displacement current term in the second Maxwell equation correctly accounts for the current in an air region between two coaxial cylinders which are placed in a circuit.

13. A long and cylindrical conductor with infinite conductivity and with radius R carries a current $I = I_0 \sin \omega t$. As a function of radius r (for $r < R$ and $r > R$) find (a) the conduction current density; (b) the displacement current density; and (c) the magnetic flux density.

14. Two flat circular plates of radius a form a parallel plate capacitor. The capacitor is connected to an alternating source, so that the charge on the plates is $Q = Q_0 \sin \omega t$. Neglecting edge effects, find the magnetic field between the plates in terms of the distance r from the axis.

15. A coaxial capacitor is formed by two cylinders of radii a and b and of length l. It is connected to an alternating current source. Show that the total displacement current flowing across any cylindrical radius r ($a < r < b$) is independent of radius. Assume that the variation of electric field with radius is the same as the electrostatic case.

16. Starting from the second Maxwell equation, derive the electromagnetic wave equation for the magnetic flux density, $\boldsymbol{B}$.

17. Derive the electromagnetic wave equations for the general case of arbitrary time variations (i.e. not sinusoidal waves).

18. Expand the $\boldsymbol{E}$ field electromagnetic wave equation into its rectangular components for E_x, E_y and E_z.

19. Find the power flow in a square copper rod of side 3 mm carrying a current of 100 mA.

20. Confirm that the power density for a complex wave with electric and magnetic field amplitudes of E and H, respectively, is given by $P = \frac{1}{2}\mathrm{Re}(\boldsymbol{E} \times \boldsymbol{H}^*)$.

21. If the Earth has an electric field of 100 mV m^{-1} at the equator which points into the Earth and a magnetic field of 40 mA m^{-1} which points to the north, find the Poynting vector.

22. The Earth receives 0.14 W/cm^2 sunlight on the ground. (a) What is the Poynting vector in W/m^2? (b) What is the power output of the sunlight from the Sun, assuming that the Sun radiates uniformly in all directions?

4 Boundary conditions and quasi-static fields

4.1 INTRODUCTION

Maxwell's equations and the electromagnetic wave equation are valid at all frequencies and in all regions of space. They provide general solutions for the behaviour of fields and waves. In the case of waves which travel in the atmosphere or space, no additional information is needed and the solutions to the electromagnetic wave equation completely define the behaviour of the electromagnetic waves. If, however, the wave impinges onto an object, for instance a lens, or if the wave is guided along a transmission line, the situation is different. In this case the electromagnetic wave equation still provides a general solution, but to find a *unique* solution additional information must be provided. The *boundary conditions* provide this additional information and enable the precise form of the electromagnetic fields in any region to be determined. An example of a situation where the boundary conditions are needed is shown in Figure 4.1 where a dielectric slab lies on top of a conducting sheet and is surrounded by another dielectric medium. The electromagnetic wave equation will give general solutions for the regions marked R_1 and R_2. These solutions must satisfy certain conditions on the metal surface and the amplitudes of the fields in regions R_1 and R_2 will be determined by the boundary conditions across three boundaries: the air–dielectric interface, the air–conductor interface and the dielectric–conductor interface.

There are four boundary conditions, and these will be derived for dielectric boundaries and conducting boundaries in the next two sections. The rest of the chapter deals with cases where the electromagnetic fields can be assumed to be static in space. A simplified form of the wave equation for scalar fields, called Laplace's equation, is derived and solutions are obtained using analytical and numerical techniques. The main merit of the quasi-static approximation is its simplicity though it is an assumption which is valid in quite a number of microwave and optical situations. It is particularly useful for gaining an insight into the form of the electric fields and equipotentials. Practical techniques will be described for this purpose.

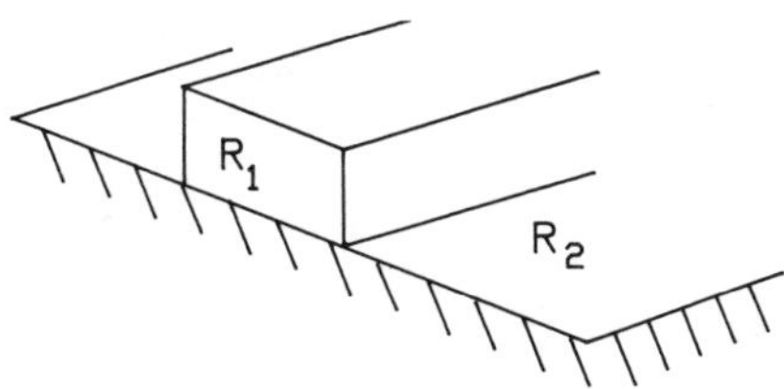

Figure 4.1. Dielectric slab on conducting sheet

4.2 BOUNDARY CONDITIONS BETWEEN TWO DIELECTRIC MEDIA

The integral form of Maxwell's equations (equations (3.31) to (3.34)) can conveniently be used to formulate the boundary conditions. They will be applied to small regions between two dielectrics, one of which could be air or free space.

TANGENTIAL ELECTRIC FIELDS AT A BOUNDARY

Consider the situation in Figure 4.2(a) which shows the boundary between two regions R_1 and R_2 with an imaginary contour, or closed line, crossing the boundary. This has a length δl and height δh, both of which are small but δl is much greater than δh. Assuming that there is no charge on the surface, which is the usual case for a non-conducting medium, the first Maxwell equation in integral form is

$$\int \boldsymbol{E}\cdot d\boldsymbol{l} = -j\omega \int \boldsymbol{B}\cdot d\boldsymbol{s} \tag{4.1}$$

For a small distances, the line integral on the left-hand side is the sum of the electric field times the distance along the four sides of the contour. If the height δh approaches zero, then the only contributions will come from the two sides of length δl. The electric field along each side will be the components of the total electric field which are tangential to the surface. Designating these as E_{t1} and E_{t2}, equation (4.1) becomes

$$\lim_{\delta l \to 0} \int \boldsymbol{E}\cdot d\boldsymbol{l} = E_{t1}\delta l - E_{t2}\delta l \tag{4.2}$$

The right-hand side of equation (4.1) is the magnetic flux density times the area enclosed by the contour. This will tend to zero as δh tends to zero. Hence equation (4.2) is equal to zero and

$$E_{t1} = E_{t2} \tag{4.3}$$

This is the first boundary condition and states that the tengential components of the electric field must be continuous across the boundary. It is the most important of the boundary conditions because the tangential electric field at a boundary is usually easier to derive than either the normal components of the electric fields or the magnetic fields.

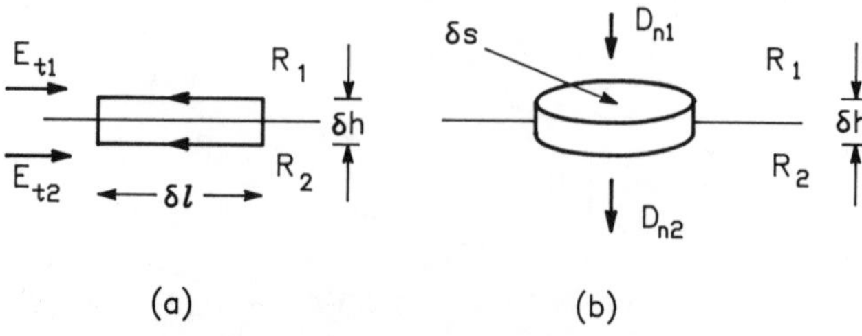

Figure 4.2. Geometry for boundary conditions at a dielectric boundary. (a) Tangential components of electric fields. (b) Normal components of electric flux density

In some cases, more than one boundary condition is needed, in which case the choice must be made from the remaining three cases.

TANGENTIAL MAGNETIC FIELDS AT A BOUNDARY

The boundary condition for the tangential magnetic fields can be obtained in a similar way to the tangential electric fields by using the second Maxwell equation,

$$\int \boldsymbol{H}\cdot d\boldsymbol{l} = \int \mathrm{j}\omega \boldsymbol{D}\cdot d\boldsymbol{s} + \int \boldsymbol{J}\cdot d\boldsymbol{s}$$

The current density, $\boldsymbol{J}$, is equal to zero, because no currents can flow on the surface.

Using the same contour as in the first case but now with the tangential components of the magnetic field, H_{t1} and H_{t2}, and applying the same approximations, leads to

$$\lim_{\delta l \to 0} \int \boldsymbol{H}\cdot d\boldsymbol{l}(= H_{t1}\delta l - H_{t2}\delta l) = \lim_{\delta l \to 0} \int \mathrm{j}\omega \boldsymbol{D}\cdot d\boldsymbol{s}(=0) \tag{4.4}$$

This simplifies to

$$H_{t1} = H_{t2} \tag{4.5}$$

In words, this states that the tangential components of the magnetic field are continuous across the boundary.

NORMAL ELECTRIC FIELDS AT A BOUNDARY

Relations which link the normal components of the electric and magnetic field on either side of a boundary can be obtained from the remaining two Maxwell's equations. Consider a very small box across the boundary as shown in Figure 4.2(b). The area of the surface is δs and the height is δh. The components of the electric flux density normal to the surface are designated D_{n1} and D_{n2}. The third Maxwell equation in integral form is

$$\int \boldsymbol{D}\cdot d\boldsymbol{s} = \int \rho \, \mathrm{d}v$$

for the situation where there are no charges, the right-hand side is zero so this gives directly

$$\lim_{\delta l \to 0} \int \boldsymbol{D}\cdot d\boldsymbol{s} = D_{n1}\delta s - D_{n2}\delta s = 0 \tag{4.6}$$

or

$$D_{n1} = D_{n2} \tag{4.7}$$

This states that the normal components of the electric flux densities are equal across the boundary. It does not mean that the normal components of the electric field are equal across the boundary, because the permittivity on either side of the boundary must have a different value. Equation (4.7) can be expressed in terms of normal components of the electric field as

$$\varepsilon_1 \varepsilon_0 E_{n1} = \varepsilon_2 \varepsilon_0 E_{n2}$$

NORMAL MAGNETIC FIELDS AT A BOUNDARY

Similar reasoning shows that the fourth Maxwell equation will lead to a relation for the magnetic flux density,

$$B_{n1} = B_{n2} \tag{4.8}$$

This states that the normal components of the magnetic flux densities are continuous across the boundary.

Equations (4.3), (4.5), (4.7) and (4.8) are the four boundary conditions and may be summarised as follows

> The tangential components of electric and magnetic fields are equal across a dielectric boundary.
> The normal components of electric and magnetic flux densities are equal across a dielectric boundary.

The boundary conditions can be expressed in vector notation by expressing the electric and magnetic fields as vectors of the form $\boldsymbol{E} = E_t\hat{\boldsymbol{t}} + E_n\hat{\boldsymbol{n}}$ where $\hat{\boldsymbol{t}}$ is a unit vector tangential to the surface and $\hat{\boldsymbol{n}}$ is a unit vector normal to the surface. Then using the relations between unit vectors with dot products and cross products gives for equations (4.3), (4.5), (4.7) and (4.8), respectively

$$\hat{\boldsymbol{n}} \times (\boldsymbol{E}_1 - \boldsymbol{E}_2) = 0 \tag{4.9}$$

$$\hat{\boldsymbol{n}} \times (\boldsymbol{H}_1 - \boldsymbol{H}_2) = 0 \tag{4.10}$$

$$\hat{\boldsymbol{n}} \cdot (\boldsymbol{D}_1 - \boldsymbol{D}_2) = 0 \tag{4.11}$$

$$\hat{\boldsymbol{n}} \cdot (\boldsymbol{B}_1 - \boldsymbol{B}_2) = 0 \tag{4.12}$$

In most practical cases only two boundary conditions are needed because the tangential and normal field components are related to each other.

Example 4.1 Boundary conditions

Find the electric field on the lower side of the boundary shown in Figure 4.3 where the electric field in the upper region is $\boldsymbol{E}_1 = -E_1\hat{\boldsymbol{y}}$. The electric field must be resolved

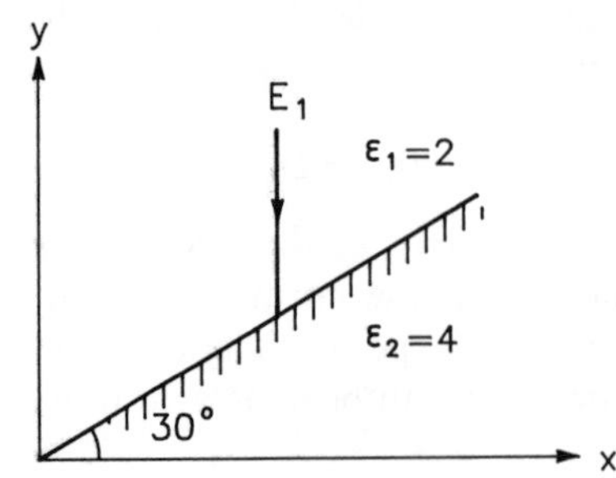

Figure 4.3. Electric field on a boundary

into components tangential to the surface and normal to the surface. These are

$$E_{t1} = -\sin(30°)E_1 = -0.5E_1$$

$$E_{n1} = -\cos(30°)E_1 = -0.866E_1$$

Two boundary conditions are needed because there are two components. Applying boundary condition, equation (4.3), gives $E_{t2} = -0.5E_1$ and applying boundary condition, equation (4.7) with $\varepsilon_1 = 2$ and $\varepsilon_2 = 4$ gives $E_{n2} = \varepsilon_1/\varepsilon_2 E_{n1} = -0.433E_1$. The electric field in the lower region has an amplitude $E_2 = -0.66E_1$ and is at an angle of $\tan^{-1}(E_{n2}/E_{t2}) = 40.9°$ to the surface.

This example shows the general behaviour of a static field at a dielectric boundary. It does not apply to a travelling wave because part of the electric field would be reflected at the boundary and part transmitted. The boundary conditions still apply for a travelling wave but it is the total field on each side of the boundary which satisfies the conditions. This will be demonstrated in Chapter 6 when the properties of optical waves and microwaves at boundaries will be studied.

4.3 BOUNDARY CONDITIONS AT A PERFECT CONDUCTOR

When one of the media at a boundary is a perfect conductor, Figure 4.4, the boundary conditions must be modified in two ways. Firstly there are no electric or magnetic fields inside a perfect conductor and secondly a charge density and current density can flow on the surface of the conductor. Zero fields inside a perfect conductor immediately leads to a simplification of the boundary conditions for tangential electric fields and normal magnetic flux densities as follows

$$E_{t1} = 0 \tag{4.13}$$

$$B_{n1} = 0 \tag{4.14}$$

The other two boundary conditions involve an extra term because the presence of charges and sources means that the second and third Maxwell's equations must contain $\boldsymbol{J}$ and ρ, respectively. Equations (4.5) and (4.7) need modifying. For the tangential magnetic field components, the second Maxwell equation gives

$$\lim_{\delta l \to 0}\left(\int \boldsymbol{H}\cdot d\boldsymbol{l}(=H_{t1}\delta l) = \int j\omega \boldsymbol{D}\cdot d\boldsymbol{s}(=0) + \int \boldsymbol{J}\cdot d\boldsymbol{s}(=J\delta h\delta l)\right) \tag{4.15}$$

where the contour of the line integral is shown in Figure 4.4. As δh tends to zero the

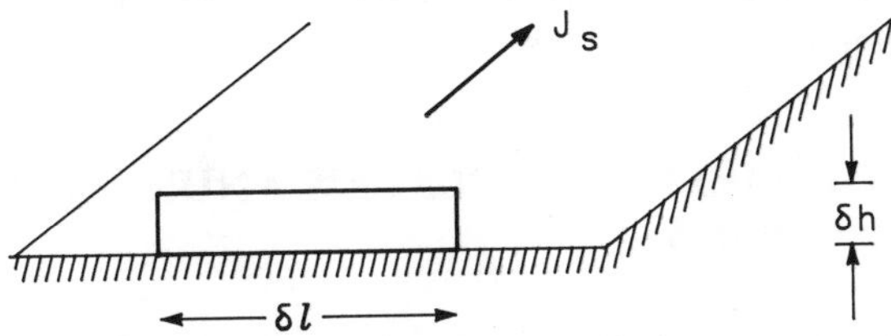

Figure 4.4. Boundary conditions at a conductor

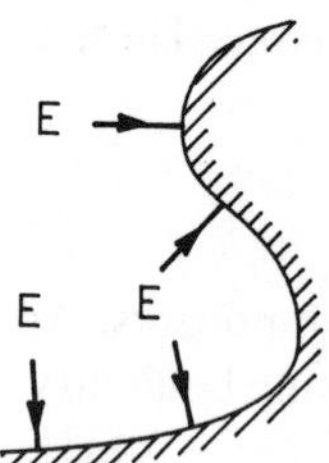

Figure 4.5. Electric field at a conductor is normal to the surface

term $J\delta h$ tends to the surface current density, J_s, which simplifies equation (4.15) to

$$H_{t1} = J_s \tag{4.16}$$

Associated with the current density is a charge density ρ_s and the third Maxwell equation then simplifies to

$$D_{n1} = \rho_s \tag{4.17}$$

The four boundary conditions at a conductor can be expressed in words as

> The tangential components of the electric field and the normal component of the magnetic flux density are zero at a perfectly conducting surface.
>
> The tangential component of the magnetic field is equal to the current density flowing on the surface.
>
> The normal component of the electric flux density is equal to the charge density flowing on the surface.

The two tangential equations are the boundary conditions which are usually used in practical problems. The fact that the tangential components of the electric field are zero at a perfect conductor is a particularly simple condition to apply for cases where only the fields in the medium are required. An alternative expression of this boundary condition is that the electric fields at a perfect conductor must be normal to the surface. This is shown in Figure 4.5 and makes the plotting of electric fields at conductors simple. It was shown in Chapter 2 that the potential at a point in an electric field is the gradient of the electric field at that point, or the potential is orthogonal to the electric field. Thus at a conductor the potential must be zero.

The boundary conditions are always needed when the fields in two or more media are required. They will be used in later chapters to find the characteristics of microwaves when they are reflected by metal surfaces and by microwaves or optical waves when they are reflected and refracted by two dielectrics. The solution of the field equations for all guiding transmission lines such as coaxial lines and optical fibres depends on a knowledge of the boundary conditions.

4.4 QUASI-STATIC FIELDS—LAPLACE'S AND POISSON'S EQUATIONS

There are many practical examples where the variation of the electromagnetic fields with time are very small. This obviously applies to static problems where by definition the time variations are zero, but it also applies in cases at high frequencies where the time varying fields across an object change by only a small amount. Examples are objects which are small in wavelength so that the effect of the changing fields are not apparent as far as the object is concerned. The fields are then said to be *quasi-static.*

The electromagnetic wave equation simplifies for quasi-static fields. The wave equation for electric fields as derived in Chapter 3 is

$$\nabla^2 \boldsymbol{E} + \omega^2 \mu_0 \mu_r \varepsilon_0 \varepsilon_r \boldsymbol{E} = 0 \tag{4.18}$$

For static or quasi-static fields, ω tends to zero so the second term tends to zero and

$$\nabla^2 \boldsymbol{E} = 0 \tag{4.19}$$

The electric field is a vector and it is usually more convenient in quasi-static problems to work with the scalar potential, V, which was defined in Section 2.5 as

$$\boldsymbol{E} = -\nabla V \tag{4.20}$$

A general equation describing the behaviour of V when a charge density ρ is present can be obtained by using Gauss's law,

$$\nabla \cdot (\varepsilon_1 \varepsilon_0 \boldsymbol{E}) = \rho \tag{4.21}$$

and inserting equation (4.20) to give $\nabla \cdot (\nabla V) = -\rho/\varepsilon_1 \varepsilon_0$. The vector operation $\nabla \cdot (\nabla V)$, or (div grad V), is written as $\nabla^2 V$. This equation is called *Poisson's* equation, named after Siméon-Denis Poisson (1781–1840),

$$\nabla^2 V = \rho/\varepsilon_0 \varepsilon_r \tag{4.22}$$

In most practical problems the source can be isolated from the problem of finding the potential in a region. If there are no sources, the Poisson's equation becomes *Laplace's* equation, named after Pierre Simon Laplace (1749–1827),

$$\nabla^2 V = 0 \tag{4.23}$$

Laplace's equation is the most widely used equation is quasi-static electromagnetic fields. It is similar in form to equation (4.19) but deals only with a scalar variable. The vector operation ∇^2 is called the *Laplacian* operator.

The rest of this chapter is devoted to analytical, experimental and numerical solutions to Laplace's equation.

4.5 ANALYTICAL SOLUTIONS TO LAPLACE'S EQUATION

Analytical solutions to Laplace's equation use the equation in component form. In rectangular coordinates Laplace's equation is

$$\frac{\partial^2 V}{\partial x^2} + \frac{\partial^2 V}{\partial y^2} + \frac{\partial^2 V}{\partial z^2} = 0 \tag{4.24}$$

This is a second-order differential equation which can be solved analytically for problems with rectangular boundaries and simple boundary conditions. The equivalent form in cylindrical coordinates is more complicated, but applicable to many problems involving cylindrical geometries

$$\frac{1}{r}\frac{\partial}{\partial r}\left(\frac{r\partial V}{\partial r}\right)+\frac{1}{r^2}\left(\frac{\partial^2 V}{\partial \phi^2}\right)+\frac{\partial^2 V}{\partial z^2}=0 \tag{4.25}$$

Examples of the analytical solution of Laplace's equation will be demonstrated by solving for the potential between parallel plates and in a coaxial cylinder.

Example 4.2 Potential and electric field between parallel plates

Find the potential and electric field between two parallel plates which are infinite in the y and z directions, Figure 4.6, and spaced d apart in the x direction. There is a voltage of V_1 V on the top plate and 0 V on the bottom plate.

The process of solution is in two parts. First, the general form of the potential between the plates is obtained from Laplace's equation. Second, the boundary conditions on the plates are used to find the precise form of the potential.

It will be assumed that there are no variations of the fields in the y and z directions, so $\partial V/\partial y = \partial V/\partial z = 0$, and equation (4.24) becomes

$$\frac{\partial^2 V}{\partial x^2}=0 \tag{4.26}$$

Integrating twice gives as a solution

$$V(x) = Ax + B \tag{4.27}$$

where A and B are constants to be determined by applying the boundary conditions. This clearly demonstrates that the wave equation or Laplace's equation gives the general form of the fields in the region of interest, but the quantitative values must be obtained from a knowledge of the fields at the boundaries to the region.

The boundary condition to use in this case is the continuity of potential. At $x=0$, $V=0$, so equation (4.27) gives $0=B$. At $x=d$, $V=V_1$, and the result is $A=V_1/d$, thus the complete solution is

$$V(x)=\frac{x}{d}V_1 \tag{4.28}$$

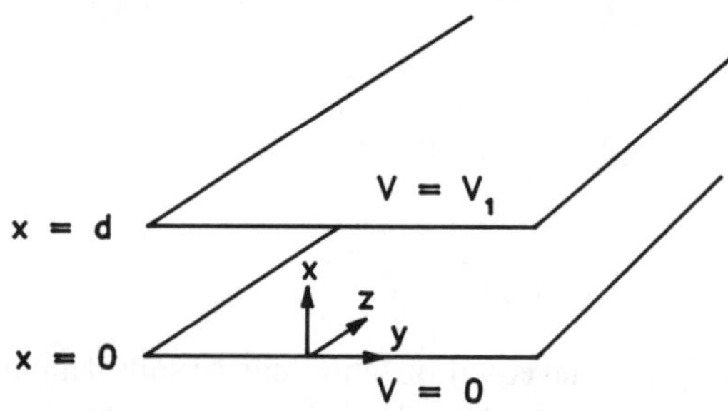

Figure 4.6. Parallel plates

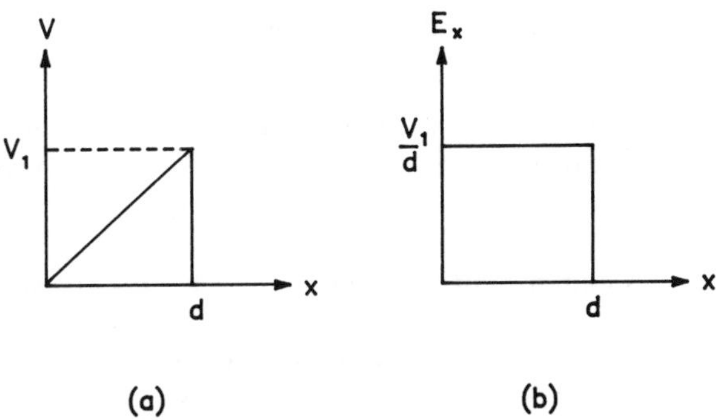

Figure 4.7. (a) Potential and (b) electric field between parallel plates

The electric field is given by

$$E = -\nabla V = -\frac{\partial V}{\partial x}\hat{x} = -\frac{V_1}{d}\hat{x} \tag{4.29}$$

This is constant with the value V_1/d in the x direction. The potential and field between the plates are sketched in Figure 4.7.

Example 4.3 Potential between coaxial cylinders

Find the variation of potential in between coaxial cylinders with inner radius a and outer radius b. There is a potential difference of 100 mV between the earthed inner and outer.

This case is the cylindrical version of the previous example and the solution follows the same procedure using cylindrical coordinates. There are no variations in the fields in the z direction and the fields do not vary around the cylinders, thus $\partial/\partial z = \partial/\partial\phi = 0$ and Laplace's equation is

$$\frac{1}{r}\frac{\partial}{\partial r}\left(\frac{r\partial V}{\partial r}\right) = 0 \tag{4.30}$$

The radius, r, lies between $r = a$ and $r = b$, so there is no possibility of $r = 0$. This needs to be ascertained so that a pole does not occur in the equation. Multiplying by r and integrating twice gives

$$V = A\ln(r) + B \tag{4.31}$$

where A and B are constants to be determined by the boundary conditions. Using the condition that $V = 0$ at $r = a$ and $V = V_1$ at $r = b$ gives $B = -A\ \ln(b)$ and $A = V_1/\ln(b/a)$. The variation of potential is

$$V = \frac{V_1}{\ln(b/a)}\ln(r/a) \tag{4.32}$$

This agrees with the variation of potential found in Example 2.8 (equation (2.23)) which used Gauss's law. The multiplying factors are different because this example uses the potential between the conductors whilst the earlier example used the charge between the conductors.

The procedure used in the above examples to find the solution to Laplace's equation applies to all bounded field problems which are amenable to analysis. The procedure can be summarized as

1. Simplify Laplace's equation in component form for the specified geometry.

2. Find the general solution containing unknown constants.

3. Apply the boundary conditions to find the values for the constants.

This procedure will also be shown to be the basis for the solution of guided wave problems studied in later chapters.

4.6 FIELD MAPPING TECHNIQUES

The behaviour of the electromagnetic field in a region which is bounded by conductors depends on the shape of the conducting surfaces. In many practical cases, the shape of the conductors is such that the electric field and potential takes on a complicated form which cannot be easily predicted by analytical techniques. If the parallel plates in Figure 4.6 had sides which were at a different potential from the top and bottom plates then this would be an example for which there would be no analytic solution. In such cases the form of the fields can be studied either by measuring the field in an experiment or by attempting to numerically predict the fields. Numerical solution will be described in the next section. Here, an experimental method is described and aids to drawing field patterns explained.

Of the two quantities involved in electric fields, the potential, V, is the easier to measure because it is a scalar voltage. A map of the equipotential lines or surfaces are also the most meaningful from the point of view of understanding the field behaviour. It is unnecessary to also measure the electric field intensity, $\boldsymbol{E}$, because it can be predicted, given the equipotential lines or surfaces and some basic knowledge of the relation between the two quantities.

Practical electromagnetic field structures are invariably three-dimensional devices. They have length, width and height. The measurement of fields in a three-dimensional model can be accomplished by setting up a model in an electrolytic tank and probing the volume to establish the potentials. This is, however, difficult to set up and tricky to obtain reliable measurements. Fortunately, many practical structures can be simulated with a two-dimensional model because the variation of the fields in the third dimension are small and stay relatively constant. The field mapping technique to be described will therefore be two dimensional.

EQUIPOTENTIALS BY RESISTANCE PAPER ANALOGY

A two-dimensional technique which is easy to implement and reveals the behaviour of the electric fields with relatively little effort is to measure the potential using a resistance paper analogy. This depends on the analogy between the electric flux, $\boldsymbol{D}$, and the current density, $\boldsymbol{J}$, as seen in the two relationships

$$\boldsymbol{D} = \varepsilon \boldsymbol{E} \qquad (4.33)$$
$$\boldsymbol{J} = \sigma \boldsymbol{E}$$

If the conductivity, σ, can be made to simulate the permittivity, ε, then the current density will be the analogy of the flux density and the behaviour of the current density will mirror the behaviour of the flux density. In two dimensions, $\boldsymbol{J} = \sigma \boldsymbol{E}$ becomes $I = V/R_s$. R_s is the *sheet resistance* and equals length/(width × thickness × conductivity). It has units of Ω per square (not Ω/m^2, because this is a two-dimensional sheet). Measuring the voltages and currents in a two-dimensional analogy will produce plots which mirror the behaviour of the fields in an electromagnetic device. The analogy can be extended to other field quantities, including the capacitance.

The currents and voltages which exist at any point on the surface of carbon impregnated paper, or resistance paper, satisfy the requirements for the two-dimensional analogy. The experimental set-up is shown in Figure 4.8, and consists of a d.c. voltages source, a probe and a voltmeter. The conducting surfaces are carefully painted onto the resistance paper using a conducting paint. The source is connected to the conducting surfaces and the resistance paper probed to measure the voltages. It is easiest to track a line of constant voltage which is an equipotential line. The probe can be pushed into the surface to make a mark so that the resistance paper is the record of the equipotential lines.

Figure 4.9 shows the equipotential lines for a pair of parallel plates of finite length and Figure 4.10 shows the equipotential lines for a conductor near an array of wires, both obtained by this technique. The parallel plates demonstrates the fringing field which exists at the edge of the plates. The analytical solution for parallel plates was studied in the previous section. In order to make the analytical solution feasible the assumption had to be made that the plates were infinitely long. The resistance paper analogy is much nearer to reality. The example in Figure 4.10 has a geometry which is much too complicated to be solved analytically. It is relatively easy to map the equipotentials using this technique. The plot demonstrates how an array of wires

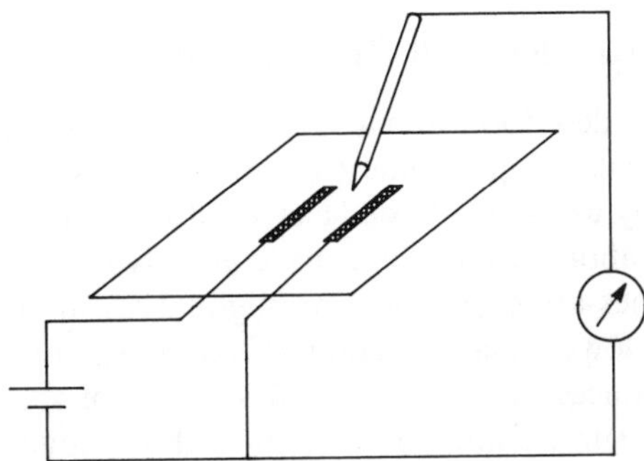

Figure 4.8. Experimental set-up for resistance paper analogy

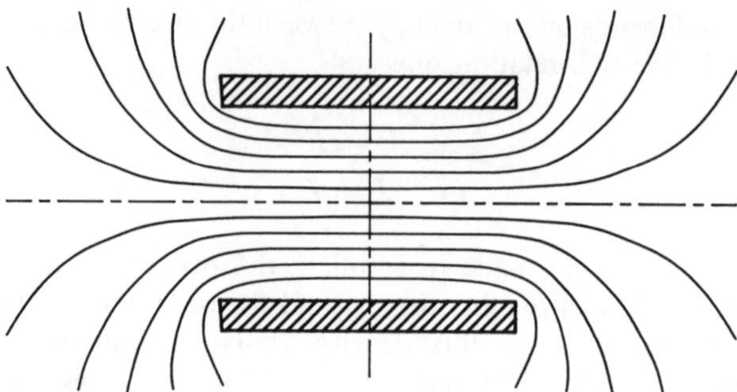

Figure 4.9. Equipotential lines for a pair of parallel plates

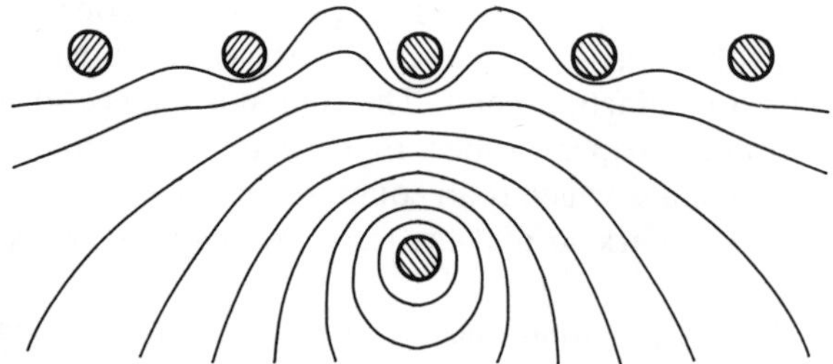

Figure 4.10. Equipotential lines for a charged conductor and an array of wires

effectively acts as a conducting shield or screen. The screen approximates to a conducting surface. The concept is widely used at all frequencies up to microwaves as a way of creating a conducting sheet without the weight or wind loading of a solid sheet. Thus the simple analogy give a useful insight into a practical situation.

ELECTRIC FIELD LINES BY CURVILINEAR SQUARES

The analogy technique described above plots the equipotential lines. A complete description of the fields also requires the electric field lines. These can be drawn over the equipotential lines by using two basic facts and the technique known as curvilinear squares. The basic facts about electric fields have already been studied. These are firstly, that the electric field lines are everywhere orthogonal, or at right angles, to the equipotential lines. This follows from the gradient relation between $\boldsymbol{E}$ and V. Secondly that the conducting surfaces are equipotential surfaces of zero potential so that the electric field lines at the surface must be normal to the conducting surface.

The concepts of curvilinear squares helps to draw the correct number of electric field lines. It can be explained by considering the equipotential lines shown in Figure 4.11.

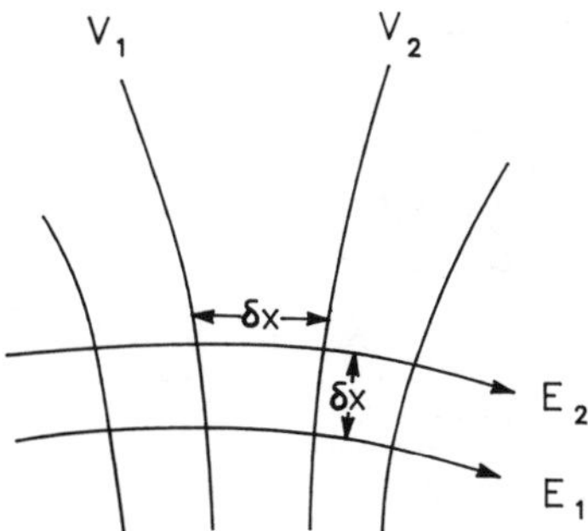

Figure 4.11. Fields for curvilinear squares

V_1 and V_2 are two equipotentials and E_1 and E_2 are the electric field lines at the edges of a bar of electric flux. In three dimensions this would be a flux tube. If the distances apart of the lines are small, the flux between E_1 and E_2 is

$$\delta\psi = \varepsilon(E_1 - E_2)\delta x \tag{4.34}$$

The corresponding potential difference is

$$\delta V = V_1 - V_2 = (E_1 - E_2)\delta x \tag{4.35}$$

Dividing these two equations gives

$$\frac{\delta\psi}{\delta V} = \varepsilon \tag{4.36}$$

The ratio of the change in flux to the change in potential is a constant, assuming that the box is square. The fields do not, however, follow straight lines so that the box can only be approximately square and is called a curvilinear square. This property will be true at all points in the electric field. The electric field lines can therefore be drawn over the equipotential lines by drawing the electric field lines to form curvilinear squares, and ensuring that the electric field lines always cross the equipotentials at right angles. This is shown in Figure 4.12 for the parallel plates of Figure 4.9. The result can only be drawn approximately but does give a good picture of the behaviour of the electric fields.

SYMMETRY AND IMAGES

The amount of effort needed to map the field can often be reduced by making use of symmetries in the fields. For instance the equipotentials and fields of the parallel plates in Figure 4.9 have two axes of symmetry so only a quarter needs to be shown in Figure 4.12. It is therefore only necessary to map the field in one quadrant. The other quadrants are the mirror images of the mapped quadrant. This saves time as well as effort.

Images also occur when conductors are present and sometimes this can be useful. Consider the situation in Figure 4.13(a) showing a charge $+q$ near to a conducting sheet.

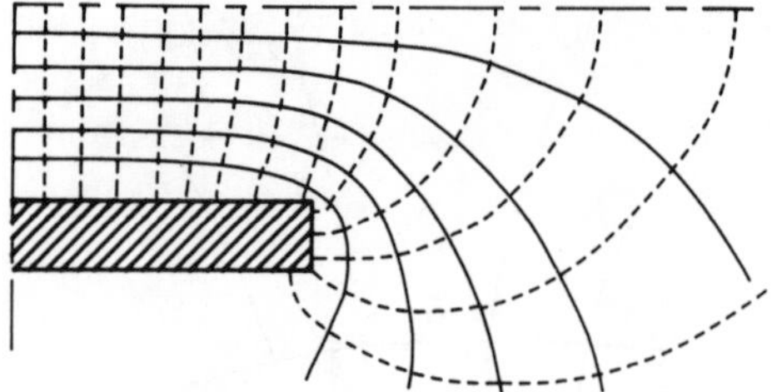

Figure 4.12. Equipotentials (———) and electric field (··········) between part of a pair of parallel plates

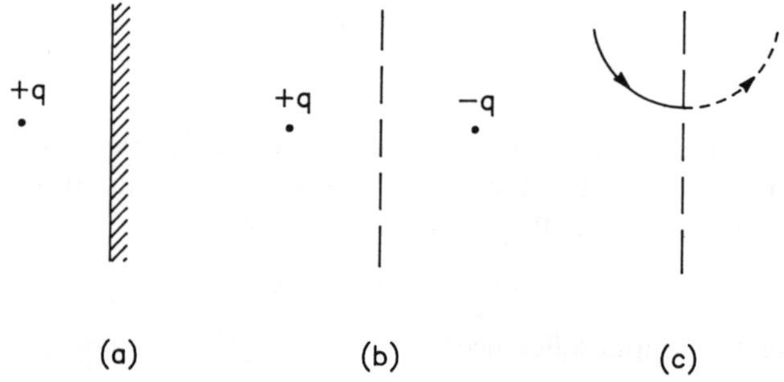

Figure 4.13. Images near conductors. (a) Charge near a conductor. (b) Mirror image for charge near conductor. (c) Image of electric field line

The charge is *mirrored* in the conductor and exactly the same field pattern would be created by two charges $+q$ and $-q$ without a conducting sheet, Figure 4.13(b). The conducting sheet is a zero equipotential surface which is created half way between the two charges in the image model. Similarly the electric field line in Figure 4.13(c) has an image as shown. The direction of the image electric field line is opposite to the real line so that if the conductor is removed, the fields are continuous.

4.7 NUMERICAL SOLUTION OF LAPLACE'S EQUATION

There are many practical problems where the fields can be found by numerical solution of Laplace's equation. This will be demonstrated by using an iterative technique that is very well suited to implementation on a computer. A computer-based technique can use the powerful capability of graphics to quickly plot the equipotentials or electric fields and provide visual insight into the behaviour of electromagnetic fields. In this section the technique will be explained by hand so that the principles of the numerical solution can be demonstrated.

There are a number of techniques available for solving Laplace's equation numerically. Here a finite difference technique will be used. In this method the fields are found at specific points by using a simplified version of Laplace's equation which applies to slowly varying fields together with a knowledge of the boundary conditions. The method will be explained for a two-dimensional geometry, but it could be extended to three dimensions.

The region of the electric field in which a solution is required is divided into a grid of equally spaced points. Figure 4.14 shows five points separated by δx and δy. It is assumed that the potential at the four outer points, V_1 to V_4, are known and the aim is to find to potential at the centre point. The fields between the points are assumed to change linearly which will be valid if the points are close enough together. Laplace's equation in two dimensions is

$$\frac{\partial^2 V}{\partial x^2} + \frac{\partial^2 V}{\partial y^2} = 0 \tag{4.37}$$

The first term is the rate of change of the rate of change of V with respect to x. The second term is the same with respect to y. The rate of change of the potential between the points 1 and 0 is $\partial V_{10}/\partial x$ which is approximately the difference in potential divided by the distance, or

$$\frac{\partial V_{10}}{\partial x} \approx \frac{V_1 - V_0}{\delta x} \tag{4.38}$$

Similarly the rate of change of the potential between points 0 and 2 is

$$\frac{\partial V_{02}}{\partial x} \approx \frac{V_0 - V_2}{\delta x} \tag{4.39}$$

The second differential, $\partial^2 V/\partial x^2$, can be found by similar reasoning to be given approximately by

$$\frac{\partial^2 V}{\partial x^2} \approx \frac{\left(\dfrac{\partial V_{10}}{\partial x}\right) - \left(\dfrac{\partial V_{02}}{\partial x}\right)}{\delta x} \tag{4.40}$$

The second differential in the y direction can be found in the same manner by using points 3 and 4. Inserting equation (4.40) into Laplace's equation, equation (4.37) gives

$$\frac{\left(\dfrac{V_1 - V_0}{\delta x}\right) - \left(\dfrac{V_0 - V_2}{\delta x}\right)}{\delta x} + \frac{\left(\dfrac{V_3 - V_0}{\delta y}\right) - \left(\dfrac{V_0 - V_4}{\delta y}\right)}{\delta y} = 0 \tag{4.41}$$

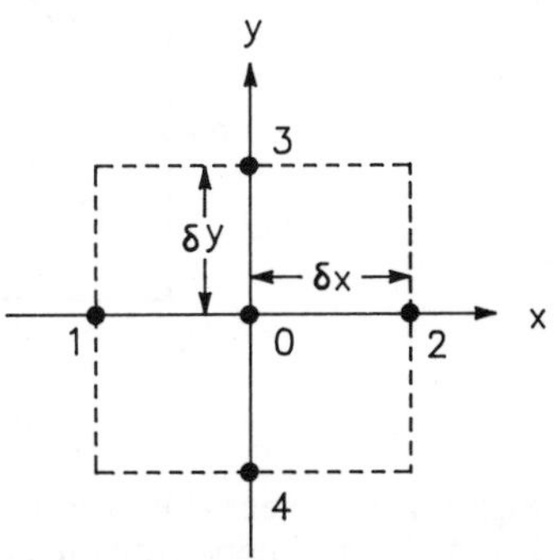

Figure 4.14. Numerical solution of Laplace's equation showing fields around a point

If the x and y spacings are equal, δx will equal δy and equation (4.41) has the following simple result

$$V_0 = \tfrac{1}{4}(V_1 + V_2 + V_3 + V_4) \qquad (4.42)$$

This states that the potential at the point 0 is the average of the potentials at the four surrounding equidistant points. This equation is particularly suited for use on a computer because the arithmetic manipulations are trivial and can be carried out very quickly.

The practical implementation of equation (4.42) will be illustrated with three examples.

Example 4.4 Potential in a trough guide

Find the 5 V, 10 V and 20 V equipotentials for the trough guide shown in Figure 4.15(a), where the bottom and two sides are at zero potential and the isolated top plate is at a potential of 30 V.

The potential at the top corners can be assumed to be the average of the potentials on the top and side plates. i.e. 20 V. A relatively course grid of equally spaced points is shown though for detailed mapping a fine grid of any density could be used. The letters on the grid points show the order in which they will be calculated. The potential at the centre point A is found by using the potentials at the four sides as the grid points, Figure 4.15(b). Equation (4.42) gives

$$V_A = \tfrac{1}{4}(40 + 0 + 0 + 0) = 10\text{ V}$$

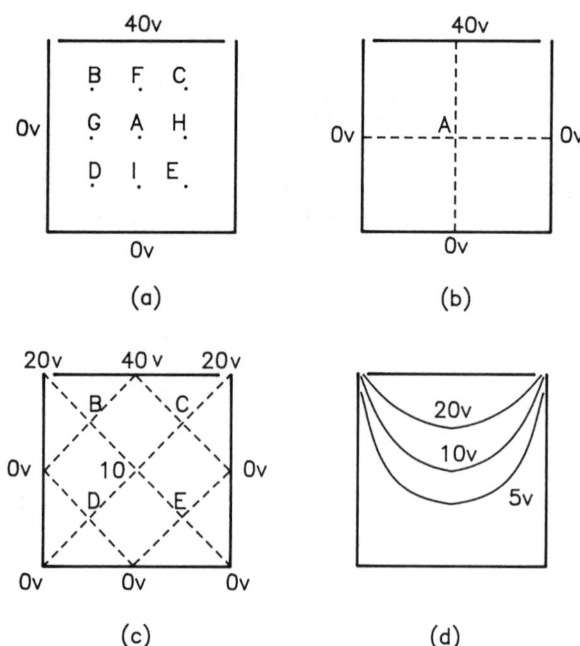

Figure 4.15. Potential in a trough guide. (a) Geometry and points. (b) First point A. (c) Points B, C, D, E. (d) Equipotential lines

Next the potential at the points B, C, D and E are found by using a grid which is rotated by 45° with respect to the first grid, Figure 4.15(c) and using the potential at the gap as one end of the grid. This gives $V_B = V_C = 17.5$ V and $V_D = V_E = 2.5$ V. The potentials at the other points in Figure 4.15(a) can be found by returning to the grid which is parallel with the sides to give $V_F = 21.1$ V, $V_G = V_H = 7.5$ V and $V_I = 3.8$ V. With more points, a dense grid of potentials is obtained which then enables the equipotential contours to sketched as shown in Figure 4.15(d).

A few comments can be made from this example. The first point, A, used a very coarse grid so that the assumption behind equation (4.42) of a small grid would appear to be violated. This is strictly true, but in this example, the fields decay monotonically from the top plate so the assumption is not bad. In general the first few points can only be found approximately because the initial information is on the boundaries which are some distance from the first points. A way around the error introduced by the large grid is to repeat the calculation, as will be demonstrated in the next example. The solution was helped by using two grids which were at 45° to each other. This enabled the potential at more points to be found with the same boundary voltages. This procedure can be implemented with any size of grid. If the process is done by hand, relatively few points are needed in order to sketch the equipotentials. On a computer a denser grid would be required, because the graphics package cannot visualize the whole plot and is not able to make assumptions about the variation in field outside the small region. In many problems, geometrical symmetry can be used to reduce the number of potentials which need to be calculated. This is the case for the example above. There is a line of symmetry through the points F, A and I so it is unnecessary to calculate the potentials on both sides of this line.

Example 4.5 Equipotential lines in a deep trough guide

Find, by iteration, the potentials at the points shown in Figure 4.16, using a coarse grid.

The process of iterating towards the precise value for the potential field can be demonstrated by considering a trough guide which is twice as tall as the previous example. The potential at the point A can be found by using a grid at 45° and will have the initial value of 10 V. The potential at point B must be found from all four

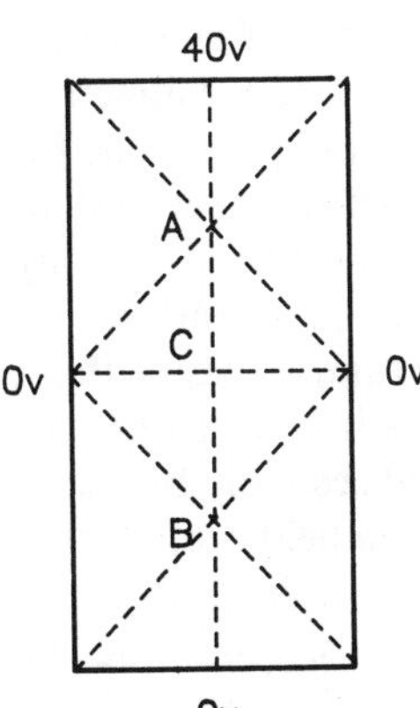

Figure 4.16. Potential in a deep trough guide

sides which have zero potential. This gives an initial value of 0 V which is clearly wrong since there must be finite fields inside the region. The way around this difficulty is to use A, B and the two sides to find the finite potential at C (2.5 V) and then to use C and three sides at 0 V to find a finite value for the potential at B (0.625 V) which replaces the initial value. This iterative loop can be repeated to obtain a better estimate for C and B but the loop should also include A since a finite value for C will modify A and so on.

The above procedure will work but is not very elegant. A more systematic procedure is to use only the square grid and to assign initial values of 0 V to the points inside the trough guide. Then work consistently down the guide in the order A, C, B and keep repeating the calculations until a specified accuracy is reached. The results of this process are

	A (V)	B (V)	C(V)
Loop 1	10.0	2.50	0.625
Loop 2	10.625	2.656	0.664
Loop 3	10.664	2.666	0.666

Convergence is surprisingly rapid. The difference between the values on the third and second loops is only 0.4%.

It is a characteristic of numerical techniques that the precise answer is not usually computable. Nor is it actually needed, an approximate answer is acceptable for most engineering applications. The iterative process is stopped when the specified accuracy is reached. This is found by comparing the difference between the values computed on the last loop with the values computed on the next to the last loop, against the desired accuracy. In most cases an accuracy of between 0.1% and 1% is adequate.

Example 4.6 Equipotential lines in a ridged guide

Sketch the equipotential lines in the ridged guide shown in Figure 4.17(a), which is assumed to be infinite at the left and right.

In this example symmetry means that only half the structure needs to be used for calculating the potentials, Figure 4.17(b). The equipotential lines between the plates will vary linearly between the plates except for the region around the step in the bottom plate. This is the only region in which the fields need to be mapped but near the step the potentials will change rapidly. The solution procedure is the same as in the last example except that effort can be saved firstly by assigning initial values to the grid points which are linear functions of distance between the top and bottom plates, and secondly by using a non-uniform grid. The non-uniform grid ensures that the required field around the step is found without unnecessary work away from the step. A suitable grid is shown in Figure 4.17(b) together with the computed equipotentials. Since the aim is to sketch the equipotentials, only approximate estimates of the potentials need by computed.

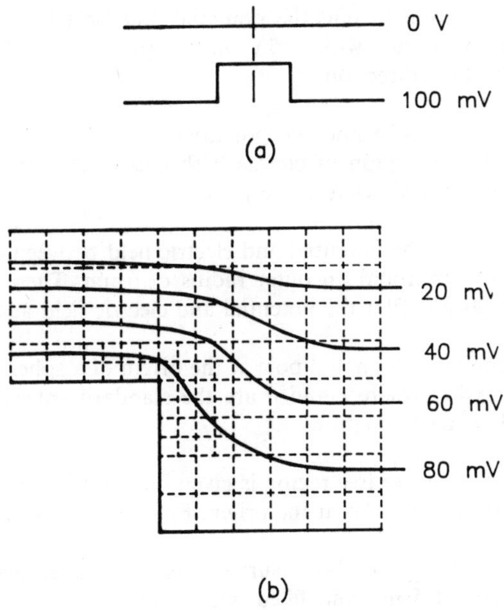

Figure 4.17. Ridged guide. (a) Geometry of conductors. (b) Equipotential lines and grid

PROBLEMS

1. A groove is cut in a block of dielectric with relative permittivity ε_r. If an electric field of flux density $\boldsymbol{D}$ is present in the dielectric on one side of the groove. What is the value of the electric field in the groove (a) if $\boldsymbol{D}$ is perpendicular to the groove and (b) if $\boldsymbol{D}$ is parallel to the groove?

2. A dielectric plate of permittivity ε_2 is placed in an infinite media of permittivity ε_1 in which there is a uniform electric field of E_1 V/m. The angle between the normal to the plate and the direction of E_1 is θ_1, and the angle between the normal to the plate and the direction inside the plate is θ_2. Show that the relationship between θ_1 and θ_2, is $\tan\theta_1/\tan\theta_2 = \varepsilon_1/\varepsilon_2$.

3. Repeat Example 4.1 but using a magnetic field in the upper region given by $\boldsymbol{H}_1 = -H_1\hat{\boldsymbol{y}}$, in order to find the direction of the magnetic field in the dielectric.

4. Derive the vector form of the boundary conditions, as given in equations (4.9) to (4.12).

5. The magnetic field in air just above a perfect conducting sheet, which is in the xy plane, is given by $\boldsymbol{H} = \hat{\boldsymbol{x}} + \hat{\boldsymbol{y}}$. Find the current density flowing on the surface.

6. Show that if a steady current flows across a plane boundary between two conductors with different conductivities, a charge appears at the boundary. Find an expression for the charge in terms of the current.

7. Solve Laplace's equation to find the potential and electric field between the plates of a parallel plate capacitor with $+5\,\text{V}$ on one plate and $-5\,\text{V}$ on the other plate. Sketch the electric field direction.

8. Solve Poisson's equation to find the potential between parallel plates separated by 3 mm, which enclose a region of charge with charge density $3\,\mu\text{C/m}^3$ and $\varepsilon_r = 1.1$. There is a potential of 2 V between the plates.

9. Obtain expressions for the potential and electric field between coaxial cylinders with an inner radius of 2 mm and an outer radius of 6 mm. There is a voltage between the cylinders of 100 mV. Plot the potential and electric field against radius.

10. The electric potential for a small dipole at the origin of a spherical coordinate system is given by $V \alpha \cos\theta/r^2$ where r and θ are the standard spherical parameters. Show that V satisfies Laplace's equation.

11. A potential field in a charge free region is given by $V = x^3 - 5y^2 + f(x)$. The potential V and electric field E_x are zero at the origin. Find an expression for the factor $f(x)$.

12. Find the charge density on both surfaces of a parallel plate capacitor with a separation distance of 5 mm and filled with dielectric of $\varepsilon_r = 3$. The upper plate is at $+60\,\text{V}$ and the lower plate at $+30\,\text{V}$.

13. Draw the electric field lines on the equipotential contours shown in Figure 4.10, using the method of curvilinear squares.

14. Show that the capacitance of a system of cuvilinear squares between two conductors is given by $C = \varepsilon N/M$ where N is the number of flux tubes between the conductors and M the number of potential tubes.

15. Draw a map of the curvilinear squares between an air-filled coaxial capacitor of inner radius 5 mm and outer radius 15 mm. Compute the capacitance per metre using the equation given in Problem 14. Check the result using the exact formula.

16. A point electric charge of value $3 \times 10^{-9}\,\text{C/m}$ is 3 cm above a conducting sheet. Use the method of images to find the electric field on the surface 10 cm away from the point immediately below the charge. Hence deduce that the electric field is everywhere normal to the surface.

17. Find the distribution of potential in the box shown in the figure which has sides $2a$ by a.

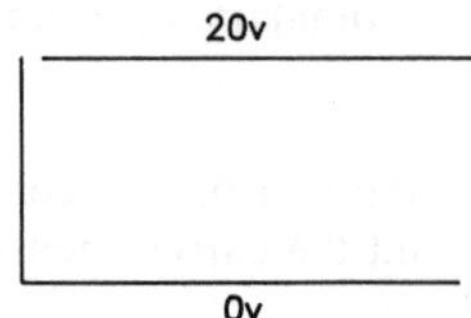

18. Find and sketch the equipotential lines in the guide shown in the figure, which has

sides $3a$ by $2a$. Confirm that the potentials at the points X and Y are 1.0 V and 0.25 V, respectively.

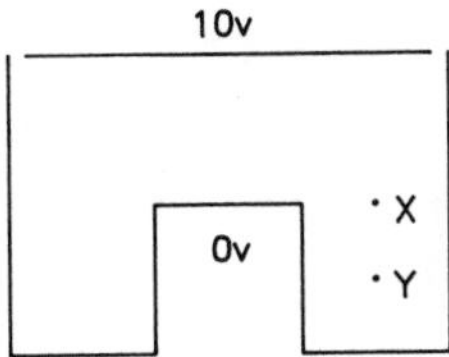

19. The figure shows a square coaxial system of conductors with + 10 V on the central conductor and 0 V on the outer conductor. Draw the lines of symmetry for the system and sketch the 2, 4, 6 and 8 V equipotential contours.

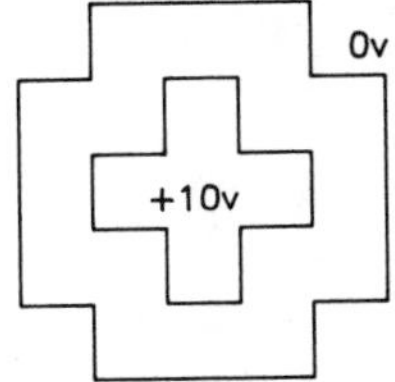

20. Sketch the distribution of potential in the system of conductors shown in the figure. Symmetry means that only one quarter of the system is needed to give the complete equipotential pattern.

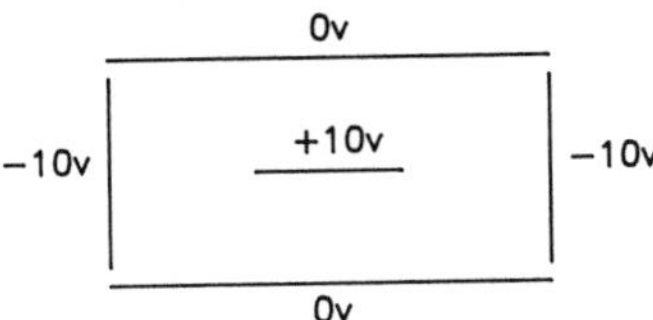

5 Plane waves

5.1 INTRODUCTION

A plane electromagnetic wave is one where there are spatial variations only in the direction of travel. In directions transverse to the direction of travel the wave is exactly the same at all points in space. This is shown in Figure 5.1, where the lines below the sine wave represent the wavefronts or points of constant amplitude on the wave. There is one wavefront per wavelength. The direction of travel is the z direction. This will be assumed to be the direction of travel of all waves described in this book. The wave is sometimes said to *propagate* in the z direction.

In practice radio, microwave or optical waves are generated by a source which must be at a point. A wave in free space therefore spreads out from the source and the wavefronts are spherical. A cross-section of the wavefronts in a plane along the axis of propagation is sketched in Figure 5.2. The wavefronts are shown in the figure as portions of circles of ever increasing diameter. In reality the wavefronts are portions of a sphere. A long distance away from the source, the arc of the circle shown in Figure 5.2 has become almost flat, or plane. This is the situation for most practical application because the receiver, or observer, is a long distance from the source of microwaves or optical waves. Then the section of spherical waves which is detected is essentially a plane wave, thus plane waves are a good description of most non-guided waves. All radio and microwave systems used for communications, radar and broadcasting fall into this category and so do all free space optical systems. A laser beam is a good example of this later case as will be described later in the chapter. The laser beam is confined in the transverse plane but the transverse width in wavelengths is very large.

The theory for the plane wave is much simpler than that for spherical wave so if the true spherical wave can be approximated by a plane wave, it is well worth doing. The question naturally arises as to how close the receiver can be to the source before the approximation of a plane wave breaks down. There is no fixed answer to the question, but a commonly accepted criteria is to say that if the path length difference between the plane wave and the spherical wave is less than $\lambda/16$, the result is acceptable. This distance is called the *far-field distance* and is derived by the geometry shown in Figure 5.3. It depends on the width, D, of the detector or antenna, and can be expressed as

$$R = \frac{2D^2}{\lambda} \tag{5.1}$$

5.2 WAVE AND RAY REPRESENTATION OF PLANE WAVES

Plane waves can be represented in diagrams either as waves or as rays. In the case of the wave representation, the wave is drawn as a series of lines showing points of constant

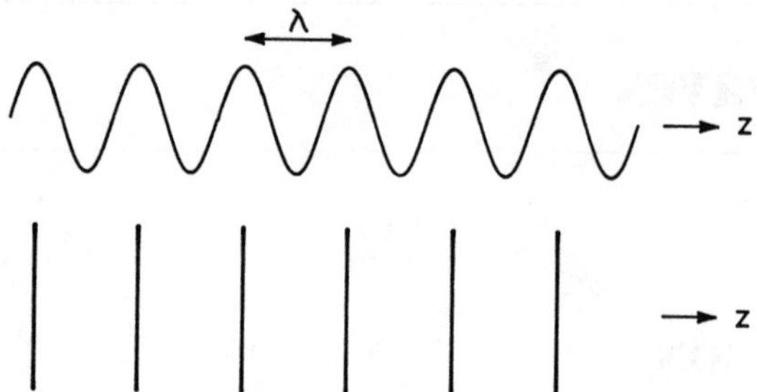

Figure 5.1. Plane wave showing sinusoidal wave and plane wavefronts

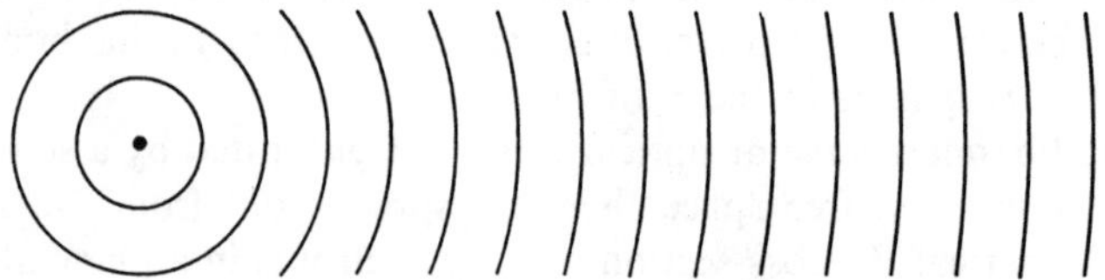

Figure 5.2. Spherical wave from a source

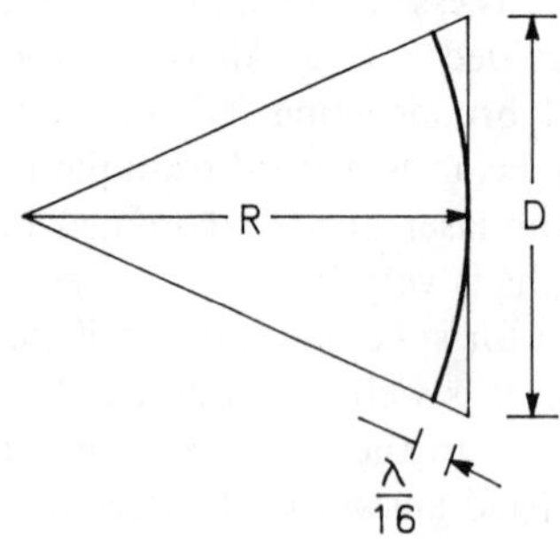

Figure 5.3. Far field distance showing difference between spherical and plane wavefronts

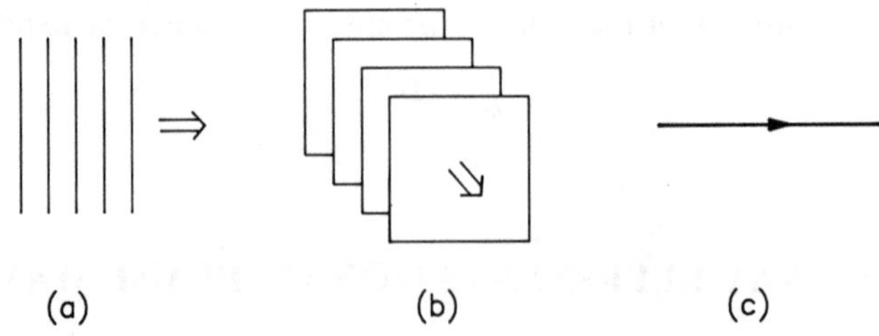

Figure 5.4. Representation of plane waves (a) wavefronts in two dimensions, (b) wavefronts in three dimensions, (c) ray representation

phase. These are at right angles to the direction of travel as shown schematically in Figure 5.4(a) in two-dimensional form and Figure 5.4(b) in three-dimensional form. For the purposes of drawing the wave, it has been truncated in the transverse direction, whereas by definition it is infinite in the transverse direction, thus the sketch is approximate.

A *ray* is a line drawn on paper or on a screen which *depicts* a *ray* of light or a *ray* of a microwave beam. The ray representation does not suffer from the approximation inherent when drawing a wavefront, but neither does it represent anything physical. The line is drawn to correspond to the direction of travel of the plane wave, as shown in Figure 5.4(c). It is a mathematical device without physical reality. The nearest practical realization is a narrow laser beam, though as has been discussed above, a laser beam is actually very wide if measured in wavelengths. The ray is easy to draw and visualize; thus it is the method of representation of plane waves which is most used. A ray is everywhere orthogonal to a wavefront, so the wavefront can be easily visualized.

5.3 BASIC PROPERTIES OF PLANE WAVES

The theory of plane waves will be developed in this section starting from Maxwell's equations and the electromagnetic wave equation. The equations obtained will enable the characteristics of plane waves to be deduced. The initial assumption is made that the plane wave is in free space so that the medium has a relative permittivity of unity and a relative permeability of unity ($\varepsilon_r = \mu_r = 1$). The assumption will also be made that the region of interest is sufficiently far from the source of the waves so that there are no sources or charges influencing the wave. This implies that the current density, $\boldsymbol{J} = 0$ and the charge density, $\rho = 0$.

As well as being a good approximation to real systems, a plane wave is a very useful mathematical description because it simplifies the analysis of electromagnetic waves. The basic characteristic of a plane wave is that there is no variation in two directions of a rectangular coordinate system. These will be taken to be the x and y directions, Figure 5.5. This is expressed mathematically by the fact that the partial derivatives in the x and y directions are zero, i.e. $\partial/\partial x = 0$ and $\partial/\partial y = 0$. When these constraints are applied to Maxwell's equations, a number of important deductions can be made about the basic properties of plane waves.

The first Maxwell equation is

$$\nabla \times \boldsymbol{E} = -\mathrm{j}\omega\mu_0 \boldsymbol{H} \tag{5.2}$$

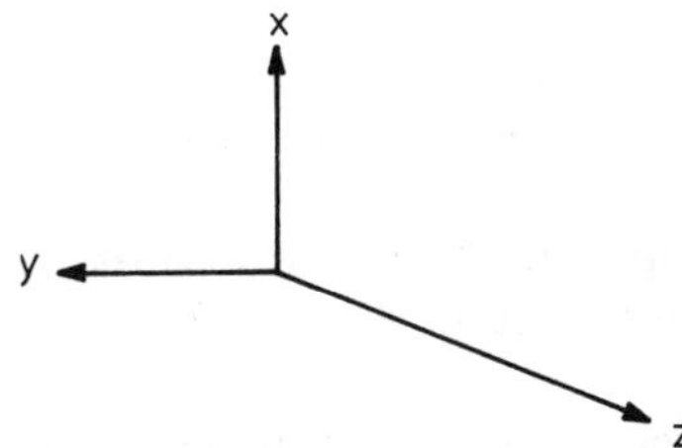

Figure 5.5. Rectangular coordinate system with x and y transverse to the wave direction

It can be expanded in component form as

$$\left(\frac{\partial E_z}{\partial y}-\frac{\partial E_y}{\partial z}\right)\hat{x}+\left(\frac{\partial E_x}{\partial_z}-\frac{\partial E_z}{\partial_x}\right)\hat{y}+\left(\frac{\partial E_y}{\partial x}-\frac{\partial E_x}{\partial y}\right)\hat{z}=-\mathrm{j}\omega\mu_0(H_x\hat{x}+H_y\hat{y}+H_z\hat{z}) \quad (5.3)$$

The first, fourth, fifth and sixth parts of the left-hand side of equation (5.2) are zero because the partial derivatives are zero. Equating the x, y and z components separately gives:

In the x direction

$$\frac{\partial E_y}{\partial z}=\mathrm{j}\omega\mu_0 H_x \quad (5.4)$$

In the y direction

$$\frac{\partial E_x}{\partial z}=-\mathrm{j}\omega\mu_0 H_y \quad (5.5)$$

In the z direction

$$0=\mathrm{j}\omega\mu_0 H_z \quad (5.6)$$

A similar set of equations can be written starting from the second Maxwell equation with $\boldsymbol{J}=0$,

$$\nabla\times\boldsymbol{H}=\mathrm{j}\omega\varepsilon_0\boldsymbol{E} \quad (5.7)$$

In the x direction

$$\frac{\partial H_y}{\partial z}=-\mathrm{j}\omega\varepsilon_0 E_x \quad (5.8)$$

In the y direction

$$\frac{\partial H_x}{\partial z}=\mathrm{j}\omega\varepsilon_0 E_y \quad (5.9)$$

In the z direction

$$0=\mathrm{j}\omega\mu_0 E_z \quad (5.10)$$

Examination of equations (5.4) to (5.6) and (5.8) to (5.10) enables the following deductions to be made:

1. H_z and E_z are both zero. The field components of the plane wave are E_y, H_x, E_x and H_y which are electric and magnetic field components transverse to the z direction. There are no components in the z direction.

2. If the electric field component, E_y exists, so does the magnetic field component, H_x. This follows from equations (5.4) and (5.9).

3. If the electric field component, E_x exists, so does the magnetic field component, H_y. This follows from equations (5.5) and (5.8).

4. There is no mathematical connection between the E_y/H_x solution and the E_x/H_y solution so both sets can exist independently.

5. The equations derived from the first and second Maxwell's equations are equivalent. Only one of the equations need be used.

The implications of these deductions are of considerable importance for understanding how microwaves and optical waves travel in free space. The first deduction implies that the wave can be described as a transverse wave. The electric and magnetic fields are in a plane which is transverse to the direction in which the wave is moving. As the wave moves through space, there are thus no magnetic and electric field components along the direction of travel. The picture of the wavefronts shown in Figure 5.4(b), also applies if the planes are assumed to contain the electric and magnetic fields. This property of plane waves considerably simplifies the design of antennas which are needed to capture a portion of the wave.

The second deduction comes from the relationships between electric and magnetic fields in Maxwell's equations. It states that there are only two field components necessary for a valid plane wave, one electric field component, E_y and an associated magnetic field component, H_x. The interlinking of the electric and magnetic fields is a basic property of an electromagnetic wave which is manifest here for these two particular field components. The third deduction states the same principle as the second deduction, but says that it also applies to the E_x electric field component and the H_y magnetic field component.

The fourth deduction comes from examining the equations and noting that none of the equations containing the transverse components show any linkage between E_x and E_y. Thus, mathematically, the E_y and H_x wave is completely independent of the E_x H_y wave. This must also be true physically and means that a basic property of plane waves is that there can be two orthogonal plane waves at the same frequency, each with two field components. Both the plane waves can exist simultaneously and they will not interact with each other. This has many significant practical applications. For instance it means that two communication signals can be carried simultaneously at the same frequency and no transfer of power will take place. Only one set of antennas, transmitters and receivers are needed for two independent communication channels. The actual wave that is sent out from a transmitter will have field components that are determined by the physical design of the transmitting antenna, or in the case of an optical wave, by the lens. It is the designers task to ensure that the antenna or lens is such that the correct field component (E_x or E_y) is excited. The property of two waves to exist simultaneously will be found later to be valid in a more general sense for waves in waveguides.

The fifth and final deduction is to note that there was no extra information provided by expanding the second Maxwell equation into equations (5.8) to (5.10). This is due to the absence of sources in the analysis which means that there are redundant terms in Maxwell's equations.

5.4 PLANE WAVE EQUATION AND ITS SOLUTION

The properties of the plane wave which were deduced above set the scene for deriving the electric and magnetic field components of a plane wave. In order to find the form of the fields, the electromagnetic wave equation must be solved. This was derived in

Chapter 2 in vector differential form as

$$\nabla^2 \boldsymbol{E} = -\omega^2 \mu_0 \varepsilon_0 \boldsymbol{E} \tag{5.11}$$

In full differential form it becomes

$$\frac{\partial^2 \boldsymbol{E}}{\partial x^2} + \frac{\partial^2 \boldsymbol{E}}{\partial y^2} + \frac{\partial^2 \boldsymbol{E}}{\partial z^2} = -\omega^2 \mu_0 \varepsilon_0 (E_x \hat{\boldsymbol{x}} + E_y \hat{\boldsymbol{y}} + E_z \hat{\boldsymbol{z}}) \tag{5.12}$$

This can be simplified for plane waves by applying the restriction that $\partial/\partial x = \partial/\partial y = 0$ and $E_z = 0$, then

$$\frac{\partial^2 E_x}{\partial z^2} \hat{\boldsymbol{x}} + \frac{\partial^2 E_y}{\partial z^2} \hat{\boldsymbol{y}} = -\omega^2 \mu_0 \varepsilon_0 (E_x \hat{\boldsymbol{x}} + E_y \hat{\boldsymbol{y}})$$

By equating the x and y components, two separate second-order differential equations result. One for the E_x wave and one for the E_y wave.

$$\frac{\partial^2 E_x}{\partial z^2} = -\omega^2 \mu_0 \varepsilon_0 E_x \tag{5.13}$$

and

$$\frac{\partial^2 E_y}{\partial z^2} = -\omega^2 \mu_0 \varepsilon_0 E_y \tag{5.14}$$

The two equations have identical forms showing that the characteristics of the E_x and E_y waves will be the same. The solution to equations (5.13) and (5.14), together with equations (5.4) and (5.5), completely describe the characteristics of plane waves in free space.

It would have been possible to have started with the magnetic field wave equation and obtained the same type of relationships as equations (5.13) and (5.14) but with E_x replaced by H_x and E_y by H_y. Only one wave equation is needed since Maxwell's equations allow the other equations to be derived. It is usual to start with the electric field wave equation as has been done above. The next stage is to solve the plane wave equation in order to derive a mathematical description of a plane wave. This will then be interpreted physically in order to demonstrate the characteristics of plane waves travelling in free space. The solution for the E_y equation will be developed but the form of the E_x wave will be identical.

The wave equation in any of the forms above, equations (5.11) to (5.14) implicitly contains a factor $\exp(j\omega t)$ which expresses the fact that the wave equation is for a wave made up of sinusoidal variations. This was inserted when the electromagnetic wave equation was derived in Chapter 3. Thus the electric field E_y in equation (5.14) is actually $E_y e^{j\omega t}$. The equation does not need modifying because the exponential factor appears on both sides of the wave equation and therefore cancels. In order to discuss the physical meaning of the solution to the wave equation, the sinusoidal factor needs to be included in the solution. The remainder of the chapter will therefore use a description of plane waves which include the time dependence term to show the full properties of the wave. In later chapters, the time dependence will only be included where it is pertinent to the physical understanding. This is common practice in microwaves and optics.

The time dependence can be written either using the phasor form of $e^{j\omega t}$ or $\sin(\omega t)$ $[= \mathrm{Im}(e^{j\omega t})]$ or $\cos(\omega t)$ $[= \mathrm{Re}(e^{j\omega t})]$. The phasor form is most general, but all forms contain a full description of the time dependence. There is no difference between using sine or cosine because $\cos(\omega t) = \sin(\omega t + 90°)$ and a 90° time factor, or quarter of a time period, has no significance as time continuously changes. The solutions to the wave equation will be written using the sine form because it is easier to visualise the wave.

The wave equation for the E_y wave, equation (5.14), is a second-order differential equation which will be given a trial solution of the form

$$E_y = \sin \omega \left(t + \frac{z}{c} \right) \tag{5.15}$$

where c is a constant to be determined by inserting equation (5.19) into equation (5.14). This gives

$$-\frac{\omega^2}{c^2} \sin \omega \left(t + \frac{z}{c} \right) = -\mu_0 \varepsilon_0 \omega^2 \sin \omega \left(t + \frac{z}{c} \right) \tag{5.16}$$

which is satisfied if $c^2 = 1/(\mu_0 \varepsilon_0)$. The absolute permeability, μ_0, has the units of inductance per metre or the dimensions of $ML\,I^{-2}T^{-2}$. The absolute permittivity ε_0 has the units of capacitance per metre or the dimensions of $I^2 T^4 M^{-1} L^{-3}$. Thus the dimensions of $1/(\mu_0 \varepsilon_0)$ are $L^2\,T^{-2}$ or velocity squared. Hence c is a velocity with a value given by

$$c = \frac{1}{(\varepsilon_0 \mu_0)^{1/2}} \tag{5.17}$$

or

$$c = 1/(4\pi \times 10^{-7} \times 8.854 \times 10^{-12})^{1/2} = 2.998 \times 10^8 \text{ m/s}$$

This is called the *velocity of light* but is the velocity of all electromagnetic waves. The velocity of light is so fundamental to many aspects of physics and engineering that an international value was agreed in 1983 of exactly $2.997\,924\,58 \times 10^8$ m/s. In all engineering applications the velocity of light can be taken to have the convenient rounded value of 3×10^8 m/s. The difference between this value and the internationally agreed value is only 0.069% which is much less than the tolerances on components or other electrical quantities.

The solution to the wave equation, equation (5.15), can be written in a modified form which is better for physical interpretation. This is done by taking the ω term into the bracket and replacing ω/c by $2\pi f / f \lambda$ or $2\pi/\lambda$, where λ is the wavelength of the wave. Then

$$E_y = \sin(\omega t + \beta z) \tag{5.18}$$

where $\beta = 2\pi/\lambda$. This is called the *phase constant* or *propagation constant* of the wave because it describes how the phase of the wave behaves in the direction of travel. A phase constant will be found to appear in all descriptions of electromagnetic waves but only in free space does it depend on the reciprocal of the wavelength. In other medium it is modified by the parameters of the medium, as described below.

Equation (5.17) is one form for the solution but it is not the only form of solution to the wave equation. Other possible solutions are

$$E_y = E_0 \sin(\omega t - \beta z) \tag{5.19}$$

$$E_y = E_0 \cos(\omega t \pm \beta z) \tag{5.20}$$

$$E_y = E_0 \exp[\mathrm{j}(\omega t \pm \beta z)] \tag{5.21}$$

An amplitude factor, E_0 has been added in front of all the solutions. The significance of the sign in front of βz will be seen in the next section. The exponential form in equation (5.21) is the most general description of a plane wave.

5.5 TRAVELLING WAVES

The solutions to the wave equations represent travelling waves. This can be demonstrated by plotting the basic plane wave solution, equation (5.18), $E_y = \sin(\omega t + \beta z)$, against the distance parameter βz for three instances of time, Figure 5.6. When $t = 0$, $E_y = \sin \beta z$ and curve A results. A quarter of a time period later, $t = T/4$, and $\omega t = 2\pi f\, t = 2\pi t/T = \pi/2$. This is shown in curve B. Similarly after half a time period, $\omega t = \pi$ and the situation is as shown in curve C.

If a point, P, on the crest of the wave is considered and its behaviour monitored, Figure 5.6, it is seen that as time progresses, the point P moves to the *left*. The plane wave solution thus represents a wave *travelling* to the left, that is, in the negative z direction.

Now consider the alternate form of solution, $E_y = \sin(\omega t - \beta z)$. If this is plotted in the same way, then a point on the crest of the wave moves to the *right* and the solution represents a wave travelling in the positive z direction. This is the natural direction to use as a reference direction, so the form of solution of the plane wave which will be used from now on will contain the negative sign in front of the space factor, that is

$$E_y = \sin(\omega t - \beta z) \quad \text{or} \quad E_y = \exp[\mathrm{j}(\omega t - \beta z)]$$

The argument of the sine or exponential term is proportional to frequency, as

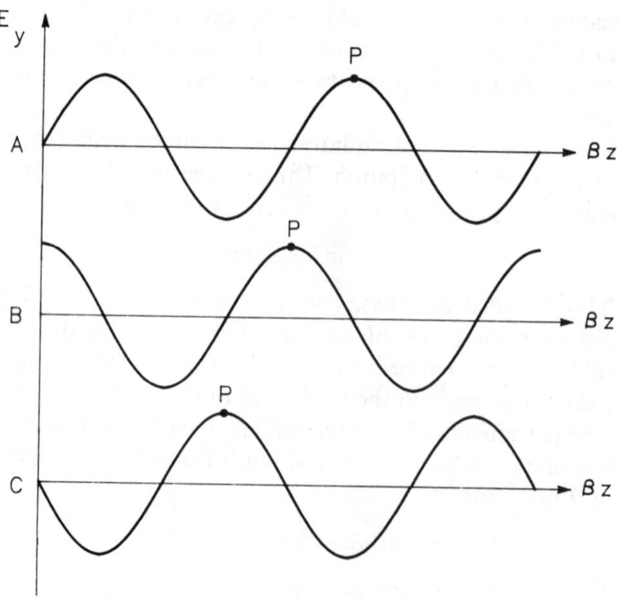

Figure 5.6. Travelling wave at $t = 0$ (curve A); $t = T/4$ (curve B); $t = T/2$ (curve C)

equation (5.15) shows thus the frequency of the wave is crucial in determining the period of the plane wave in both the time domain and the space domain. The argument $(\omega t - \beta z)$ can be written as $2\pi(ft - z/\lambda)$ which shows that the frequency directly controls the time dependent part of the argument, whilst the spatially dependent part is determined by distance divided by the wavelength. This distance is called the *normalised* distance because it is a function of the wavelength. It is the normalised distance which appears in most equations concerned with electromagnetic waves, rather than the absolute distance. This is important because the spatial part of the argument will have the same value whether the distance is 10 mm with a wavelength of 30 mm or a distance of 0.5 μm with a wavelength of 1.5 μm.

5.6 PHASE VELOCITY

It has already been established that the plane wave travels along at the velocity of light. This can be derived in another way by working out the speed of movement of the point P in Figure 5.6. The point P is a point on the wavefront of constant phase and is characterized by the condition

$$\omega t - \beta z = \text{constant}$$

Differentiating this condition with respect to time gives

$$\omega - \beta \frac{\mathrm{d}z}{\mathrm{d}t} = 0$$

or

$$\frac{\mathrm{d}z}{\mathrm{d}t} = +\frac{\omega}{\beta} = +c \tag{5.22}$$

This is the velocity of a constant phase point on the wave and is called the *phase velocity*. The phase velocity is positive, confirming that the wave is travelling in the positive z direction. In the case of a plane wave in free space, the phase velocity is the velocity of light. This is not, however, true for waves that are enclosed in waveguides. It will be shown in later chapters that the phase velocity for guided waves can differ markedly from the velocity of light.

5.7 TRAVELLING WAVES IN A DIELECTRIC

The discussion so far has considered only the case of plane waves in free-space, of which the atmosphere is a good approximation. A more general situation is when the plane wave is travelling in a homogeneous dielectric with a relative permittivity that is greater than one, i.e. $\varepsilon_r \geqslant 1$. The wave equation will have the same form as that derived in Section 5.3 except that ε_0 will be replaced by $\varepsilon_0\varepsilon_r$. Consequently the form of the solution for a plane wave will be similar to that in Section 5.4 except that the velocity of light, c, will be replaced by a velocity v, so that

$$E_y = E_0 \sin \omega \left(t - \frac{z}{v} \right) \tag{5.23}$$

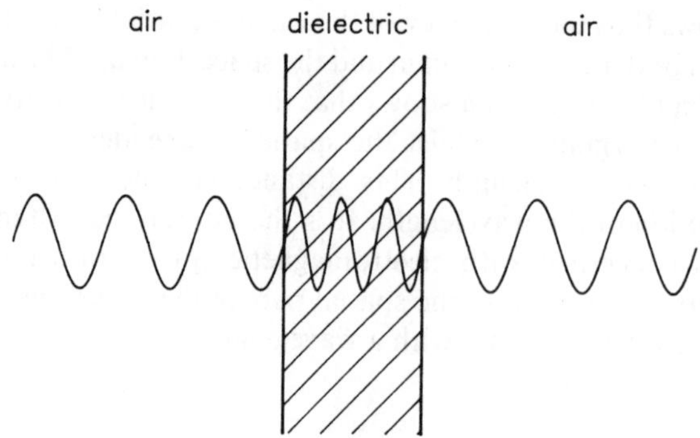

Figure 5.7. Plane wave passing through a slab of dielectric

where

$$v = \frac{1}{(\varepsilon_0 \varepsilon_r \mu_0)^{1/2}} = \frac{c}{\varepsilon_r^{1/2}} \tag{5.24}$$

This shows that the wave is slowed down in the dielectric, by comparison with the velocity in free space. The phase constant, β, is now equal to ω/v. It will increase by comparison with free space and will have the value $2\pi\varepsilon_r^{1/2}/\lambda$ where λ is the wavelength in free space. It can also be written as $\beta = 2\pi/\lambda'$ where $\lambda' = \lambda/\varepsilon_r^{1/2}$ and λ' is the apparent wavelength of the wave in the dielectric. This is always shorter than the free space wavelength. This is sketched in Figure 5.7 which shows a plane wave passing through a slab of dielectric. The wave is bunched in the dielectric as it slows down and the wavelength decreases.

Example 5.1 Velocity and wavelength in glass

Find the velocity and apparent wavelength of an optical wave at 0.7 μm in air and in glass with a relative permittivity of 3.1.

The values are calculated by using the above relationships. The velocity in air is 3×10^8 m/s and in glass it is $3 \times 10^8/\sqrt{3.1}$ or 1.70×10^8 m/s. The wavelength in air is 0.7 μm, and in glass it is $0.7/\sqrt{3.1}$ or 0.40 μm.

5.8 INDEX OF REFRACTION IN GENERAL MEDIA

A general medium could have both a non-unity relative permittivity and a non-unity relative permeability, that is $\varepsilon_r \neq 1$ and $\mu_r \neq 1$. Extrapolating the discussion of the last section indicates that the phase velocity of a plane wave in a general medium will be

$$v = \frac{1}{(\varepsilon_0 \varepsilon_r \mu_0 \mu_r)^{1/2}} \tag{5.25}$$

The *index of refraction, n*, is defined as the reciprocal of the relative phase velocity

$$n = \frac{c}{v} = (\mu_r \varepsilon_r)^{1/2} \tag{5.26}$$

There are very few practical media where the permeability is not nearly unity, so the index of refraction is usually expressed as the square root of the relative permittivity, or

$$n = \varepsilon_r^{1/2} \tag{5.27}$$

If a material such as glass has $\varepsilon_r = 3$, then $n = 1.73$. The index of refraction is merely an alternative way of expressing the electrical parameters of a material or gaseous medium. Historically it was common practice for physicists to use the index of refraction when working with optics and for engineers to use the relative permittivity when working with microwave systems. This practice has stuck and so medium in which microwave signals travel is usually described by its relative permittivity and medium in which optical waves travel is usually described by its refractive index.

5.9 MAGNETIC FIELD OF THE PLANE WAVE

The characteristics of the plane wave described in Sections 5.4 to 5.7, were deduced from the solution to the electric field wave equation and the E_y electric field component. The associated magnetic field, H_x, can be obtained from the appropriate part of Maxwell's equations. In this case, this means equation (5.4), but because the solution for E_y uses the sine form for the time dependence, the full time differential must be included, so equation (5.4) for a plane wave in a general medium becomes

$$\mu_0\mu_r\frac{\partial H_x}{\partial t} = \frac{\partial E_y}{\partial z} \tag{5.28}$$

(Note that if the exponential form, $\exp[\mathrm{j}(\omega t - \beta z)]$ had been used, then equation (5.4) could be used directly.) Substituting $E_y = E_0 \sin(\omega t - \beta z)$ into equation (5.28) and evaluating the differential with respect to z and the integral with respect to time gives

$$H_x = -\frac{\beta E_0}{\omega\mu_0\mu_r}\sin(\omega t - \beta z)$$

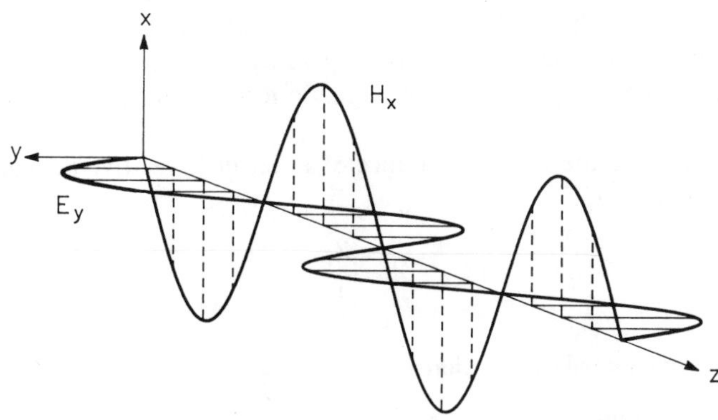

Figure 5.8. Electric and magnetic fields of the E_y/H_x plane wave

or by replacing ω/β by the velocity and expanding

$$H_x = -\left(\frac{\varepsilon_0 \varepsilon_r}{\mu_0 \mu_r}\right)^{1/2} E_0 \sin(\omega t - \beta z) \tag{5.29}$$

Comparing this equation for the H_x magnetic field with that for the E_y electric field indicates that the phase factor is the same and the only the amplitude factor is different.

The electric and magnetic fields of a plane wave are shown schematically in space and at one instant of time in Figure 5.8. The negative sign in front of the amplitude of H_x determines that the magnetic field lies in the negative xz plane. The electric and magnetic components of the plane wave are in synchronism at all points in time and space and differ only by the orthogonal planes in which they oscillate.

5.10 WAVE IMPEDANCE IN A GENERAL MEDIA

The ratio of the electric field and the magnetic field of the plane wave has the dimensions of resistance since the units of the electric field are V/m and of the magnetic field are A/m. Dividing E_y by H_x gives

$$Z = \frac{E_y}{H_x} = \left(\frac{\mu_0 \mu_r}{\varepsilon_0 \varepsilon_r}\right)^{1/2} \Omega \tag{5.30}$$

where Z is called the *wave impedance* (or sometimes the *intrinsic* impedance) of the medium through which the plane waves are travelling. The wave impedance is independent of any of the parameters of the plane wave, that is the frequency and the amplitude of the wave. It is a characteristic only of the medium containing the plane waves, in the same way as a resistor does not depend on the actual voltage across it.

If the medium through which plane waves are travelling is free space, the wave impedance becomes

$$Z_0 = \left(\frac{\mu_0}{\varepsilon_0}\right)^{1/2} \tag{5.31}$$

This has a value of 376.731 Ω and is usually referred to as the impedance of free space. For engineering purposes it can be taken to have a value of $Z_0 = 377\,\Omega$ or $120\pi\,\Omega$. It may seem strange that the air has an impedance, but the air, or free space, is only a particular medium which has characteristics give by the absolute permittivity, ε_0, and the absolute permeability, μ_0. In electromagnetic terms there is nothing special about free space.

It is useful to express the wave impedance of a dielectric relative to that of free space, in which case equation (5.30) becomes

$$Z = \frac{Z_0}{\varepsilon_r^{1/2}} \tag{5.32}$$

Example 5.2 Wave impedance in materials

Calculate the wave impedance of a plane wave travelling in (a) water with $\varepsilon_r = 80$ and in (b) polyethylene with $\varepsilon_r = 2.25$.

direction in a rectangular coordinate system. Two orthogonal waves have been derived. In a practical system, such as a communication channel in free space, there may be a combination of waves and there may be no reference coordinate system. A way is needed of describing the orientation of the plane waves in space. This is done with the *polarisation* of the plane wave. The polarisation describes the way in which the shape and orientation of the locus of the ends of the electric field vector in a fixed plane in space varies with time. The fixed plane is transverse to the direction of travel, meaning the xy plane. The polarisation of a wave is subdivided into a number of categories.

LINEAR POLARISED PLANE WAVES

If a transverse plane is placed at any point along the z axis of Figure 5.8, then the electric field of the E_y wave will move along the line in the transverse plane shown in Figure 5.10(a). This wave is said to be *linearly polarised* because the electric vector varies only in one plane which in this case is the yz plane.

When the electric field vector is horizontal to the surface of the Earth, as implied in Figure 5.10(a), the wave is also said to be *horizontally polarised*. This notation is widely used in broadcasting to describe the orientation of a plane wave broadcast by a transmitting antenna. By analogy a *vertically polarised* wave is a wave with an electric field vector which is normal to the surface of the Earth. This is the case for the E_x wave using the same coordinate system, Figure 5.10(b), (corresponding to Figure 5.9). A linear polarised wave does not have to aligned with one of the axes of the coordinate system, it could be a line at an arbitrary angle to the x axes, Figure 5.10(c).

It has already been established mathematically that an E_y wave and an E_x wave can exist simultaneously. This is also physically observed by comparing Figures 5.10(a) and 5.10(b). If they are superimposed, both E_y and E_x could exist simultaneously and there would be no interaction between the two components and both can exist without cross-coupling. They are called orthogonally polarised waves and the situation is shown schematically in Figure 5.11 which plots just the electric field components of the two plane waves. This ability for two plane waves to exist simultaneously is put to good use in some advanced satellite and radio relay communication systems where two orthogonal linear polarised signals using the same microwave frequency are transmitted through one transmitting antenna and received by one receiving antenna. This enables the effective communication channel capacity to be doubled. There is a penalty in that the antennas and other components at either end of the link have to be of high quality

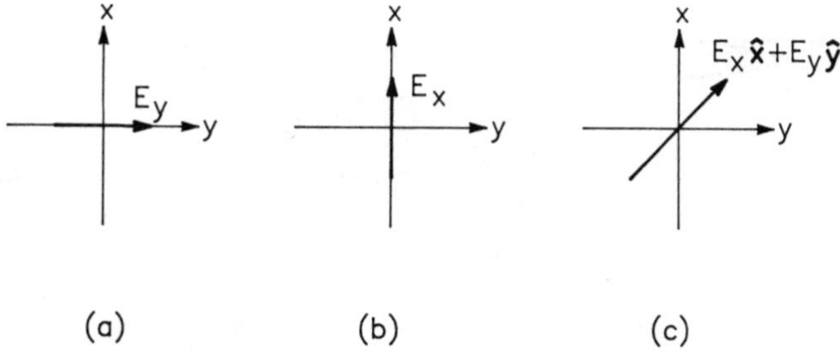

Figure 5.10. Electric vector for linear polarised wave. (a) horizontal polarisation, (b) vertical polarisation, (c) general linear polarisation

Inserting the values into equation (5.32) gives $Z = 42.1\,\Omega$ for water and $Z = 251\,\Omega$ for polyethylene. The higher the relative permittivity, the lower the wave impedance. It is sometimes useful to express the wave impedance as a dimensionless quantity normalised to the impedance of free space. Equation (5.32) shows that the normalised impedance is $1/\varepsilon_r^{1/2}$ or the reciprocal of the refractive index. For the values in this example, the normalised impedance is 8.94 for water and 1.5 for polyethylene.

5.11 THE ORTHOGONAL ELECTRIC FIELD SOLUTION

It was shown in Section 5.3 that there are two independent solutions to the wave equation and consequently two independent plane waves can exist. The other electric field solution is that of the E_x wave. This is orthogonal to the E_y wave which has just been studied. It is described by an identical equation to the E_y wave as can be seen by examination of equations (5.14) and (5.13). The associated magnetic field is the H_y wave and is given by equation (5.5). The only difference between the E_x and H_y wave and the E_y and H_x wave is a sign change for H_y. The fields are given by

$$E_x = E_0 \sin(\omega t - \beta z) \tag{5.33}$$

$$H_y = \frac{E_0}{Z} \sin(\omega t - \beta z) \tag{5.34}$$

The schematic representation of this wave is shown in Figure 5.9. If this is compared with Figure 5.8, it is seen that the E_x/H_y wave is just the E_y/H_x wave rotated through 90°, thus, all the properties of the E_x/H_y wave are identical to those of the E_y/H_x wave.

5.12 POLARISATION OF PLANE WAVES

The description of plane waves in the previous sections has dealt with them by mathematical analysis which expressed the field components of the plane wave by their

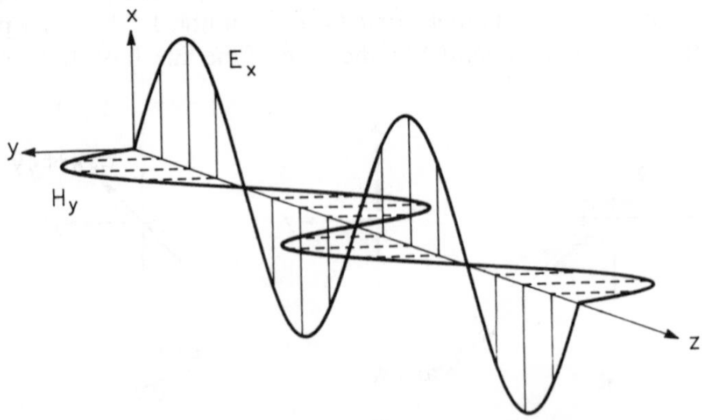

Figure 5.9. Electric and magnetic fields of the E_x/H_y plane wave

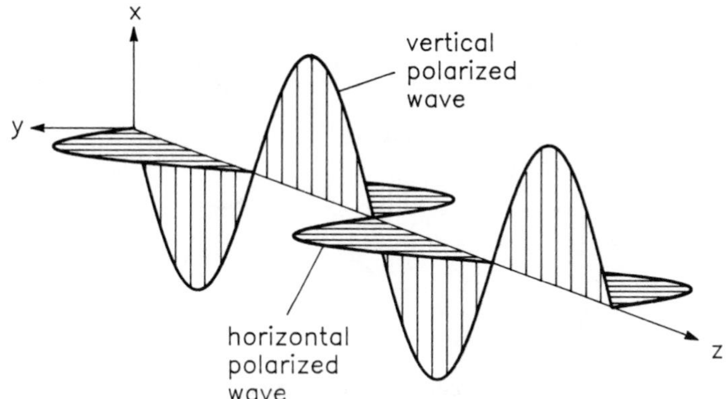

Figure 5.11. Two orthogonal linear polarised plane waves

so that they do not couple any signal into the orthogonal channel. The quality of the antennas are measured by the cross-polarisation which is the amount of cross coupling between the two linear polarisations.

CIRCULAR POLARISED PLANE WAVES

A linear polarised plane wave has one vector in a single plane transverse to the direction of propagation. It is also possible to have a combination of two orthogonal waves which do not exist independent of each other. The most useful is called *circular polarisation*. A circular polarised plane wave is one in which there are two orthogonal components of equal amplitude and with a quarter of a time period between the components. This can be demonstrated by writing down the field components of a general plane wave,

$$\boldsymbol{E} = E_x\hat{\boldsymbol{x}} + E_y\hat{\boldsymbol{y}}$$

If $E_x = E_1 \sin(\omega t - \beta z)$ and $E_y = E_2 \sin(\omega t - \beta z + \theta)$ where θ is the phase difference between E_y and E_x, then

$$\boldsymbol{E} = E_1 \sin(\omega t - \beta z)\hat{\boldsymbol{x}} + E_2 \sin(\omega t - \beta z + \theta)\hat{\boldsymbol{y}} \tag{5.35}$$

When $E_1 = E_2$ and a quarter of time period exists between the two components which is expressed as $\theta = \omega t = \pm\pi/2$, equation (5.35) becomes

$$E_x\hat{\boldsymbol{x}} + E_y\hat{\boldsymbol{y}} = E_1[\sin(\omega t - \beta z)\hat{\boldsymbol{x}} \pm \cos(\omega t - \beta z)\hat{\boldsymbol{y}}] \tag{5.36}$$

To find the magnitude of this equation, the sum of the squares is found

$$\begin{aligned} E_x^2 + E_y^2 &= E_1^2[\sin^2(\omega t - \beta z) + \cos^2(\omega t - \beta z)] \\ &= E_1^2 \end{aligned}$$

This is the equation of a circle in the transverse xy plane with radius E_1. It is plotted in Figure 5.12(a). In three-dimensional space the locus of the electric field vector traces out a helix, Figure 5.13. The rotational direction of the helix, or of the vector in Figure 5.12(a), is determined by whether $\theta = +\pi/2$ or $\theta = -\pi/2$. It is important to be consistent in the definition of the rotational direction because the transmitter and receiver

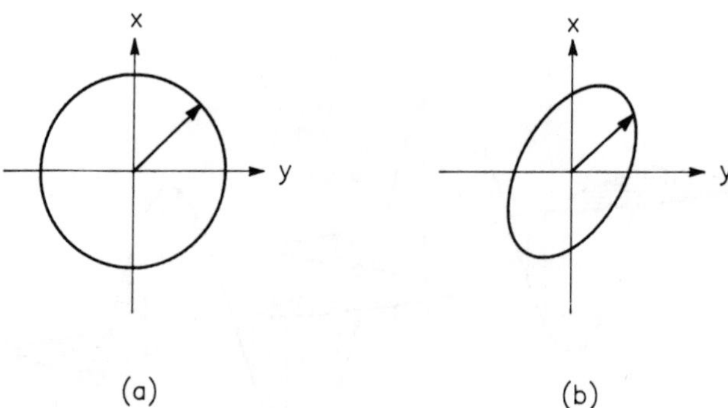

Figure 5.12. Electric vector of (a) a circular polarised plane wave, (b) an elliptical polarised plane wave

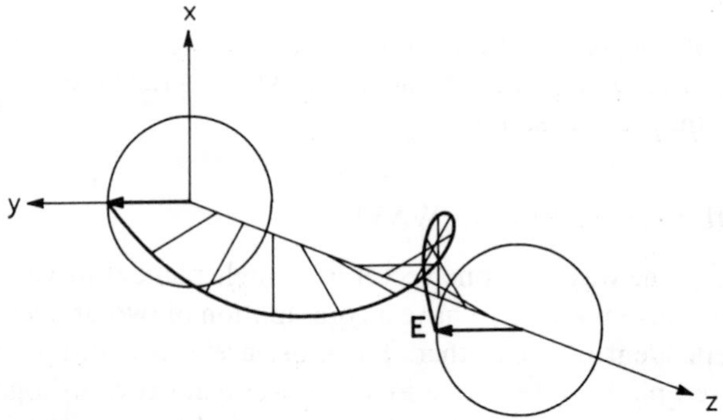

Figure 5.13. Locus of electric vector of a circular polarised wave

in a system using circular polarisation can be a long way apart. The IEEE definition defines the rotational direction by the *hand* of polarisation *as seen by an observer looking in the direction of travel of the wave.* A *left-hand* polarised wave is one in which the tip of the electric vector appears to move in a left-hand direction, looking along the direction of travel. This corresponds to $\theta = +\pi/2$ so that the E_y wave is advanced by quarter of a time period.

If the E_y field had been delayed in phase by a quarter of a time period then the electric vector would circle in the right-hand direction and be called a *right-hand circular polarised* wave. A right-hand circular polarised wave and a left-hand circular polarised wave do not interact with each other and so both can exist simultaneously, just as with two orthogonal linear polarised waves.

Circular polarisation is used in some communication and radar systems even though it is more complicated to transmit and receive. The advantages of circular polarisation are that at some frequencies it is less affected by the atmosphere and the receiving antenna does not have to be carefully aligned to the electric vector. It is used in some satellite communication systems and in direct broadcasting satellites. In radar it is used because when a circular polarised wave hits a metal target it is reflected with the opposite polarisation. This then gives a method of distinguishing metal targets (e.g. aircraft) from clouds and clutter.

The circular polarised wave is generated in practice by a *polariser* which produces a circular polarised wave from a linear polarised wave. The polariser first splits the linear polarised wave into two orthogonal waves and then delays one of the waves by quarter of a time period. A quarter of a time period is equivalent to a phase delay of 90° and this can be achieved in practice by making one of the waves travel through a piece of dielectric of the correct length so that it has been delayed in phase by 90°. A polariser will work equally well in the reverse direction. If a circular polarised wave is incident on the polariser, a single linear polariser wave will be generated at the output.

ELLIPTICAL POLARISED PLANE WAVES

If the amplitudes E_1 and E_2 are unequal and the phase difference θ is also arbitrary, then equation (5.35) describes an elliptical polarised plane wave. As the name implies, the tip of the electric field vector traces out an ellipse in a transverse plane, Figure 5.12(b). The ratio of the length of the axes of the ellipse define the *axial ratio*. An axial ratio of one is circular polarisation. Elliptical polarisation is the most general form of polarisation, but also in practice the least desirable because it is the most complicated. Waves will usually be transmitted in circular polarisation, but the influence of a non-perfect transmission medium, such as gases in the atmosphere, will distort the wave so that at the receiver it is elliptically polarised. If the receiver is expecting a circular polarised wave, power will be lost because of the mismatch when the elliptically polarised wave arrives.

UNPOLARISED PLANE WAVES

An unpolarised plane wave is one where the phase and amplitude of the plane wave varied randomly and rapidly with time. The polarisation changes in a completely unpredictable fashion. Sunlight is an example of unpolarised plane waves. It is generated by a large number of random atomic emitters in the Sun. Each one radiates a polarised plane wave which lasts for no more than about 10^{-8} s, to be followed by other waves with random orientations. Another example occurs when a radio wave propagates through the ionosphere. The electron density in the ionosphere changes rapidly and causes the radio wave to change its polarisation.

5.13 POWER FLOW IN A PLANE WAVE

The power flow was derived for a general time harmonic electromagnetic wave in Section 3.10. The power density is given by the Poynting vector, $\boldsymbol{S} = \boldsymbol{E} \times \boldsymbol{H}$. This will now be derived for the case of a linear polarised plane wave in free space. Since a plane wave only has one electric field component, which will be taken as E_y and one magnetic field component, H_x, the complex Poynting vector has only one component in the z direction,

$$\boldsymbol{S} = S_z = \boldsymbol{E} \times \boldsymbol{H} = E_y H_x \hat{z} \tag{5.37}$$

As expected this expression states that the direction of energy flow is the z direction which is the direction of travel of the wave. Inserting the equations for E_y and H_x gives

$$S_z = \frac{E_0^2}{Z} \sin^2(\omega t - \beta z) \tag{5.38}$$

The time average power per unit area is given by $P_z = \frac{1}{2}\mathrm{Re}[\boldsymbol{E} \times \boldsymbol{H}^*]$ where $\boldsymbol{E}$ and $\boldsymbol{H}$ are the peak values of the time harmonic wave. For the plane wave there are no complex terms so

$$P_z = \frac{E_0^2}{2Z} (\mathrm{W/m^2}) \tag{5.39}$$

This is the power in a plane wave. It is analogous to the V^2/R power in a electric circuit.

Example 5.3 Power density in a radio wave

Calculate the power density in plane radio wave in free space which has an electric field amplitude of $E_0 = 1\,\mathrm{mV/m}$. Also calculate the power received by a receiving antenna with a circular cross-section of effective diameter 30 mm.

The average power density is given by equation (5.39) with $Z = Z_0 = 377\,\Omega$, i.e. $1.33 \times 10^{-9}\,\mathrm{W/m^2}$. This is a very small power, partly because it is spread over a square metre. The area of the receiving antenna is $\pi D^2/4$ or $7.07 \times 10^{-4}\,\mathrm{m^2}$ so the amount of power picked up (assuming 100% efficiency) is $9.4 \times 10^{-13}\,\mathrm{W}$.

5.14 LASER BEAM AS A PLANE WAVE

A laser beam provides a good practical example of a plane wave in free space. The beam has a fixed width so appears to violate the fundamental properties of plane waves that it is infinite in the transverse plane. The width of a typical laser beam measured in wavelengths is, however, very large. For instance a helium–neon laser emits light at a wavelength of 0.6328 μm, so a beam with a nominal diameter of 2 mm is 3160 wavelengths across. The beam will have an amplitude in the radial direction which decays away gradually from the centre. The centre of the beam will, however, have both an approximately constant amplitude and wavefronts which are very nearly flat. A typical cross-section is shown in Figure 5.14 and follows a Gaussian profile. This means that the electric field is approximated by

$$E_y = A \exp(-r^2/w^2) \sin(\omega t - \beta z) \tag{5.40}$$

where A is the amplitude on axis and w is called the effective beam radius. It is the radius where the electric field has fallen to 1/e of the value on axis. The Gaussian profile is caused by the shape of the lens or aperture at the end of the laser. It occurs in many optical systems and also microwave systems.

The average power density for the laser beam can be evaluated by using the formula in the last section. From equation (5.39)

$$P_z = \frac{1}{2}\frac{A^2}{Z_0} \exp(-2r^2/w^2)\,\mathrm{W/m^2} \tag{5.41}$$

The total power carried by the laser beam is the integral of the power density taken

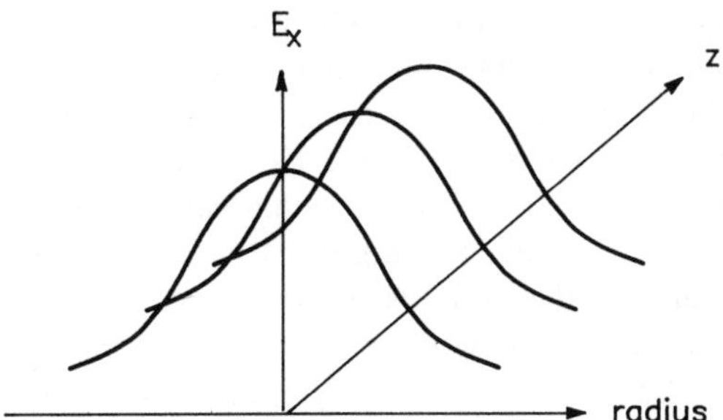

Figure 5.14. Cross-section of electric field of a laser beam

over the circular cross-section of the beam.

$$P_T = \int_0^{2\pi}\int_0^{\infty} P_z r\,dr\,d\phi = \frac{1}{2}\frac{A^2}{Z_0}\int_0^{2\pi}\int_0^{\infty} \exp(-2r^2/w^2) r\,dr\,d\phi \tag{5.42}$$

$$= \frac{\pi w^2}{4}\frac{A^2}{Z_0}$$

The power density at the centre of the beam is given by equation (5.41) at $r = 0$, or $P_0 = A^2/2Z_0$ W/m^2. The total power can be expressed in terms of the power in the centre of the beam as

$$P_T = \frac{\pi w^2}{2} P_0 \tag{5.43}$$

The effective cross-sectional area of the beam is πw^2 so the total power is the axial power density times half the cross-sectional area.

Example 5.4 Power in a laser beam

Calculate the power density at the centre of a typical laboratory helium–neon laser with a total power, P_T, of 5 mW and an effective radius, w, of 0.4 mm.

The power density at the centre of the beam, P_0 using equation (5.42) works out at 19.9 kW/m^2. For comparison the power density of bright sunlight on the Earth's surface, near the equator, is about 1.2 kW/m^2. Thus the power density in a laser beam is almost seventeen times greater than sunlight. This is one reason why it is dangerous to look into a laser. A second reason is that the laser beam consists of plane waves which are all in phase, whereas light is a random collection of waves. The result is that the laser light is more easily focussed onto the retina of the eye by the lens in the eye. The laser beam is said to be collimated or coherent and the sunlight is incoherent.

5.15 VELOCITY OF ENERGY TRAVEL AND GROUP VELOCITY

Section 5.6 defined a phase velocity as the speed at which a point on the wave travelled. For a single plane wave at one frequency this is also the velocity at which energy travels, but there are relatively few practical cases where there is just a single plane wave. For instance, if information is to be communicated by a microwave or optical wave, then the information must be modulated onto a carrier wave. This will produce a multiplicity of plane waves. Another example is when a wave of arbitrary shape is being transmitted. Fourier analysis indicates that any periodic signal can always be expressed as a summation of sinusoidal waves, so each component can be treated as a separate plane wave. However, the energy in the signal is associated with the complete group of plane waves and not the separate components. A new velocity of energy travel needs to be defined which describes the speed of travel of the signal. This is called the *group velocity*.

The group velocity can be derived by considering the case of an amplitude modulated signal described by

$$E = E_0 \cos(\delta\omega t - \delta\beta z) \sin(\omega t - \beta z) \tag{5.44}$$

where $\delta\omega$ and $\delta\beta$ are small deviations in ω and β. The signal has the form shown in Figure 5.15. The sine part of equation (5.44) is the standard plane wave term. The velocity of energy travel is the rate at which the amplitude of the signal moves along, as was found for the power flow of a plane wave in Section 5.13. The amplitude is $E_0 \cos(\delta\omega t - \delta\beta z)$ and this is also the envelope of the signal which carries the information. The group velocity, v_g, is obtained by keeping the phase $(\delta\omega t - \delta\beta z)$ constant. In the limit as $\delta\omega$ and $\delta\beta$ approach zero, the group velocity is thus

$$v_g = \frac{d\omega}{d\beta} \tag{5.45}$$

This is the definition of group velocity. For a single plane wave equation (5.44) shows that $v_g = \omega/\beta$, thus confirming that the group velocity and phase velocities are equal in this case.

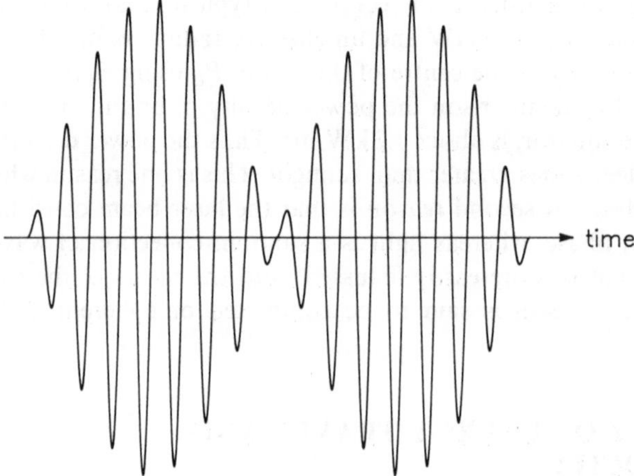

Figure 5.15. Amplitude modulated wave showing the envelope of the wave which moves at the group velocity

PROBLEMS

1. Show that $\boldsymbol{E}=\exp[j(6.28\times10^8t-2.1z)]$ is a solution to the wave equation and that it represents a travelling wave in the z direction. Find the wavelength.

2. Prove that the addition of an arbitrary phase factor ψ to $\sin(\omega t-\beta z)$ is still a valid solution to the wave equation.

3. Any general function of the form $E(x,y,z,t)=f(\omega t-ax-by-cz)$, where a,b,c are coefficients, represents as plane wave in an arbitrary direction. Verify that this is true by inserting E into the wave equation.

4. The amplitude of the electric field of a plane wave has a complex value $(8.09-j5.88)$. Show that this corresponds to a phase delay of 1/10 of a wavelength.

5. A wave has an electric field at $z=0$ given by $E_x=\cos(\omega t)+0.5\cos(3\omega t)$. Plot E_x against distance for a few points in free space. Repeat for a dielectric in which the phase velocity at angular frequency 3ω is 1/3 of that at angular frequency ω.

6. A phase shifter is a device added to a microwave system to introduce a phase delay. It contains a piece of dielectric which reduces the velocity of the waves. If $\varepsilon_r=2$, plot a graph of phase delay in degrees against thickness of dielectric.

7. A plane wave of frequency 3 GHz is incident normally on a large sheet of dielectric with $\varepsilon_r=2.7$. How thick must the sheet be to retard the wave in phase by 180° with respect to a wave which travels through a large hole in the sheet?

8. A plane wave is given by $E_x=4\cos[2\pi\times10^{15}(t-z/c)+\pi/2]$. Find the frequency, wavelength, phase angle at $z=0$, and the magnitude of the magnetic field.

9. The magnetic field of a plane wave is given by $H_x=5\times10^{-7}\cos(10^8\pi t-\beta z+\pi/4)$. Find the phase constant β and the location along the axis where H_x vanishes at $t=3$ ns.

10. A plane wave has a peak electric field of $E_0=6$ V/m. If the medium is lossless with $\mu_r=1$ and $\varepsilon_r=3$, find the velocity of the wave, the impedance of the medium and the peak value of the magnetic field H.

11. At 100 kHz a bath of distilled water has constants $\mu_r=1$ and $\varepsilon_r=81$. Find (a) Z/Zo, the ratio of the impedance of the medium to that of free space; (b) λ/λ_0, the ratio of the wavelength in the medium to that in free space; (c) v/v_0, the ratio of the phase velocity in the medium to that in free space and (d) the index of refraction.

12. Write down the equations for the electric and magnetic fields of a linear polarised plane wave where the electric vector is orientated at an angle of 60° to the x axis.

13. The electric field of a wave is given by $\boldsymbol{E}=\hat{x}\cos(10^8t-z/\sqrt{4})-\hat{y}2\sin(10^8t-z/\sqrt{4})$ Vm^{-1}. Find the frequency, the wavelength, the relative permittivity of the dielectric, the magnetic field and the polarisation.

14. Show that $\boldsymbol{E}=\hat{x}E_a\sin(\omega t-\beta z)+\hat{y}E_b\sin(\omega t-\beta z+\pi/2)$ is an elliptically polarised wave.

15. A light wave with a peak electric field of 5 V/m travels through glass with refractive index 1.5. Find the amplitude of the magnetic field and the value of the Poynting vector.

16. A travelling wave has a peak electric field with an amplitude of $E_0 = 6$ V/m. Find the average Poynting vector and the stored electric and magnetic energy.

17. A spacecraft at lunar distance from Earth transmits plane waves. If a power of 10 W is radiated isotropically, find (a) the average Poynting vector at Earth, (b) the rms electric field at Earth, (c) the time taken for the radio waves to travel from the space craft to Earth. (Take the Earth–Moon distance as 380 Mm.)

18. Find the power density emitted by a 1 mW laser beam which has an effective radius of 1 mm. Plot the distribution of power with radius.

19. Repeat Problem 18 for the case where the power in the laser beam is diverging at the rate of $1/(R/\lambda)^2$ from the source. Find the power density at 10 m if the source has a wavelength of 0.6328 μm.

20. Show that the group velocity, v_g, can be expressed as $v_g = v_p - \beta(\mathrm{d}v_p/\mathrm{d}\beta)$ where v_p is the phase velocity.

6 Reflection and refraction of plane waves

6.1 INTRODUCTION

The last chapter analysed and described the behaviour of plane waves which travel in a homogeneous medium. In practice this means microwave and optical transmissions through space, air and a dielectric. The microwave or optical waves will probably encounter obstacles along their transmission path. The obstacles will be a medium with finite boundaries which cause a perturbation and change in direction of the waves. This chapter is concerned with cases where a plane wave is incident upon a boundary between two media. The incident plane wave travels in the first medium, Figure 6.1, and then hits the boundary of a second medium. If the second medium is a perfect conductor then all the energy in the plane wave is reflected. This widely occurs and is the basis of reflector antennas and mirrors. It is also used in radar which detects a portion of the transmitted signal that is reflected back to the receiver.

If the second medium is a dielectric, part of the wave is reflected and part transmitted. The transmitted part can alter its direction of travel and is then said to be *refracted.* It is very widely used at optics and to a lesser extent at microwaves. This is the basis of lenses, prisms and some types of optical processing components. An important special case of reflection and refraction at a dielectric boundary is when the incident plane wave hits a boundary where the relative permittivity of the second medium is less than that of the first medium. Then under certain conditions, the wave can be totally reflected. This is the basis of the guiding action in optical and dielectric waveguides.

In this chapter the physical characteristics of the reflected and refracted waves will be studied and expressions developed which quantify the behaviour. In order to use the concept of plane waves incident on a boundary it is assumed that the boundary is large in wavelengths so that the fields over the local region where the wave hits the boundary can be considered plane. The general name for this assumption is *geometric optics.* As the name implies it is almost always true for optical systems where most objects are many thousands of wavelengths across. It is less true at microwave frequencies where most objects are comparable in size to the wavelength of the plane wave, but experience shows that the results provide a good estimate of the physical behaviour.

The total field in a region will often be the sum of more than one plane wave. For instance the field adjacent to a conductor will be the sum of the incident plane wave and the reflected plane wave. It will be assumed that the medium through which the waves are travelling is homogeneous and isotropic so that the resultant wave is the linear sum of the incident and reflected waves. In a general case with a number of plane waves at one frequency, the principle of superposition applies. This means that the resultant wave is the linear sum of any number of individual plane waves, and is itself a plane wave.

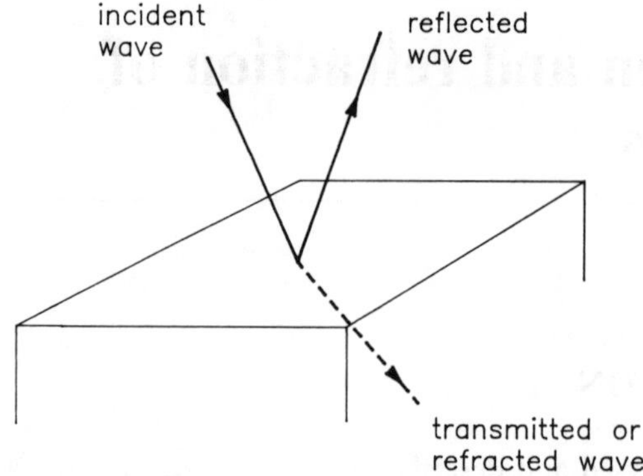

Figure 6.1. Reflection and refraction at a boundary

6.2 PLANE WAVES NORMALLY INCIDENT ON A DIELECTRIC

The first case to be considered is the simple situation of a plane wave incident at right angles to the boundary of a dielectric, Figure 6.2. A linear polarised plane wave with its electric vector along the y axis will be used. This is the E_y plane wave of the last chapter, but the results in this section will apply equally well to the E_x plane wave. This is because the boundary is normal to the z direction of travel and there are no factors which lead to a preferential polarisation. The general case will be studied where the incident plane wave is in a dielectric medium with a relative permittivity ε_1 and the second dielectric medium has a different relative permittivity of ε_2. Neither ε_1 nor ε_2 are in general equal to unity, however one of the most common situations is when one of the media is air. In this case either ε_1 or ε_2 will be equal to unity. This is a special case for the results which follow.

There are three plane waves to consider, the incident wave, E_{yi}, the reflected wave, E_{yr}, and the transmitted wave, E_{yt}. The incident wave will be given by

$$E_{yi} = E_i \sin(\omega t - \beta z) \tag{6.1}$$

where E_i is the amplitude of the wave. The reflected and transmitted waves will similarly

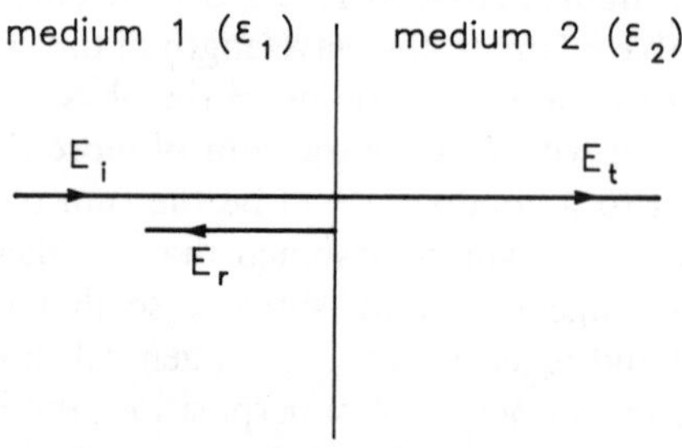

Figure 6.2. Plane wave normally incident on a dielectric

be given by

$$E_{yr} = E_r \sin(\omega t + \beta z) \tag{6.2}$$

$$E_{yt} = E_t \sin(\omega t - \beta z) \tag{6.3}$$

where E_r and E_t are the amplitudes of the reflected and transmitted waves. These are in general complex and contain the information about the result of the plane wave hitting the boundary. Note that the sign of the βz term in equation (6.2) is positive to indicate that the wave is travelling in the reverse direction to the other two components. The phase factor, $\beta(=2\pi\sqrt{\varepsilon_r}/\lambda)$, describes the speed and relative wavelength of the plane wave.

The value of the amplitude of the reflected and transmitted waves needs to be related to the amplitude of the incident wave. This is done by (a) applying a boundary condition and (b) applying the principle that the total power must be conserved across the boundary. This procedure is used in all the cases considered in this chapter.

The boundary condition which is used is that the total tangential electric field on one side of the boundary must equal the total tangential electric field on the other side of the boundary. This is straightforward to apply in this case because the E_y wave is completely tangential to the boundary. If the boundary is at $z=0$, then the sinusoidal part of the fields in equations (6.1) to (6.3) are equal and the total electric field on the incident side of the boundary is the sum of the incident electric field and the reflected electric field, or $E_i + E_r$. The total electric field on the transmitted side of the boundary is just E_t, therefore the boundary condition gives

$$E_i + E_r = E_t \tag{6.4}$$

Power conservation states that the total power is constant as the wave is incident on the boundary. The average power in the incident plane wave is E_i^2/Z_1 (see Section 5.13), where Z_1 is the wave impedance in medium 1. At the boundary, the incident power will be split between the power in the reflected wave and the power in the transmitted wave, or

$$\frac{E_i^2}{Z_1} = \frac{E_r^2}{Z_1} + \frac{E_t^2}{Z_2} \tag{6.5}$$

where Z_2 is the wave impedance in the second medium. Inserting equation (6.4) into equation (6.5) and solving gives

$$\frac{E_r}{E_i} = \frac{Z_2 - Z_1}{Z_2 + Z_1} \tag{6.6}$$

This relates the amplitudes of the waves to the impedance of the waves. It is convenient to express the ratio in terms of the relative permittivities. This can be done by using $Z_1 = Z_0/\varepsilon_1^{1/2}$ and $Z_2 = Z_0/\varepsilon_2^{1/2}$ where Z_0 is the impedance of free space, so

$$\frac{E_r}{E_i} = \frac{\varepsilon_1^{1/2} - \varepsilon_2^{1/2}}{\varepsilon_1^{1/2} + \varepsilon_2^{1/2}} = \frac{1 - \varepsilon_r^{1/2}}{1 + \varepsilon_r^{1/2}} \tag{6.7}$$

where $\varepsilon_r = \varepsilon_2/\varepsilon_1$.

The ratio E_r/E_i is an important quantity because it does not depend on the actual power in the plane wave. It can be measured in a practical system so it is given the name of the *reflection coefficient*. In general it is defined as

$$\rho = \frac{E_{yr}}{E_{yi}} \tag{6.8}$$

Although less used, than the reflection coefficient, a *transmission coefficient* can also be defined as the ratio E_{yt}/E_{yi}. This can be obtained directly from equation (6.4).

Example 6.1 Reflection and transmission coefficients for a wave incident normally on a dielectric

Find the reflection and tramsmission coefficients at a boundary for a plane wave incident from both sides of a dielectrc with $\varepsilon_1 = 1$ and $\varepsilon_2 = 2.5$.

Inserting the values into equation (6.7) for a wave incident in ε_1 gives $\rho = -0.23$. This means that 23% of the electric field is reflected. Since the power ratio is the square of the electric field ratio, this also means that 5% of the incident power is reflected. The value of ρ can be written as $0.23e^{j180°}$ so the negative sign means that the phase of the reflected field is changed by 180° relative to the incident field. The reflection coefficient is often expressed in decibels, given by $20\log_{10}(|\rho|)$, which in this case gives -12.8 dB. The transmission coefficient is $1 + \rho$ or 0.77.

If the wave is incident from the other side, then the permittivities need reversing around so that $\varepsilon_1 > \varepsilon_2$. The magnitude of the reflection coefficient is the same but the sign of the reflected electric field is positive so the phase of the reflected electric field is the same as the incident field.

Figure 6.3 shows a graph of how the reflection coefficient varies as the ratio ε_r changes. When the incident wave is in the more dense medium, the values are the same, but the

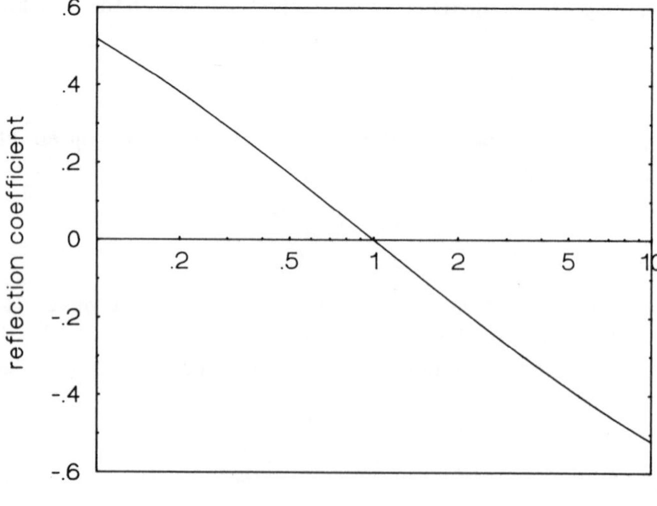

Figure 6.3. Reflection coefficient for normal incidence as a function of relative permittivity

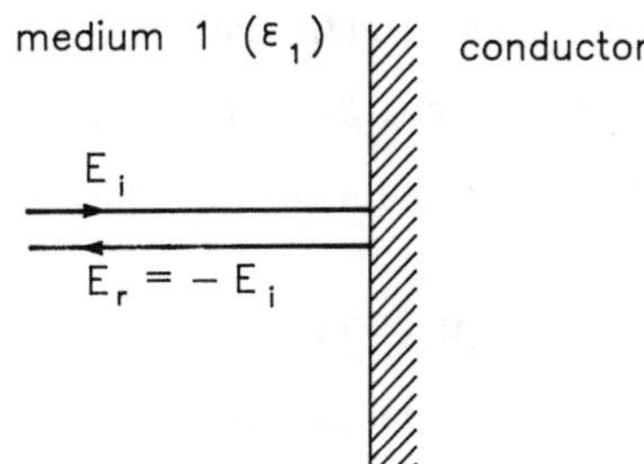

Figure 6.4. Plane wave normally incident on a conductor

sign of the reflection coefficient is positive. In many practical situation where a plane wave is incident on a dielectric, the dielectric is being used as part of the transmission system, as in a lens. Ideally all the incident power would go into the dielectric and the fact that some of the incident power is reflected is an inconvenience which lowers the efficiency of the system, for example the lens on a camera, thus the design aim is often to reduce the reflection coefficient to as near unity as possible. Equation (6.7) seems to indicate that this can only be done by using dielectrics with a lower relative permittivity, such as a foam plastic. This is not practicable at optical frequencies, however there is a way of reducing the reflection coefficient by using multiple layers of dielectric, and this is described later in the chapter.

PLANE WAVE INCIDENT NORMALLY ONTO A CONDUCTOR

If the second medium is a perfect conductor then there can be no electric field inside the conductor so there is no transmitted electric field and $E_t = 0$, Figure 6.4. In this special case the boundary condition of equation (6.4) shows that $E_r = -E_i$ and the reflection coefficient is $\rho = -1$. This shows that when a plane wave is reflected by a metal surface, the phase is changed by 180°.

An interesting feature of reflections from a conductor comes if a circular polarised wave (Section 5.12) is incident on the conductor. In this case the hand of polarisation, or direction in which the electric vector rotates, is reversed by the reflection. If a right-hand circular polarised wave is incident on the conductor, a left-hand circular polarised wave is reflected. This feature of reflections from a conductor has been implemented in some radar systems.

6.3 STANDING WAVES

The wave which is reflected normally from the surface and was studied in the last section will interact with the incident wave. An observer has no means of distinguishing the incident and reflected waves. The resulting field in space forms a wave which appears to be stationary in space and is called a *standing wave*. The mathematical form of the standing wave can be derived by summing together the incident wave and the reflected wave.

$$E_y = E_{yi} + E_{yr} = E_i \sin(\omega t - \beta z) + E_r \sin(\omega t + \beta z) \tag{6.9}$$

Expanding this equation with a trigonometric identity, gives

$$E_y = (E_r + E_i)\sin(\omega t)\cos(\beta z) + (E_r - E_i)\cos(\omega t)\sin(\beta z) \tag{6.10}$$

The second medium can be either a conductor or a dielectric.

STANDING WAVES AT A CONDUCTOR

If the second medium is a perfect conductor, then it was found above that $E_r = -E_i$. Equation (6.10) then simplifies to

$$E_y = 2E_i \sin(\beta z)\cos(\omega t) \tag{6.11}$$

The equation contains $\sin(\beta z)$ as a multiplying factor which means that the amplitude of the wave depends on the position along the z axis. At some points the incident and reflected waves add together and at other points they subtract. The equation thus represents a wave which is stationary in space and it is a standing wave. This fact can be confirmed by sketching the wave against βz for different instances of time, Figure 6.5. Curve A shows the case for $t = 0$ when the wave is a sine wave of amplitude $\pm 2E_1$. Curve B is for one eighth of a time period later when the amplitude is $\pm 1.41E_i$. Curve C is for one quarter of a time period after $t = 0$ when the amplitude is zero. If a point of constant phase on the wave, P, is tracked as time changes, it is observed to remain fixed at one value of βz. The peak amplitude of the standing wave is $\pm 2E_i$.

Reflections from a conductor occur widely in microwaves. It can be demonstrated in the laboratory with the set-up shown in Figure 6.6. A source of microwaves is reflected off a metal sheet and the field in the intervening space is sampled with a small metal probe. The probe records the field according to equation (6.11). If the source is modulated with a low frequency and the probe output detected with a tuned low frequency receiver, only the envelope of the standing wave will be recorded. This is because the wave oscillates at the basic frequency of the wave which is very high for a microwave signal and not detected by the tuned low frequency receiver. The envelope of the wave is stationary in space and therefore detected.

The envelope is given by the magnitude of equation (6.11), that is $|2E_1 \sin(\beta z)|$. This is plotted in Figure 6.7 against normalised distance. The detected signal will be zero at

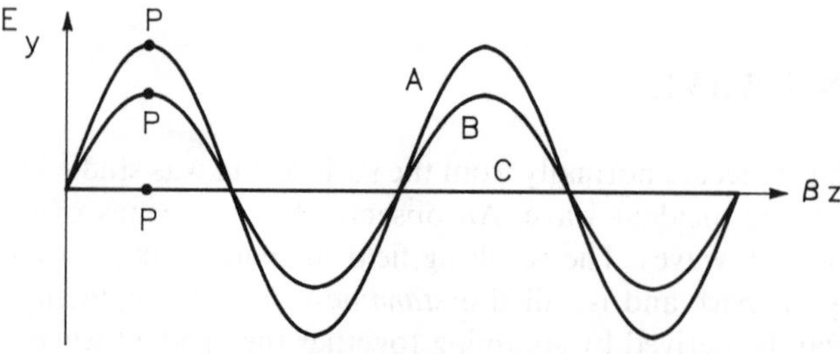

Figure 6.5. Standing waves at a conductor. Curve A is for $t = 0$, curve B for $t = T/8$ and line C for $t = T/4$

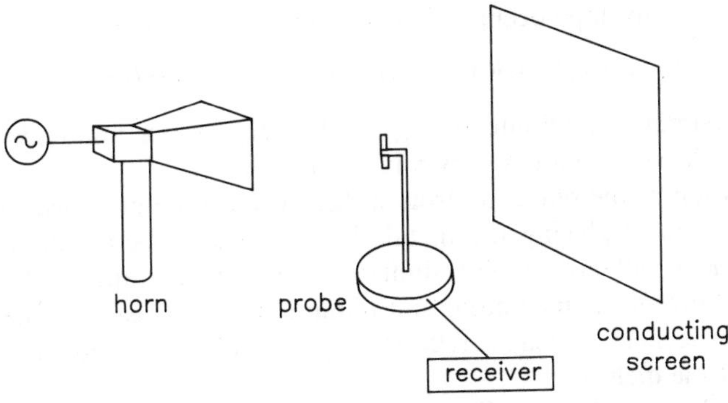

Figure 6.6. Demonstration of standing waves in a laboratory

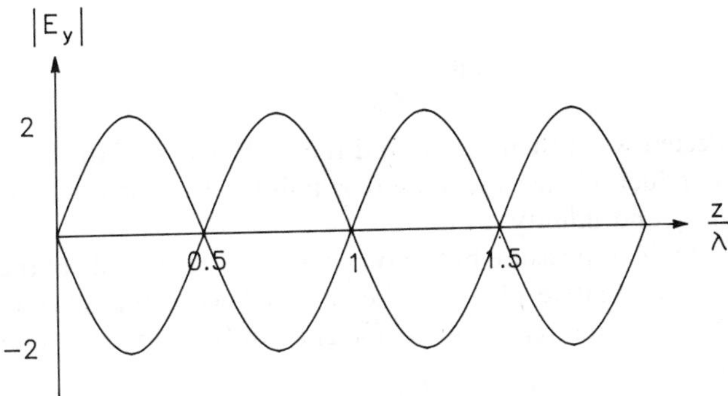

Figure 6.7. Standing wave envelope at a conductor

$\beta z = \pi, 2\pi, 3\pi$ etc. Since $\beta = 2\pi/\lambda$ this means that the signal will be zero at

$$z = \frac{\lambda}{2}, \lambda, \frac{3\lambda}{2}, \ldots, \frac{n\lambda}{2}$$

or every half wavelength apart. This gives a very simple and powerful method of measuring the wavelength of a microwave signal. It can be demonstrated using the simple set up in Figure 6.6. The method would also work in principle at optical frequencies, but the wavelength is too short to be easily measurable.

STANDING WAVES AT A DIELECTRIC

The other case is where the second medium is a dielectric. The incident and reflected waves will interact in the same way for a conductor but the amplitudes of the incident and reflected waves are different. This means that the interaction of the waves gives only partial addition and subtraction. An equation for the envelope of the standing wave can be obtained from equation (6.10) by trigonometric manipulation as

$$E_y = [E_r + E_i)^2 \cos^2(\beta z) + (E_r - E_i)^2 \sin^2(\beta z)]^{1/2} \sin(\omega t - \beta z) \tag{6.12}$$

The standing wave envelope occurs when $\sin(\omega t - \beta z) = \pm 1$ or

$$E_y = \pm[(E_r + E_i)^2 \cos^2(\beta z) + (E_r - E_i)^2 \sin^2(\beta z)]^{1/2} \tag{6.13}$$

This has a maximum amplitude of $E_{max} = (E_i + E_r)$ and a minimum amplitude of $E_{min} = (E_i - E_r)$. A typical plot is shown in Figure 6.8.

A demonstration of the reflection from a dielectric can be provided by using the set up of Figure 6.6 but replacing the metal sheet with a dielectric sheet. Probing and recording the field will give a plot similar to the positive amplitude in Figure 6.8. Measuring the ratio of the maximum and minimum fields is easily done and gives the amount of electric field which is reflected by the dielectric and hence the relative permittivities of the dielectric.

The ratio of the magnitudes of the maximum to the minimum signal is called the *standing wave ratio* or SWR. The potential or voltage which is recorded by the probe will be proportional to the electric field so the SWR is also called the *voltage standing wave ratio* or VSWR,

$$\text{VSWR} = \frac{|E_{max}|}{|E_{min}|} = \frac{|E_i + E_r|}{|E_i - E_r|} \tag{6.14}$$

If there is no reflected wave then $E_r = 0$ and the VSWR = 1. If there is total reflection, as for a perfect conductor, then the VSWR is infinite. For dielectric the VSWR has a value between unity and infinity.

The VSWR is simple to measure but only records the magnitude of the standing wave ratio. To measure the relative phase of the reflected wave the reflection coefficient is needed. The VSWR and the reflection coefficient can be related through

$$\text{VSWR} = \frac{1 + |E_r/E_i|}{1 - |E_r/E_i|} = \frac{1 + |\rho|}{1 - |\rho|} \tag{6.15}$$

or

$$|\rho| = \frac{\text{VSWR} - 1}{\text{VSWR} + 1} \tag{6.16}$$

Thus knowing one quantity enables the amplitude of the other quantity to be calculated.

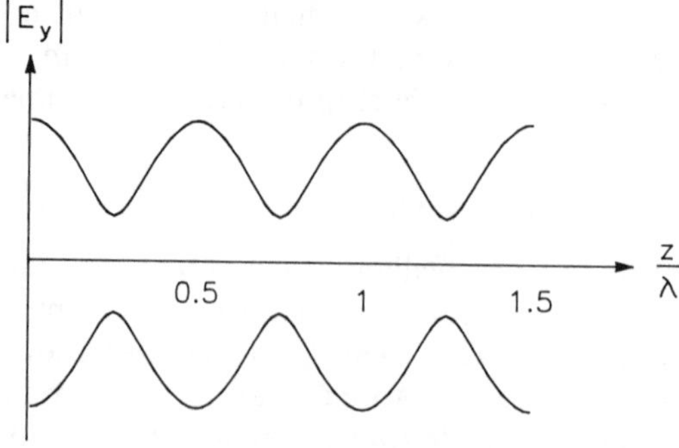

Figure 6.8. Standing wave envelope at a dielectric

The reflection coefficient at the boundary can also be expressed in terms of the relative permittivities of the medium using equation (6.7). The VSWR is a quantity which is independent of position but the reflection coefficient depends on position away from the source. Inserting this into equation (6.15) shows that the VSWR $= \varepsilon_r^{1/2}$ or VSWR $= n$, where n is the refractive index.

Example 6.2 VSWR and reflection coefficient at a dielectric

Find the VSWR and the magnitude of the reflection coefficient for a wave incident in air normally on (a) silicon ($\varepsilon_r = 12$), (b) polystyrene ($\varepsilon_r = 2.7$), (c) foam polystyrene ($\varepsilon_r = 1.1$).

The VSWR is the reciprocal of the refractive index of the medium so the three cases it is 3.46, 1.64 and 1.05, respectively. The magnitude of the corresponding reflection coefficients are, from equation (6.16), 0.55, 0.24 and 0.024. There is a considerable mismatch for the wave incident on silicon because of the relatively high permittivity, but the foam polystyrene is well matched to air.

6.4 RESONANT WAVES

If two metal plates are transverse to the direction of wave and there is a source of plane waves in between them, the waves will be continuously reflected from both plates, Figure 6.9. When the plates are an arbitrary distance apart, the total field between the plates will be a jumble of waves, but if the plates are an integer number of half wavelengths apart the situation will be as sketched in Figure 6.10. This is exactly equivalent to inserting two metal plates at any point where the total field is zero in Figure 6.7. The field between the plates will satisfy the standing wave conditions and the waves will be continually reflected back and forth by the two plates. The waves between the plates are now *resonant waves* and the system forms a resonant cavity. The amplitude of the envelope in the resonant cavity corresponds to that of standing waves.

Resonant cavities are widely used at microwave frequencies, and it can be considered to be the direct electromagnetic field analogy of an RLC circuit at low frequencies. There are two types of use. Firstly as an active resonant oscillator where the cavity is part of a microwave source. This could contain a Gunn diode or other semiconductor diode. If the length of the cavity is L, then the frequency of the oscillator will be $Nc/2L$,

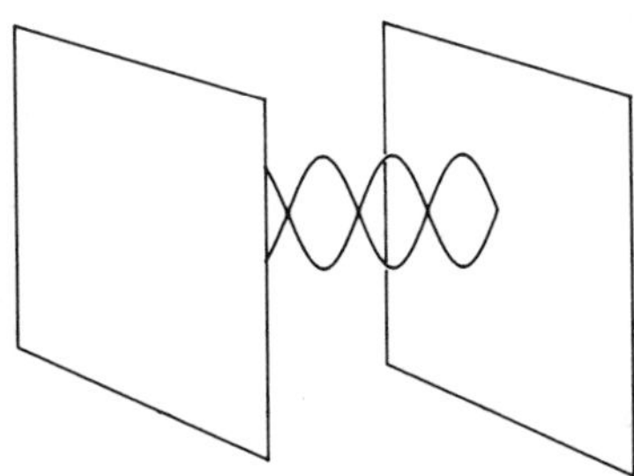

Figure 6.9. Formation of resonant waves between two plates

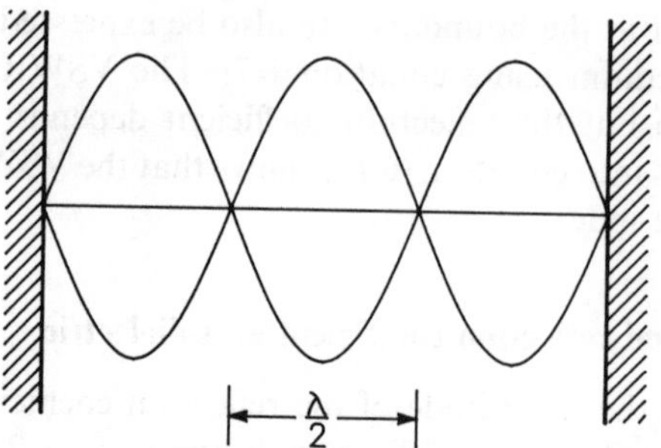

Figure 6.10. Resonant waves in a resonant cavity

where N is an integer. This second use is a passive resonant cavity where some energy is fed into the plates from outside. This can be used as a method of measuring the wavelength of a microwave signal by making the length of the cavity adjustable. The length is adjusted until resonance occurs when it will be a multiple of a half wavelength. The wavelength is thus $2L/N$. The method is simple and gives an effective wavemeter which can be calibrated directly in wavelength or frequency.

The two metal plates could be replaced by two dielectric plates and the same principle of resonance between the plates applies. Some of the energy between the plates is continuously lost due to the transmission of waves through the dielectric plates. This is the principle of a gas laser where the cavity consists of a gas-filled box containing a source of energy. Another form of the dielectric cavity is a block of dielectric. Waves inside the block will be reflected from the ends and form resonant waves if the length is correct. This is the principle of the semiconductor laser.

6.5 REFLECTION AND REFRACTION FOR ANY ANGLE OF INCIDENCE

The general case of plane waves incident on a boundary at any angle of incidence will now be studied. This is an extension of the normal incidence situation and the general case of a dielectric boundary is shown in Figure 6.11 in terms of both a ray representation

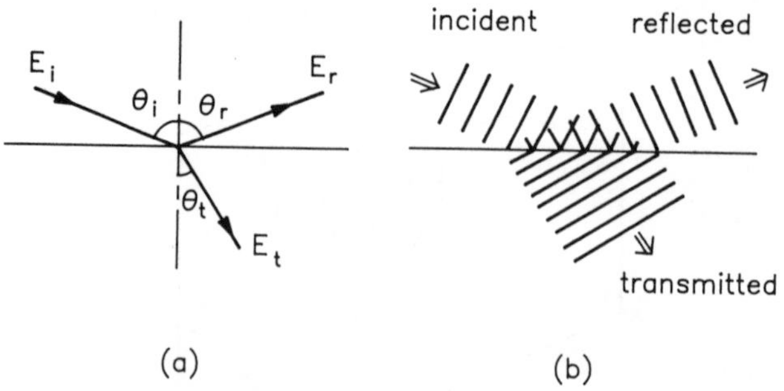

Figure 6.11. Reflection and refraction at an arbitrary angle of incidence. (a) Ray representation, (b) wavefront representation

and a wave representation. The wave is incident on the surface at an angle θ_i. Part of the incident wave is reflected at angle θ_r, and part is transmitted at angle θ_t. The angles are always measured *with respect to the normal to the surface*. The wave representation shows the interaction region at the surface. The amount of power will depend on both the strength of the incident wave and the angle of incidence. This will be derived, but first the relationship between the angles needs to be determined.

A complication for this general case of oblique incidence is that in general only part of the incident wave is tangential to the surface. The electric field can be at an arbitrary angle with respect to the plane of the paper in Figure 6.11. This problem is overcome by splitting the incident wave into two parts. One part of the electric vector is perpendicular to the plane of incidence (plane of the paper), Figure 6.12(a), and the other part of the electric vector is parallel to the plane of incidence, Figure 6.12(b). These are called the *perpendicular polarised* component and the *parallel polarised* component, respectively.

The relationship between the angles of incidence, reflection and transmission can be found by considering a tube of rays of height, AC, Figure 6.13, which is incident on the

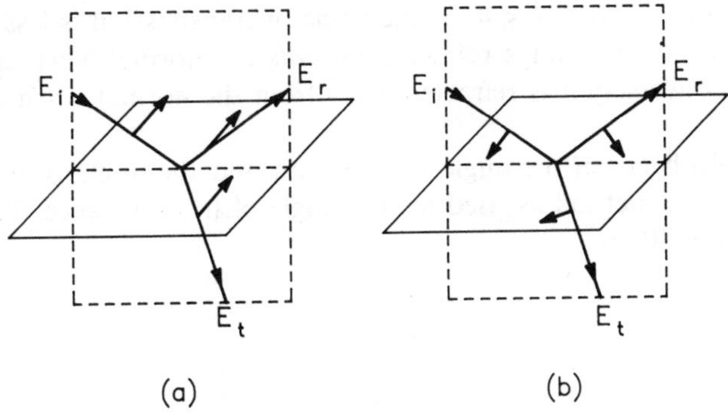

Figure 6.12. Orientation of electric vector with respect to the plane of incidence (shown dotted). (a) Perpendicular polarised wave with electric field at right angles to plane of incidence and parallel to the boundary, (b) parallel polarised wave with electric field in plane of incidence

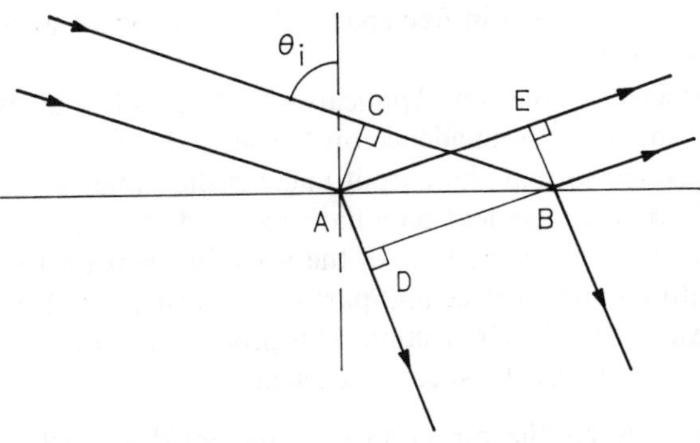

Figure 6.13. Geometry of rays incident on boundary

surface. The top edge of the incident ray travels a distance CB in the same time as the bottom edge travels either AE (for the reflected wave) or AD (for the transmitted wave). If v_1 is the velocity of the wave in the first medium with relative permittivity ε_1 and v_2 is the velocity of the wave in the second medium with relative permittivity ε_2 then

$$\frac{\mathrm{CB}}{\mathrm{AD}} = \frac{v_1}{v_2} \tag{6.17}$$

Now $\mathrm{CB} = \mathrm{AB} \sin\theta_i$ and $\mathrm{AD} = \mathrm{AB} \sin\theta_t$ so

$$\frac{\sin\theta_i}{\sin\theta_t} = \frac{v_1}{v_2} \tag{6.18}$$

but $v_1 = 1/(\varepsilon_1\varepsilon_0\mu_0)^{1/2}$ and $v_2 = 1/(\varepsilon_2\varepsilon_0\mu_0)^{1/2}$. Therefore

$$\frac{\sin\theta_i}{\sin\theta_t} = \left(\frac{\varepsilon_2}{\varepsilon_1}\right)^{1/2} = \frac{n_2}{n_1} \tag{6.19}$$

This is Snell's law of refraction, named after Willebrord Snell (1591–1626) who was a Professor at the University of Leyden. If the second medium is more dense than the first medium, so that $\varepsilon_2 > \varepsilon_1$ then $\theta_t < \theta_i$ or the angle of transmission is less than the angle of incidence. The wave is always refracted *towards* the normal when entering a more dense medium. Conversely it is refracted away from the normal when leaving a dense medium.

The relationship between the angle of reflection and the angle of incidence can be deduced by the same method as used for the angle of transmittance. The distance AE equals the distance CB so

$$\sin\theta_i = \sin\theta_r \tag{6.20}$$

or

$$\theta_i = \theta_r \tag{6.21}$$

This is the well known result that *the angle of reflection is equal to the angle of incidence.* Equations (6.19) and (6.21) are the two most important laws in the study of optics.

Example 6.3 Snell's law of refraction

Trace the ray path for a wave in free space which is incident at $\theta_i = 30°$ on a sheet of dielectric with $\varepsilon_2 = 4.5$.

Figure 6.14 shows the situation. Application of Snell's law gives $\theta_t = 13.6°$ in the dielectric. This is the angle of incidence on the second suface and equation (6.19) is then applied in reverse so that the output angle is the same as the original angle of incidence, i.e., 30°. A wave incident on a flat sheet of dielectric always emerges at the same angle as the incident angle. Part of the wave incident on the second surface is reflected back onto the first surface and part is then transmitted back into free space at the same angle as the angle of reflection. This process continues until all the incident power is used, or the end of the sheet is reached.

The relationship between the amplitudes of the incident, reflected and refracted components is found by the same procedure as used in Section 6.2 for normal incidence.

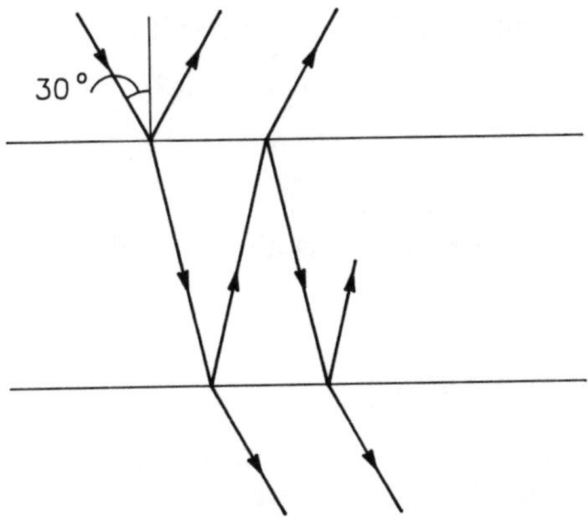

Figure 6.14. Rays incident on a sheet of dielectric showing multiple reflections and refraction

Power conservation is first used and then the boundary conditions are applied to the two polarisations.

The power density in the incident wave is E_i^2/Z_1. This is the power in a plane transverse to the direction of propagation. The component of the incident power striking the surface AB (Figure 6.13) is $(E_i^2 \cos\theta_i)/Z_1$. Similarly the component of the reflected power on the surface is $(E_r^2 \cos\theta_r)/Z_1$ and the component of the transmitted power is $(E_t^2 \cos\theta_t)/Z_2$. By conservation of power,

$$\frac{E_i^2}{Z_1}\cos\theta_i = \frac{E_r^2}{Z_1}\cos\theta_r + \frac{E_t^2}{Z_2}\cos\theta_t \tag{6.22}$$

or, substituting $Z_1 = Z_0/\varepsilon_1^{1/2}$ and $Z_2 = Z_0/\varepsilon_2^{1/2}$

$$\frac{E_r^2}{E_i^2} = 1 - \frac{\varepsilon_2^{1/2}E_t^2 \cos\theta_t}{\varepsilon_1^{1/2}E_i^2 \cos\theta_i} \tag{6.23}$$

This is the general relationship between the incident, reflected and transmitted components, E_i, E_r and E_t. In order to eliminate one of these components the boundary conditions are used, but this must be done for the two cases of perpendicular polarised waves and parallel polarised waves.

PERPENDICULAR POLARISED WAVES

If the electric field vector is perpendicular to the plane of incidence, Figure 6.12(a), the electric field is completely tangential to the surface and the boundary condition gives

$$\frac{E_t}{E_i} = \frac{E_r}{E_i} + 1 \tag{6.24}$$

Inserting this into equation (6.23) and solving for E_r/E_i gives

$$\frac{E_r}{E_i} = \frac{\varepsilon_1^{1/2}\cos\theta_i - \varepsilon_2^{1/2}\cos\theta_t}{\varepsilon_1^{1/2}\cos\theta_i + \varepsilon_2^{1/2}\cos\theta_t} \tag{6.25}$$

Then using Snell's law to eliminate θ_t, the final result is

$$\boxed{\frac{E_r}{E_i} = \frac{\cos\theta_i - (\varepsilon_2/\varepsilon_1 - \sin^2\theta_i)^{1/2}}{\cos\theta_i + (\varepsilon_2/\varepsilon_1 - \sin^2\theta_i)^{1/2}}} \tag{6.26}$$

Notice that the denominator has the same form as the numerator with a positive sign instead of a negative sign. This equation is also the reflection coefficient for arbitrary angle of incidence.

PARALLEL POLARISED WAVES

When the electric vector is parallel to the plane of incidence, Figure 6.12(b), only the tangential components of the electric fields are equal across the boundary. Hence

$$E_i \cos\theta_i - E_r \cos\theta_i = E_t \cos\theta_t \tag{6.27}$$

By the same procedure as for perpendicular polarised waves, the ratio of the reflected to incident electric fields becomes

$$\boxed{\frac{E_r}{E_i} = \frac{(\varepsilon_2/\varepsilon_1)\cos\theta_i - (\varepsilon_2/\varepsilon_1 - \sin^2\theta_i)^{1/2}}{(\varepsilon_2/\varepsilon_1)\cos\theta_i + (\varepsilon_2/\varepsilon_1 - \sin^2\theta_i)^{1/2}}} \tag{6.28}$$

This result differs from the perpendicular polarised waves by the $(\varepsilon_2/\varepsilon_1)$ factor multiplying the first term in both the numerator and denominator. This is a small but important difference because it means that under some circumstances the numerator can become zero. The angle at which this occurs is called the *Brewster angle* and is studied below.

Example 6.4 Reflection at arbitrary angles of incidence

Plot the behaviour of the reflection coefficient as a function of angle of incidence for both polarisations when a plane wave is incident on a dielectric boundary with $\varepsilon_1 = 1$ and $\varepsilon_2 = 2$ and vice versa.

Equations (6.26) and (6.28) are evaluated for a range of angles and the results are plotted in Figure 6.15(a) for the case when $\varepsilon_1 = 1$ and $\varepsilon_2 = 2$. Figure 6.15(b) shows the opposite case when $\varepsilon_1 = 2$ and $\varepsilon_2 = 1$. Other values for the relative permittivity of a

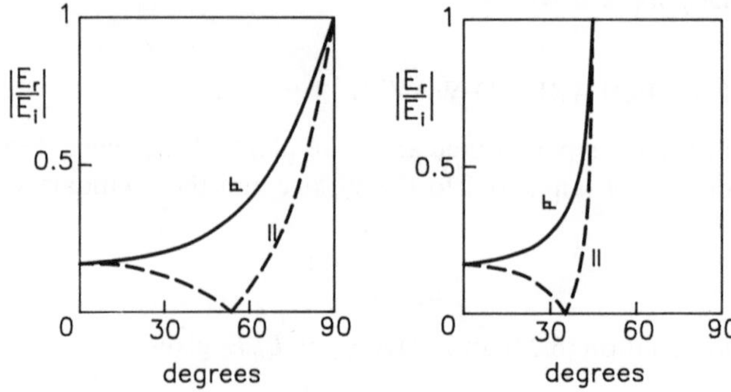

Figure 6.15. Reflection coefficient against angle for (a) $\varepsilon_1 = 1$ and $\varepsilon_2 = 2$, (b) $\varepsilon_1 = 2$, $\varepsilon_2 = 1$

dielectric will change the precise values of the reflection coefficient and angles of incidence but the general form will remain the same.

For the case of a plane wave incident on a boundary with a higher dielectric constant. Figure 6.15(a), both polarisations have the same reflection coefficient amplitude at $\theta_i = 0°$ and both incident waves are totally reflected at $\theta_i = 90°$. But for intermediate angles of incidence, the parallel polarised wave has an angle at which all the signal is transmitted and there is no reflected power ($|E_r/E_i| = 0$). This is at the *Brewster angle.*

For the opposite case of a plane wave emerging from a medium with a higher permittivity, Figure 6.15(b), shows a similar effect. However, in this case there is also an additional special feature that occurs for both polarisations. Beyond a certain angle of incidence the ratio of the reflected field to the incident field is unity which means that all the power is reflected. This is called *total internal reflection.*

6.6 TOTAL INTERNAL REFLECTION

The special feature of reflection at an oblique angle of incidence from a medium which is more dense than the second medium will now be investigated in more detail. The situation is shown in Figure 6.16 and occurs when $\varepsilon_2/\varepsilon_1 < 1$. Then the term $\varepsilon_2/\varepsilon_1 - \sin^2\theta_i$ in equation (6.26) or (6.28) is negative and its square root is imaginary which makes E_r/E_i complex and its magnitude equal to one or $|E_r/E_i| = 1$. This gives rise to what is called *total internal reflection* and the incident wave is reflected at the boundary and no wave emerges from the dense medium. It occurs when

$$\sin^2\theta_i \geqslant \varepsilon_2/\varepsilon_1 \qquad (6.29)$$

The angle in this equation is called the *critical angle*, θ_{cr}, or

$$\sin\theta_{cr} = \left(\frac{\varepsilon_2}{\varepsilon_1}\right)^{1/2} = \frac{n_2}{n_1} \qquad (6.30)$$

For waves incident on the boundary at angles greater than the critical angle all the power is internally reflected.

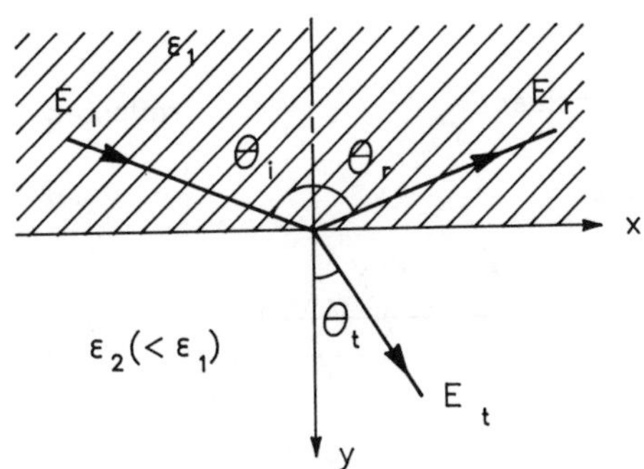

Figure 6.16. Total internal reflection. Ray picture at a dielectric boundary

Example 6.5 Total internal reflection

Find the critical angle for plane waves inside (a) glass with $\varepsilon_1 = 4.5$ and (b) water with $\varepsilon_1 = 80$ at microwaves where the second medium is air.

Applying equation (6.30) gives for the case of glass that $\theta_{cr} = 28.1°$ and for water, $\theta_{cr} = 6.4°$. As the relative permittivity increases the critical angle decreases. For all waves incident on the boundary at angles greater than 28.1° for glass or 6.4° for water, the wave will not be transmitted through the surface.

Although the above example was worked out for a wave at microwaves, the major use of internal reflection is at optical frequencies. This is because it is not easy to make a high quality flat conducting surface for optical waves. Total internal reflection can be used instead and is a good method of obtaining reflection where other methods are likely to be more lossy and more difficult to implement.

Total internal reflection is the basis of the guiding action in a dielectric or optical slab waveguide. The wave is confined between two surfaces and is reflected back and forth between the two surfaces. It will be studied in more detail in a later chapter.

The simple ray picture of total internal reflection which has just been described does not tell the complete story. There must be a field outside the dense medium as can be demonstrated from the boundary conditions. These state that the tangential components of the electric fields are equal across the boundary. Hence if there is a field inside the boundary, there must also be a field outside the boundary. To see what form this field takes a transmitted wave will be postulated and analysed for the condition of total internal reflection. Assuming that there is a plane wave incident along the direction of the transmitted ray, Figure 6.16, and that it is expressed in exponential form and resolved into components along the surface (designated the x direction in Figure 6.16) and along the normal to the surface (y direction)

$$E_t = \exp[j(\omega t - \sin\theta_t \beta x - \cos\theta_t \beta y)] \tag{6.31}$$

If the ray is emerging from a medium with relative permittivity ε_1 into air, Snell's law gives $\sin\theta_t = n_1 \sin\theta_i$. Then $\cos\theta_t = (1 - n_1^2 \sin^2\theta_i)^{1/2}$; however, the total internal reflection condition, equation (6.29), shows that $\cos\theta_t$ must be imaginary, i.e. $\cos\theta_t = -j(n_1^2 \sin^2\theta_i - 1)^{1/2}$. This means that the y dependent term in equation (6.31) is no longer an oscillatory term and E_t becomes

$$E_t = \exp[j(\omega t - \beta n_1 \sin\theta_i x)] \exp[-(n_1^2 \sin\theta_i^2 - 1)^{1/2} \beta y] \tag{6.32}$$

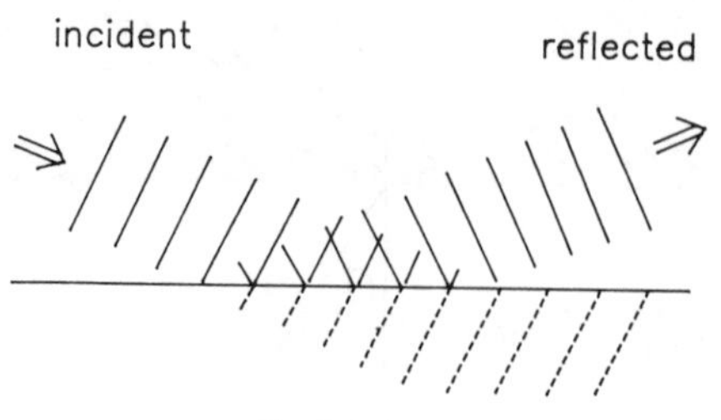

Figure 6.17. Wavefronts for wave which is total internally reflected

The exponential term for the x direction remains oscillatory whereas the exponential for the y component now represents a wave which decays with distance in the y direction. This shows that the transmitted wave in the air is a wave which apparently travels along the surface but decays exponentially in the y direction, that is away from the surface. The wave is bound to the surface and so the original assumption that all the power in the incident travelling wave is internally reflected was correct. This picture of the wave at the surface is sketched in Figure 6.17.

Example 6.6 External field due to total internal reflection

Work out the rate of decay away from the surface for a wave incident from inside a slab of glass, with a refractive index of 2 for angles of incidence of 31°, 35° and 60°.

The critical angle for $n_1 = 2$ is, from equation (6.30), 30°. Thus in all three angles of incidence, the wave is internally reflected. The attenuating part of E_t is given from equation (6.32) as $e^{-1.55y/\lambda}$, $e^{-3.53y/\lambda}$ and $e^{-8.89y/\lambda}$ for the three angles of incidence, respectively. At a distance of one wavelength from the surface the electric field has decayed to 0.21, 0.03 and 0.0001 of the value at the surface. This example shows that the rate of decay away from the surface increases very rapidly as the angle is increased above the critical angle.

In summary, when total internal reflection occurs for a wave incident on a surface, all the incident wave is reflected, but there is an attenuating or rapidly decaying wave attached to the surface which appears to travel along the surface and decays rapidly away from the surface. This phenomena is put to good use in some optoelectronic components as a means of coupling power into and out of a dielectric slab. If another slab is placed near to the boundary, as shown in Figure 6.18, then some of the attenuating wave will be coupled into the second slab. In this case *total* internal reflection will not take place because a proportion of the incident power will be coupled into the second slab. The actual proportion will depend on the relative permittivities, the angle of incidence and the gap between the two boundaries. By adjusting the width of the gap, any amount of coupling can be obtained.

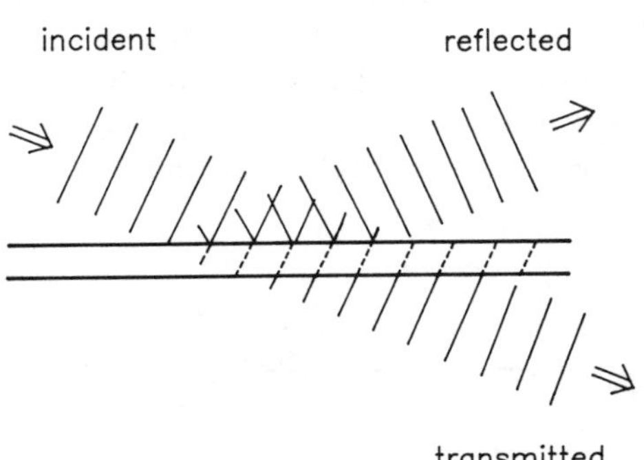

Figure 6.18. Power coupled to a second dielectric layer by total internal reflection

6.7 BREWSTER ANGLE

A parallel polarised wave incident obliquely on a surface has the property that the reflection coefficient can be zero at one particular angle. This was demonstrated with Figure 6.15 and occurs when the numerator of equation (6.28) is zero so that E_r/E_i is zero. There is no reflected wave and all the energy is transmitted. The angle at which this occurs is the *Brewster angle*, named after Sir David Brewster (1781–1868) who discovered the effect empirically. The special condition only applies for parallel polarised waves as can be seen by examining the reflection coefficient for perpendicular polarised waves, equation (6.26). In the case of perpendicular polarised waves there is no circumstance in which the numerator can be made zero.

The Brewster angle, θ_{iB}, is found by equating the numerator of equation (6.28) to zero,

$$\sin\theta_{iB} = \left(\frac{\varepsilon_2/\varepsilon_1}{1+\varepsilon_2/\varepsilon_1}\right)^{1/2} \tag{6.33}$$

or

$$\tan\theta_{iB} = \left(\frac{\varepsilon_2}{\varepsilon_1}\right)^{1/2} = \frac{n_2}{n_1} \tag{6.34}$$

If a general plane wave with both parallel and perpendicular polarisation components is incident on a boundary at the Brewster angle then the parallel component is totally transmitted but the perpendicular component is partially transmitted and partially reflected. This means that the reflected wave is now totally perpendicularly polarised and the polarisation properties of the wave have been changed. For this reason the Brewster angle is also called the *polarising angle.*

Example 6.7 Brewster angle exit from a laser

The Brewster angle concept is sometimes used on the exit of a gas laser cavity to give a polarised beam. The end of a laser cavity is tapered at the Brewster angle, Figure 6.19. Find the angles for glass with a refractive index of 1.4.

The Brewster angle is found from equation (6.34) with $n_1 = 1$ and $n_2 = 1.4$. This gives $\theta_{iB} = 54.5°$. The transmitted angle in the glass is given by Snell's law, equation (6.29), as $\theta_t = 35.6°$ and this is also the incident angle on the second surface so that the exit angle is 54.5° and the exit beam is parallel to the axis. The Brewster angle from glass to air is given by equation (6.34) with $n_1 = 1.4$ and $n_2 = 1$ which is again 35.6°. Hence the exit angle is automatically also the Brewster angle and the polarising effect takes place at both surfaces. The reflected wave is totally perpendicularly polarised but is relatively weak and is not directed along the axis. It

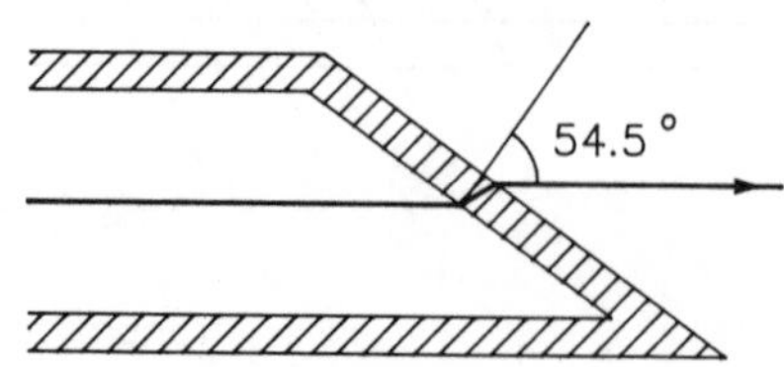

Figure 6.19. End of a cavity tapered at the Brewster angle

does not, therefore, contribute to the resonance in the cavity. On the other hand the parallel polarised component is totally transmitted and rapidly becomes the dominant polarisation state in the cavity so that the output beam is linearly polarised.

6.8 MULTIPLE REFLECTIONS

The preceding sections have dealt with reflection and refraction of plane waves at a single boundary, but what happens if there are multiple boundaries? There are many practical cases where there is more than one dielectric boundary, particularly for optical applications, the simplest is a glass window. The plane wave is incident on two or more boundaries in quick succession. The cumulative effect depends on the distance apart of the boundaries, the relative permittivities of the dielectrics and the angle of incidence. The general case is complicated to evaluate but the principles can be understand by treating the important case of two boundaries with a wave normally incident on the boundary. This is represented in practice by a sheet of dielectric between two other medium.

Figure 6.20 shows two boundaries with an single wave incident on the first boundary. At the first boundary the wave is partially reflected and partially transmitted, the same thing happens at the second boundary. A part of the wave is reflected back to the first boundary which is itself partially reflected and partially transmitted. The result is that the total field at the first boundary consists of multiple individual waves. To deal with this situation the *total* electric field amplitude is used. The total field consists of the sum of the individual components and is the quantity which would be measured in a practical system. The total field will be designed by capital subscripts, namely E_R instead of the incident component E_i. On the left-hand side of the boundary there is one incident wave and the total reflected wave

$$E_1 = E_{i1} \sin(\omega t - \beta z) + E_{R1} \sin(\omega t + \beta z) \tag{6.35}$$

In the region between the two boundaries the total field is

$$E_2 = E_{T2} \sin(\omega t - \beta z) + E_{R2} \sin(\omega t + \beta z) \tag{6.36}$$

In the third medium there is only a total transmitted wave,

$$E_3 = E_{T3} \sin(\omega t - \beta z) \tag{6.37}$$

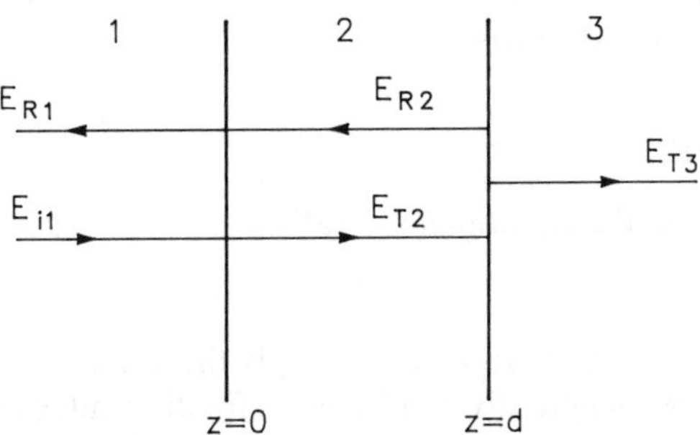

Figure 6.20. Rays at two dielectric boundaries

The solution to the problem requires that the values of E_{R1}, E_{T2}, E_{R2} and E_{T3} relative to E_{i1} be found. This requires four equations, or four boundary conditions. Equating the tangential electric fields at the two boundaries, as used previously in this chapter, will give two of the electric fields. Equating the tangential magnetic fields at the two boundaries will give the other two electric fields. The magnetic field in the first medium is given by

$$H_1 = \frac{n_1}{Z_0}[E_{i1}\sin(\omega t - \beta z) - E_{R1}\sin(\omega t + \beta z)] \tag{6.38}$$

where the refractive index, n_1, has been used to describe the dielectric properties, and Z_0 is the impedance of free space. The magnetic field for the other medium can be similarly expressed.

The solution of the boundary conditions is straightforward but will not be done here. This is because the general case is better done by using a transmission line analogy with wave impedances which is developed in Chapter 11. A particular case will be solved here. This is for the case when there is no reflected wave in the first medium from the first boundary. This is a common requirement in both microwaves and optics: for instance, that there should be no reflections from a glass lens. The procedure is called *matching* medium three to medium one. The condition requires that the reflected field, E_{R1}, should be zero so applying the boundary conditions to equations (6.35) to (6.37) and the magnetic fields gives

At $z = 0$, (Figure 6.20)

$$1 = \frac{E_{T2}}{E_{i1}} + \frac{E_{R2}}{E_{i1}} \tag{6.39}$$

$$\frac{n_1}{n_2} = \frac{E_{T2}}{E_{i1}} - \frac{E_{R2}}{E_{i1}} \tag{6.40}$$

and at $z = d$

$$\frac{E_{T2}}{E_{i1}}\sin(\omega t - \beta d) + \frac{E_{R2}}{E_{i1}}\sin(\omega t + \beta d) = \frac{E_{T3}}{E_{i1}}\sin(\omega t - \beta d) \tag{6.41}$$

$$n_2\left(\frac{E_{T2}}{E_{i1}}\sin(\omega t - \beta d) - \frac{E_{R2}}{E_{i1}}\sin(\omega t + \beta d)\right) = n_3\frac{E_{T3}}{E_{i1}}\sin(\omega t - \beta d) \tag{6.42}$$

Eliminating the electric fields between equations leads to a characteristic equation which can be satisfied by two possible cases:

HALF WAVE PLATE

If $n_1 = n_3$ and $\sin\beta d = 0$, the equations are satisfied. This means that

$$d = N\lambda_2/2 \tag{6.43}$$

where N is an integer greater than zero and λ_2 is the wavelength in medium 2 which equals the free space wavelength divided by the refractive index (λ/n_2). This means that if there is a sheet of dielectric in air, so that $n_1 = n_3 = 1$, there will be no reflections from the sheet if it is a multiple of half a wavelength thick, Figure 6.21.

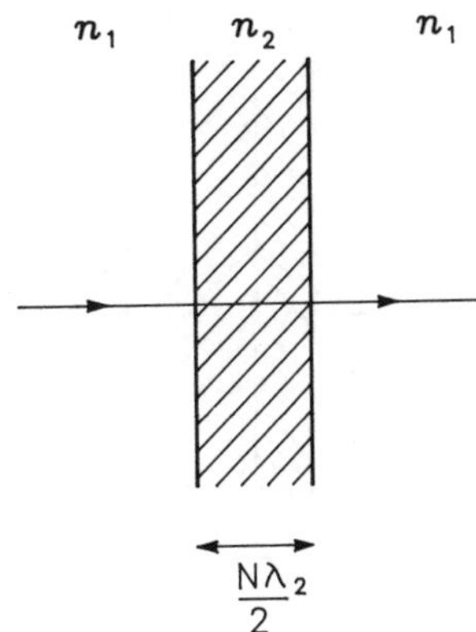

Figure 6.21. Half wavelength thick sheet of dielectric

Example 6.8 Matching with a half wave plate

Find the thickness for no reflections with a wave normally incident on a sheet of fibreglass at a frequency of 3 GHz and a sheet of glass at a wavelength of 0.6 μm. Take the refractive index as 2 for both materials.

The free space wavelength at 30 GHz is 10 mm, so the wavelength in the fibreglass is 10/2 or 5 mm. Hence the fibreglass sheet needs to be 2.5 mm or 5 mm or a multiple of 2.5 mm thick and there will be no reflections. An example of this case is a radome around a microwave radar. Note that this value only applies at the one frequency where the sheet is a multiple of half a wavelength thick, and it only applies for waves which are normally incident.

In the case of a sheet of glass at 0.6 μm (orange light), the same procedure shows that the glass must be $N \times 0.15\,\mu$m thick for no reflections. This is a very high accuracy to achieve in practice. Most glass sheets must be a few millimetres thick to be self supporting which is tens of thousands of half wavelengths thick.

QUARTER WAVE MATCHING PLATE

The other case when the condition for no reflections for normal incident waves will be satisfied is when

$$n_2^2 = n_1 n_3 \quad \text{and} \quad \sin\beta d = 1 \quad \text{or} \quad d = (2N+1)\lambda_2/4 \tag{6.44}$$

These conditions imply that if there are two media with refractive indices n_1 and n_3 then waves incident in medium 1 can be matched to medium 3, meaning that reflections can be eliminated, by using a dielectric sheet of thickness a multiple of a quarter of a wavelength and refractive index which is the square root of the product of the refractive indices of the two media to be matched, $n_2 = (n_1 n_3)^{1/2}$, Figure 6.22. This is a very useful result which is widely used at both microwaves and in optical instruments.

Example 6.9 Matching using a quarter wave plate

Find the refractive index and thickness required to match a glass lens with $n_3 = 1.9$ to air for light at a wavelength of 500 nm.

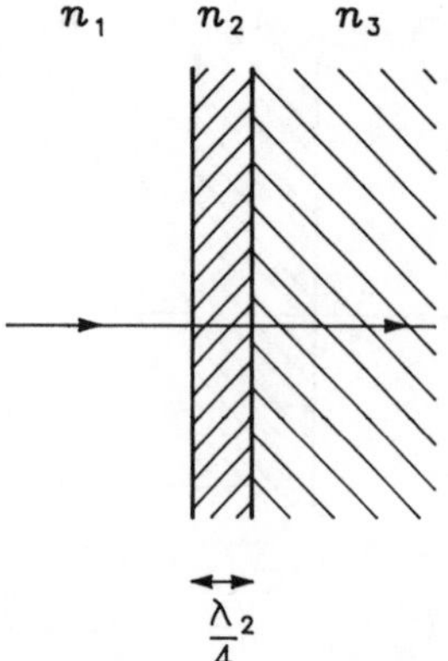

Figure 6.22. Quarter wavelength matching plate

The refractive index of the matching plate must be $(1.9)^{1/2}$ or 1.378. The wavelength in this plate is 500/1.378 or 363 nm, so the thickness should be a quarter of this wavelength or three quarters or five quarters etc., i.e. 91 nm, 273 nm, 455 nm etc. The perfect matching only occurs at the design wavelength. Light at other wavelengths will be partially reflected.

The matching can be improved by using more than a single quarter wave plate. This case is difficult to analyse with the above procedure. An easier approach is to use a transmission line analogy and wave impedances. It will be shown in Chapter 11, that this method of matching using a quarter wave plate can also by applied to transmission in waveguides.

PROBLEMS

1. A 18 GHz wave is normally incident in air on a dielectric slab with $\varepsilon_r = 1.8$. Find the amplitude and phase of the reflection coefficient and the VSWR at distances of 5 mm, 10 mm, and 20 mm from the surface of the slab.

2. A plane wave is reflected at normal incidence from a boundary. The amplitude of the electric field E_i of the incident wave is 10 mV/m and of the reflected wave is E_r. Find the VSWR when (a) $E_r = 2$ mV/m, (b) $E_r = 5$ mV/m, (c) $E_r = 8$ mV/m.

3. Compute and plot curves for the amplitude of the resultant electric field due to a standing wave in which the wave travelling in the z direction has an amplitude $E_0 = 1$ V/m and the other wave has an amplitude $E_1 = 0.4$ V/m. Plot at time $t = 0$, $t = T/8$ and $t = T/4$ and over a distance of at least 1λ in the z direction. What is the standing wave ratio and the reflection coefficient?

4. Plot the standing wave envelope for normal incidence on a dielectric of $\varepsilon_r = 2.5$.

5. A standing wave is given by $E = 10\sin(\frac{4}{3}\pi z)\cos(7\pi t)$. Determine two waves that can be superimposed to generate the standing wave.

6. An oscillator is formed by a source between two parallel conducting plates which are 60 mm apart. What will be the fundamental frequency from the oscillator?

7. An open resonant cavity is adjusted to be 100 mm long for resonance and to 105 mm long for a second resonance. Find the frequency of the waves and the number of positions along the cavity where there is zero electric field.

8. Derive expressions for the transmitted electric field for both parallel and perpendicular polarised waves incident at an oblique angle on a dielectric boundary.

9. Sketch graphs of (a) the amplitude of the ratio of transmitted electric field to incident electric field and (b) the phase of the ratio of the reflected electric field to the incident electric field against angle of incidence for the cases corresponding to Example 6.4 ($\varepsilon_1 = 1$, $\varepsilon_2 = 2$).

10. A three layer dielectric slab has relative permittivities of 2, 3, 4. A wave is incident on the sandwich at an angle of 30° to the normal. Find the output angle and sketch the path of the wave.

11. A ray is incident at an angle θ_i on a rectangular plate of thickness d. Show that the lateral displacement of the ray is given by $d \sin(\theta_i - \theta_t)/\cos\theta_i \cos\theta_t$. Find the displacement when $d = 2$ cm, $n = 1.5$ and $\theta_i = 60$.

12. Show that the reflection coefficients for oblique incidence can be expressed as

$$\rho_{\text{perp}} = -\frac{\sin(\theta_i - \theta_t)}{\sin(\theta_i + \theta_t)} \qquad \rho_{\text{para}} = \frac{\tan(\theta_i - \theta_t)}{\tan(\theta_i + \theta_t)}$$

13. Calculate the reflection coefficients from water for both perpendicular and parallel polarisation when the angles of incidence are 0°, 15°, 45° and 90°.

14. A linear polarised plane wave travelling in air is incident at an angle θ_i on the flat surface of a dielectric medium of large extent. The constants of the dielectric are $n_2 = 2.5$. Calculate the magnitude of the reflected field E_r and the transmitted field E_t relative to the incident field E_i as a function of θ_i when E_i is (a) parallel and (b) perpendicular to the plane of incidence. Draw graphs showing E_t/E_i and E_r/E_i against θ_i. What value of θ_t is the Brewster angle?

15. A beam of light is incident at angle θ_i on a cube of glass, of side 10 mm. If $n_r = 1.4$ find the value of θ_i that causes the beam to be internally reflected.

16. A small source of light is placed against the bottom face of a glass slab which is 22.5 mm thick. When viewed from above, a ring of light is observed at the bottom surface due to total internal reflection at the top surface. If the diameter of the circle is 76 mm, find the refractive index of the glass.

17. Show that a phase shift occurs at total internal reflection of $e^{-j\phi}$ where ϕ is given for perpendicular polarisation by

$$\tan\left(\frac{\phi}{2}\right) = \frac{(\sin^2\theta_i - 1/\varepsilon_1)^{1/2}}{\cos\theta_i}$$

18. An internal wave hits the surface of a dielectric of relative permittivity 2.5 at twice the critical angle for total internal reflection. Find the distance normal to a surface where the amplitude of the electric field of the attenuating wave has decayed to one tenth of the value at the surface. At what speed does the surface wave travel along the surface?

19. Two slabs of glass with refractive index 1.55 are separated by an air gap of 500 nm. A totally internally reflected wave with $\lambda = 550$ nm in one of the slabs is incident on the surface at one degree past the critical angle. How much power is transferred to the second slab of glass?

20. Determine the critical angle for total internal reflection and the Brewster angle for (a) external reflections and (b) internal reflections from a slab of dense flint glass with $n = 1.84$.

21. Find the magnitude and polarisation for a linear polarised beam which is polarised at 45° to the plane of incidence and is incident on a dielectric slab with $\varepsilon_r = 4$ at the Brewster angle.

22. The Brewster angle for a wave incident on one side of a dielectric interface is θ_{B1} and the Brewster angle for a wave incident on the other side of the interface is θ_{B2}. Show that the two Brewster angles are complementary, i.e. $\theta_{B1} + \theta_{B2} = 90°$.

23. Confirm that the conditions for no reflection with a half wavelength thick plate and a quarter wavelength thick plate (equations (6.43) and (6.44)) are given by the solution of equations (6.39) to (6.42).

24. Show that if a half wavelength layer of dielectric is placed on a thick slab of another dielectric, it has no effect on the reflection coefficient.

25. If a satellite TV signal is received at 12 GHz, how thick would a sheet of glass ($\varepsilon_r = 4$), placed normal to the wave have to be to totally transmit the signal?

26. A film is deposited onto glass of refractive index 1.50. Find the thickness and refractive index which will give no reflections for normally incident light at 500 nm.

7 Microwaves in free space

7.1 INTRODUCTION

Microwave systems which use plane wave transmission through the air or space are described in this chapter. The previous chapters have studied the theoretical background by analysing and developing an understanding of the characteristics of plane waves. This knowledge will now be used to explain the principles of systems for communication, radar and remote sensing. All of these use electromagnetic waves in the atmosphere or through space for their operation. The emphasis will be on transmission through an essentially loss less medium. The effects of a loss on the signals will be studied in the next chapter. A descriptive survey of practical systems was given in Chapter 1, so the objective is not to explain systems in detail but to put down sufficient information so that the transmission principles can be understood and appreciated. The range of applications is growing all the time as the demands for improved communications, better radar data and more data about the environment increases.

The study of microwave transmission is often concerned with the change in power due to the actions of parts of a system and not with the absolute power levels. The decibel, or dB, is usually used for the purpose of expressing the power ratio. If the input power to a part of the system is P_1, and the output power is P_2 W, then the dB ratio is defined by

$$\text{dB} = 10\log_{10}\left(\frac{P_2}{P_1}\right)\text{dB} \tag{7.1}$$

If $P_1 = E_1^2/Z_1$ and $P_2 = E_2^2/Z_2$ and $Z_1 = Z_2$, then

$$\text{dB} = 20\log_{10}\left(\frac{E_2}{E_1}\right)\text{dB} \tag{7.2}$$

Note that this produces exactly the same value of dBs and so a merit of using decibels is that potential confusion over whether power or field values are being used is avoided. Equation (7.2) is also valid for a ratio of two magnetic fields, or two voltages, or two currents. A power ratio of one half is -3 dB. A voltage ratio of one half is -6 dB. Equation (7.2) is derived from equation (7.1) and depends on the impedances at the input and output being equal. In practice this may not always be true and so allowance must be made in calculations.

Decibels can be used to express an absolute power value by making P_1 a fixed power level. There are two conventions used by microwave engineers,

$$\text{dBW when } P_1 = 1\text{ W}$$
$$\text{dBm when } P_1 = 1\text{ mW}$$

For example a power of 0.1 W is -10 dBW or 100 dBm. dBWs are often used for source calculations, whilst dBm are often used for receiver calculations.

Decibels are particularly useful when manipulating a transmission system with a number of sections because logarithms transform multiplication into addition. If $P = P_a P_b / P_c$ then in dBWs this becomes $P\,(\text{dBW}) = P_a + P_b - P_c$ since

$$10\log_{10}\left(\frac{P_a P_b}{P_c}\right) = 10\log_{10}\left(\frac{P_a}{1}\right) + 10\log_{10}\left(\frac{P_b}{1}\right) - 10\log_{10}\left(\frac{P_c}{1}\right)$$

Decibels will be used in this chapter when dealing with the gains and losses that take place in a communication system.

7.2 RADIATED POWER AND THE ROLE OF ANTENNAS

Section 5.1 explained that ideal plane waves rarely exist in practice but that they are a good approximation to the electromagnetic fields beyond a certain distance from the source of the radiation. If the source radiates uniformly in all directions then the waves will be spherical waves, and the wavefronts will be spheres with the centre of the sphere at the source. This is depicted in Figure 7.1. As the radius of the sphere gets larger, a square metre on the spherical surface becomes more nearly plane.

A source which radiates uniformly in all directions is called an *isotropic* source. If it has a trasmitted power, P_t, this power spreads out over the surface of the sphere. At a radius, R, the sphere has a surface area of $4\pi R^2$, so the power density is

$$W = \frac{P_t}{4\pi R^2}\ \text{W/m}^2 \tag{7.3}$$

An isotropic source never exists in reality because all antennas must be supported in space and this will distort the uniform radiation. This does not, however, matter because an isotropic source is not desirable. Spreading the power over all the surface of a sphere is an enormous waste of power. Electromagnetic waves are only needed over the portion of the sphere in the directions of one or more receivers. This is where the role of an antenna comes into the picture.

The job of an antenna is to concentrate the radiated waves into a specified region of space. This is shown schematically in Figure 7.2 for an antenna which concentrates the radiation into a narrow beam. The beam can be of any shape, depending on the aim

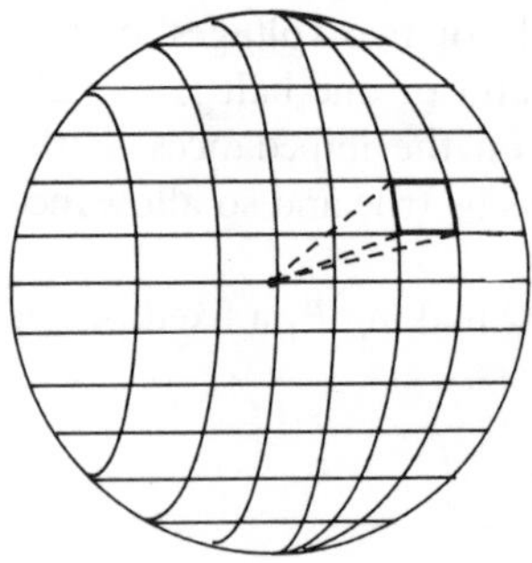

Figure 7.1. Isotropic source radiating spherical waves

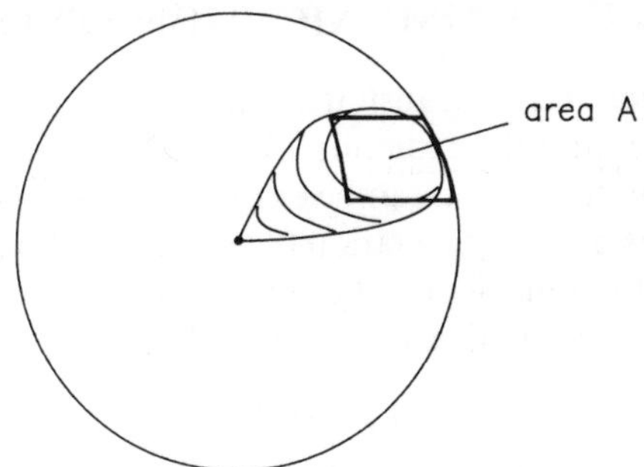

Figure 7.2. Concentration of power into a beam by an antenna

of the system. For instance a satellite Earth station antenna is required to send the signals in a very narrow beam to a satellite, whereas a broadcast antenna may be required to send signals uniformly all around it, but not up into the sky or down into the ground. An antenna does not alter the nature of the waves radiated from the source, they are still spherical waves which become plane waves after a distance.

The amount by which an antenna concentrates the power by comparison with an isotropic source is called the *gain* of the antenna. The power density at distance R, equation (7.3) then becomes

$$W = \frac{P_t G_t}{4\pi R^2} \text{ W/m}^2 \tag{7.4}$$

where G_t is the gain of the transmitter. The power density on the surface of the sphere is increased by the gain of the antenna. The gain is the amount of power radiated in a given direction so a full description is $G_t(\theta, \phi)$ where θ and ϕ are the spherical angles from the source at the origin to the portion of the surface of the sphere being observed.

The product of the transmitter power and the transmitter gain, $P_t G_t$, is often called the *effective isotropic radiated power* or EIRP. It gives the equivalent power of a source which radiates isotropically.

Example 7.1 Power and electric field radiated by an antenna

A TV transmitting antenna has an EIRP of 1000 kW, and a gain of 15 dB in the direction of the receivers. Find the power radiated and the electric field at distances of 1 km and 10 km.

A gain of 15 dB is 31.6 as a ratio so the power radiated is the EIRP divided by the gain, or $1000/31.6 = 31.6$ kW.

The power density at 1 km is obtained from equation (7.4) as $P = 10^6/(4\pi \times 10^6) = 79.6\text{ mW/m}^2$. At 10 km it has decreased 100 times to $796\ \mu\text{W/m}^2$. Assuming plane waves, the power is related to the electric field by $P = E^2/Z_0$, where Z_0 is the impedance of free space ($= 377\ \Omega$). (Note that the equation derived in Chapter 5, equation (5.39), had a factor 2 in the denominator because the peak electric field was used. In practical problem it can be assumed that the electric field is the RMS value so the 2 is omitted.) The electric field for the two distances becomes 5.5 V/m and 0.55 V/m. The electric field is inversely proportional to distance from the source.

7.3 TRANSMISSION IN COMMUNICATION SYSTEMS

A communication system has a transmitter and a receiver. The previous section considered the transmitter and the radiation from it. Now the receiver side is added to the picture so that the power received can be evaluated. The power density at a distance, R, from the transmitter is given by equation (7.4). If the receiving antenna has an area A and is ideal so that all the incident radiation is converted to power for the receiver, the power received, P_r, would be given by, Figure 7.2,

$$P_r = WA \tag{7.5}$$

Practical antennas are not perfect and only a proportion of the incident radiation is converted to power for the receiver. The radiation which is not converted is either scattered in various directions or absorbed by the antenna. If the antenna is imperfect it means that the real area, A_r, of the antenna will have to be larger than A in order to receive an amount of power, P_r. The ratio between A and A_r is called the *aperture efficiency*, η, and expressed as a percentage,

$$\eta = \frac{A}{A_r} \times 100\% \tag{7.6}$$

Typical aperture efficiencies for communication antennas are 60% upwards. It is difficult to design antennas with efficiencies above about 85% but it can be done for special purposes. It is seen from equation (7.6) that A is the effective area of the antenna, and it is sometimes so called. It is related to the gain, G, of an antenna by

$$G = \frac{4\pi A}{\lambda^2} \tag{7.7}$$

This is a basic relationship in antenna theory and enables the effective area A to be calculated because the gain of an antenna can be measured.

If the gain of the receiving antenna is G_r, then equation (7.7) can be combined with equations (7.5) and (7.4) to give

$$P_r = P_t G_t G_r \left(\frac{\lambda}{4\pi R}\right)^2 \tag{7.8}$$

This equation is called the *Friis transmission equation* and is a basic relationship for the calculation of the power received in any microwave or radio link. It appears to state that the power received is proportional to the square of the wavelength or inversely proportional to the frequency, however the gain is inversely proportional to wavelength, as equation (7.7) shows, so for fixed antennas, the power received is inversely proportional to the square of the wavelength. In practice antennas are designed for a particular band of frequencies so a higher frequency of operation would lead to a smaller antenna.

The distance appears in the Friis transmission equation through the term $(4\pi R/\lambda)^{-2}$. This term is a measure of the amount of spreading of the signal over the surface of a sphere of radius R. It is called the *path loss*, L_p, although it is not a loss in the sense that power is absorbed by a material. It is normally expressed in dB as

$$L_p = 20\log_{10}\left(\frac{\lambda}{4\pi R}\right) \tag{7.9}$$

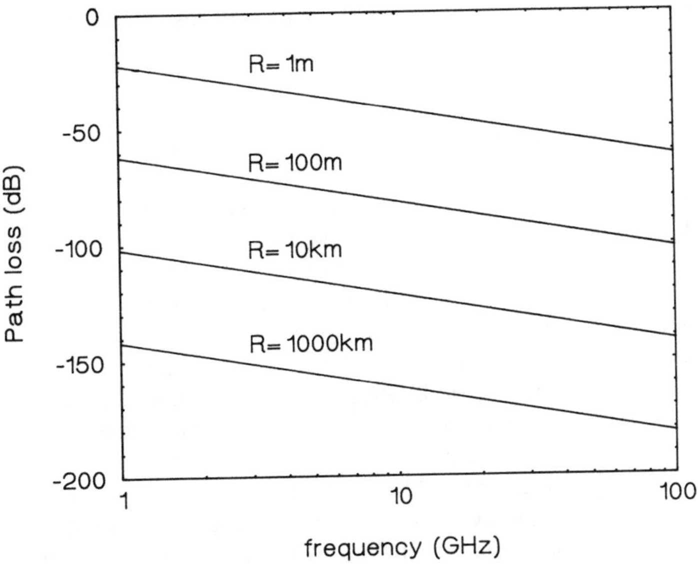

Figure 7.3. Path loss against frequency

The path loss is an important term for many communication systems and usually has a much larger value than the gains of the antennas. Figure 7.3 plots the path loss for a range of distances against frequency. The path loss increases proportional to the logarithm of the frequency and proportional to the logarithm of the inverse of the distance. This means that the same loss is obtained at a distance of 1 km away from a 1 GHz transmitter and at a distance of 10 m away from a 100 GHz transmitter. The values in Figure 7.3 demonstrate that large path losses occur, particularly at millimetrewaves.

The Friis transmission equation, equation (7.8) can be expressed in decibels as

$$P_r = P_t + G_t + G_r + L_p \quad \text{dBW} \tag{7.10}$$

where all the quantities are in dBs.

Example 7.2 Friis transmission equation

Find the power received by a 20 GHz microwave radio link with two identical antennas of effective diameter 400 mm. The distance apart of the antennas is 10 km and the transmitter power is 10 W.

The gain of the antennas is obtained from equation (7.7) with the area $A = \pi(0.4)^2/4\,\text{m}^2$ and $\lambda = 3 \times 10^8/20 \times 10^9 = 0.015\,\text{m}$ is $G = 7018$ or $G = 38.5\,\text{dB}$. The path loss from equation (7.9) is 138.5 dB. The power received in dBWs is thus

$$P_r = 10 + 38.5 + 38.5 - 138.5 = -51.5\,\text{dBW}$$

In absolute values this is 7.1×10^{-6} W. The example demonstrates why high gain antennas are necessary because they counteract to some extent the high path losses.

The version of the Friis transmission equation in equations (7.8) or (7.10) is not quite complete because there would be other losses to subtract from the right-hand side. Both the transmitting and receiving antenna will have small losses and the atmosphere will also contribute some loss.

7.4 MICROWAVE SOURCES AND DETECTORS

A complete free space microwave communication system consists of a number of sections, Figure 7.4. A transmitter is followed by waveguide conncecting to an antenna. This radiates into free space where the signals travel the free space microwave transmission path. At the reception end of the system an identical series of sections culminates in a receiver. This book is primarily concerned with the theory and practice of the guided and free space transmission sections, but the transmitter and receiver are an integral part of any system so some general knowledge of sources and detectors is helpful for a complete understanding. There are a very large variety of sources and detectors and the following short survey concentrates on the most common types.

The original microwave sources in the 1930s and 1940s were all vacuum tube devices. These were developed as highly as was possible with the technology available at the time, but were never particularly reliable. Microwave tubes were likely to be the component most in need of maintenance and careful tuning in order to work well. This is partly because tubes need high current power supplies to feed the heater circuits and stable high voltage power supplies. As soon as solid state source became available in the 1960s the vacuum devices were displaced for low power applications. Research continues with the aim of developing high power solid state devices and output powers up to 1 kW are now available. The present division between high microwave tubes and solid state sources is shown in Figure 7.5. This is a guideline and there is considerable overlap between the two types. The general conclusion is that the performance of solid state sources falls off as frequency increases. The demand for solid state sources is, however, particularly strong at millimetrewaves where one of the system drivers is the compact size of systems.

The first microwave tube to be developed was the *klystron* and this has been gradually refined over the years for high power CW microwave radar and UHF broadcast transmitter applications. An electron beam is generated by a heater which then passes through a series of resonant cavities to provide positive feedback and oscillator action. It can be tuned by adjusting the size of the resonant cavity, but this means that it is a

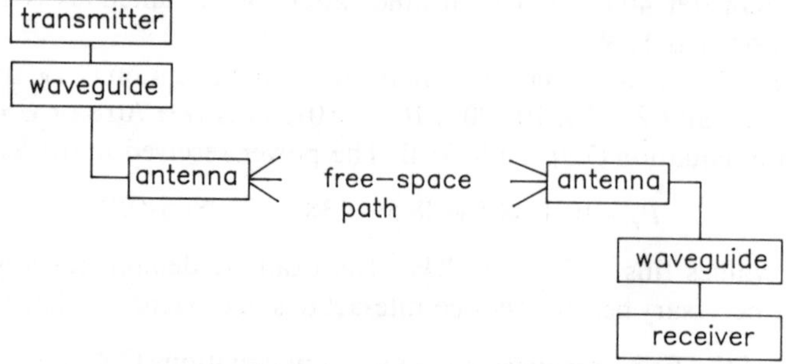

Figure 7.4. Free space microwave communication system

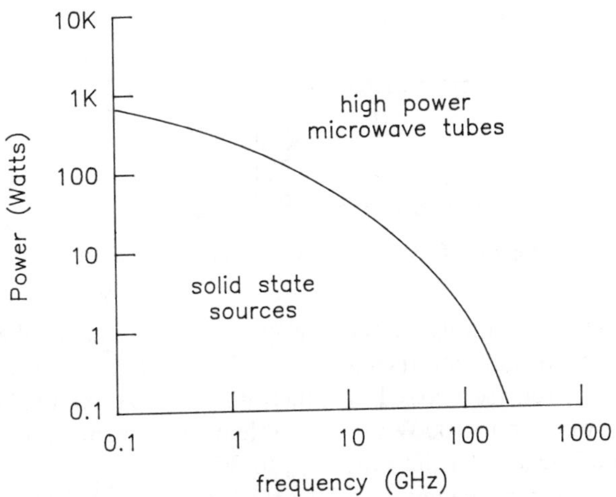

Figure 7.5. Power available from microwave solid state source and microwave tubes

narrow band device. Output powers are typically up to 1 kW. The highest powers are generated by the *magnetron.* This is a cross-field tube in which electrons interact with a magnetic field. Electron bunching transfers energy to a microwave signal. The magnetron was invented in the 1930s and was the device which enabled radar to develop rapidly. It is a mechanically simple device which can be made very reliable. Very high powers up to 10 MW are possible with high efficiencies of up to 80%. The magnetron is, however, a noisy device and for this reason is now less popular in radars which require good sensitivity.

The widest bandwidth tube available is the *Travelling Wave Tube* or TWT. This is actually an amplifier so needs to be driven by a low power solid state source. A helix produces a slow wave field which interacts with an electron beam to increases the power in a microwave signal. The bandwidth can be up to 100%, which is very wide for a microwave device. The power level depends on the bandwidth and is normally tens or hundreds of watts of CW power. The efficiency is moderate and in the range of 20 to 40%. TWTs can be made very reliable as can be seen from the fact that they are used on all high power communication satellites. They have the advantage on a satellite that the wide bandwidth can be traded for higher power, both of which are controllable remotely from the ground.

Solid state sources use either silicon or gallium arsenide (GaAs) semiconductor diodes or transistors. GaAs generally has better characteristics at microwave frequencies. The main low power source is the Gunn diode oscillator. This is a so-called transferred electron device which uses a bulk piece of GaAs semiconductor. It does not have a *pn* junction but does have a negative resistance characteristic which can be used with an external resistor to produce a stable oscillator. It is normally mounted across a piece of waveguide, Figure 7.6, and fed through a·hole in the side of the waveguide. The waveguide is partially blocked so that resonant waves are established at the wavelength determined by the length of the cavity. Gunn diodes are very simple oscillators but their efficiency if low and typically between 1% and 10%. They have been made to operate from 3 GHz to 100 GHz with CW powers up to 1 W and much higher pulse powers.

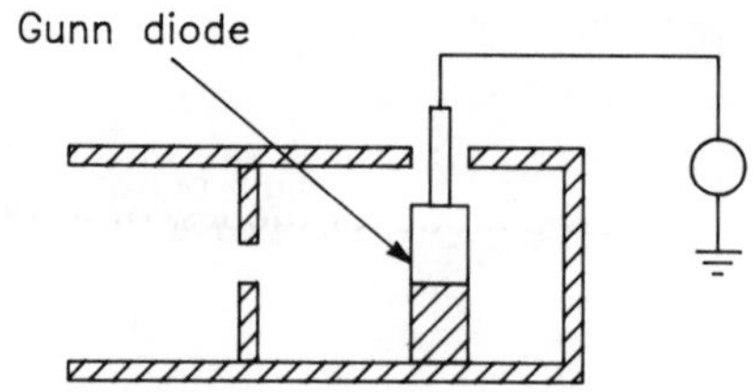

Figure 7.6. Gunn diode waveguide oscillator

The power falls off very rapidly at the top end of the band. Higher powers can be obtained from silicon or gallium arsenide IMPATT (*IMPact Avalanche Transit Time*) diodes. These have a reverse biassed *pn* junction which causes avalanche breakdown and oscillation. Powers up to 100 W are available over the same frequency range as the Gunn diode with efficiencies in the range 5% to 30%.

At the low microwave frequencies, FET transistors are used. These can be made into either oscillators or amplifiers and as well as working at low frequencies can be made to function up to 100 GHz. The output powers are low but the efficiency is reasonably high. Transistor sources have the great merit that they are very flexible and easy to integrate with other components so that *Microwave Integrated Circuits*, or MICs, can be made. The output frequency can be tuned over a wide range with either an external varactor diode or a piece of magnetized ferrite in the form of an yttrium iron garnet (YIG) sphere.

The range of microwave active components in the receiver is much more limited than in the transmitter. This is partly because the signal is converted to a lower frequency for processing as soon as possible and partly because high quality diodes and transistors are available. The main requirement for active microwave devices in the receiver is that they should introduce as little noise as possible. The noise sets the sensitivity of the receiver and is a fundamental factor in all system power level calculations, as will be described in the next section. In recent years the noise performance of microwave front-ends has improved dramatically and relatively cheap, low noise devices are available. Microwave power can be detected directly with a diode detector, Figure 7.7(a), but most microwave receivers use the superheterodyne principle, Figure 7.7(b). A local

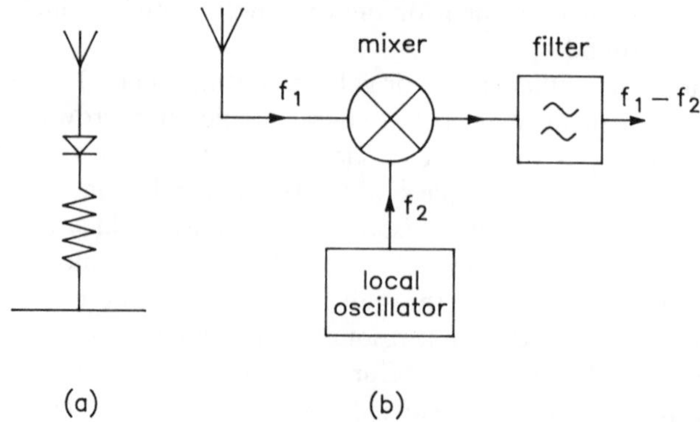

Figure 7.7. (a) Direct diode detector. (b) Superheterodyne receiver

oscillator at a frequency of the signal plus the intermediate frequency (IF), is mixed with the signal frequency to give a difference frequency which is then passed to the IF stage for amplification and demodulation. Sometimes the mixer is preceded by a *low noise amplifier*, LNA. The mixer is either a diode (or diodes) or a transistor. In the latter case the transistor can combine the functions of amplifier and mixer and is called a *low noise converter*, LNC. The favourite type of diode is a *Schottky barrier diode* made from either silicon or gallium arsenide. This behaves similar to a *pn* junction diode but operates at low d.c. bias currents and can be made to have very fast switching times as well as low noise.

7.5 NOISE

All physical objects which are at any temperature above absolute zero radiate energy in the form of electromagnetic waves. The radiation is in the form of random fluctuations and is called *noise*. It may be thought of as a collection of plane waves of random amplitudes, phase and polarisation which keep changing with time. The level of noise power radiated by natural objects is very small but it is important to understand the consequences of noise because it will always be present in any receiving or transmitting system. The low level means that it is in the receiver that the effects of noise are most significant. This is illustrated in Figure 7.8 which sketches the transfer response of a reciever to an incoming signal. The output power will normally be proportional to the input signal over a certain range of input signal levels. As the input signal level decreases, however, it will fall below the noise power and the output signal will then be determined by the noise and not by the signal. The noise therefore sets the minimum signal level of the receiver and this is often called the *sensitivity* of the receiver. The range between the maximum input signal level and the noise level that can be handled by the receiver is called the *dynamic range* of the receiver.

There are three broad categories of noise as far as microwave systems are concerned.

1. Noise due to the random motion of charge carriers. This includes thermal noise due to the temperature of an object and shot noise due to random motion of electrons in solid state or vacuum devices. This noise occurs in electronic components and is therefore added to the wanted signal by the receiver.

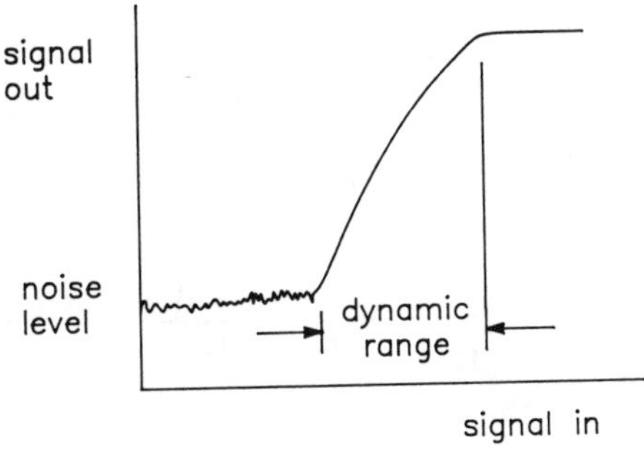

Figure 7.8. Transfer response of a receiver

2. Noise radiated by natural bodies such as the Earth, vegetation, and stars. It is the objective of remote sensing to detect and quantify the noise from the surrounding objects on the Earth. It is the objective of astronomy to detect and quantify the noise from the stars and other cosmological sources.

3. Noise caused by man-made interference. This arises unintentionally from all the free space microwave and radio wave systems which can be picked up by the system under consideration. It is a growing problem that is caused by the success of communications, broadcasting and radar. There are more and more systems so there are more and more potential source of artificial noise.

The first type of noise occurs in the receiver. The second and third type of noise are picked up by the receiving antenna and are partly influenced by the design of the antenna. A high quality antenna can eliminate most of the man-made interference. The first and second type of noise both arise from natural objects which can be considered to be perfect *blackbody* radiators. The noise power which arises from a blackbody can be measured by recording the power in an ideal resistor connected to the blackbody source of the noise, Figure 7.9. The RMS power is given by *Planck's* radiation law which was formulated by Max Planck in 1901

$$P = \frac{hfB}{\exp(hf/kT) - 1} \tag{7.11}$$

where P is the power in watts, h Planck's constant which has a value of 6.626×10^{-34} J s,

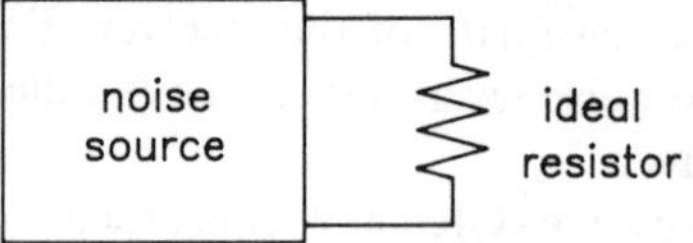

Figure 7.9. Ideal resistor connected to a noise source

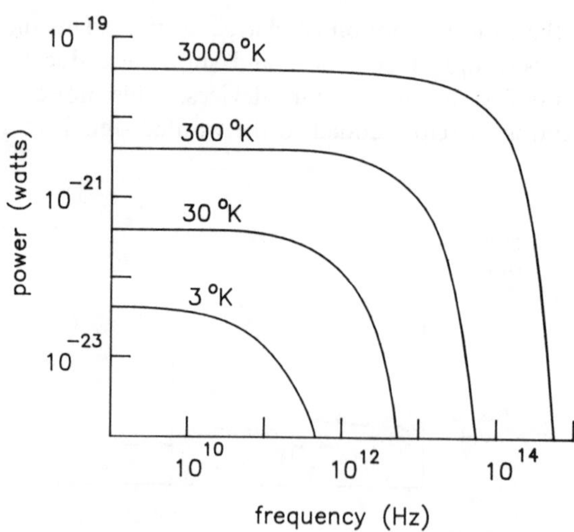

Figure 7.10. Planck's radiation law

f the frequency in Hz, B the bandwidth of the system in Hz, k Boltzmann's constant which has a value of 1.380×10^{-23} J/K and T the temperature in degrees Kelvin.

Planck's law is plotted in Figure 7.10 as power per unit bandwidth against frequency for four temperatures. From the figure it is observed that the noise power is constant as frequency is increased until a particular value after which the noise power falls sharply. The knee in the characteristics depends on temperature and occurs in the submillimetre and infrared parts of the spectrum. For temperatures near to room temperature (290 K) the noise is constant across the microwave and millimetrewave parts of the spectrum. The curve for 3 K is the base level and is important to radio astronomers because this is approximately the cosmic background temperature caused by the cosmic dust which permeates all space and matter.

Planck's law can be approximated in the flat part of the characteristics in Figure 7.10 by taking the first two terms of the series expansion of the exponential so that the denominator gives

$$\exp(hf/kT) - 1 \approx \frac{hf}{kT}$$

Equation (7.11) becomes

$$P = kTB \tag{7.12}$$

This is called the *Rayleigh–Jeans* law and applies for all RF and microwave systems. It is named after Lord Rayleigh (1824–1919) and Sir James Jeans (1877–1946) and was actually formulated before Planck's law. It was the failure of the law to explain the radiation at shorter wavelengths which prompted Planck to devise his law. The Rayleigh–Jeans law states that the noise power is proportional to temperature and to the bandwidth. The receiver noise can be reduced either by decreasing the bandwidth of the reciever, or by cooling the components. The bandwidth is usually determined by the passband of the RF or microwave system. Cooling the components is commonly done in radiometers for remote sensing by using liquid nitrogen in a dewar.

As noted above, there may be many sources of noise which will produce a total noise power, say P_t. The separate sources of noise can however all be lumped together and an equivalent *noise temperature*, T_n, defined by reversing equation (7.12):

$$T_n = P_t/kB \tag{7.13}$$

The noise temperature can then be used to specify the amount of noise which passive components or active devices produces. The external noise due to natural or man-made sources can be specified by the antenna noise temperature which is a measure of how much noise the antenna passes to the receiver.

A useful quantity in microwave system design is the ratio of the total signal power to the noise power. This is called the *signal-to-noise* ratio or in some cases the *carrier-to-noise* ratio:

$$\frac{S}{N} = \frac{P_r}{kT_nB} \tag{7.14}$$

The signal-to-noise ratio is usually specified in decibels as $10\log_{10}(S/N)$. A signal-to-noise ratio of 0 dB means that the signal is at the noise level. Each system will have a minimum signal-to-noise ratio which is necessary to receive coherent signals. The level will be determined by the data or information which is being communicated and the method of modulation. Generally digital data can withstand a lower signal to noise ratio.

7.6 SATELLITE COMMUNICATIONS

Satellites are dependent on electromagnetic waves for command, control and communication. There are a large number of different types of satellites with purposes ranging from communications to remote sensing, but all types must have some form of communication to and from the Earth. Communication satellites are specifically designed to allow the communication of information over long distances, Figure 7.11. They provide a near perfect clear communication channel because the obstructions in the path of the signals are nil. The design of the microwave parts of a communication satellite represents a good example of the use of microwaves for free space transmission.

Satellites orbiting around the Earth obey Kepler's three laws of planetary motion. The first law states that the orbit of a satellite is an ellipse with the centre of the Earth at one of the foci. The second law states that the radius vector of a satellite sweeps out equal areas in equal time, which means that the smaller the radius of the orbit, the faster the speed of the satellite. The final law states that the square of the period of revolution is proportional to the cube of the semi-major axis of the ellipse. This can be expressed as

$$T = 2\pi \frac{a^{3/2}}{\mu^{1/2}} \tag{7.15}$$

where T is the orbital period, a the semi-major axis of the ellipse and μ the product of the gravitational constant and the mass of the Earth, $\mu = 3.986 \times 10^{14}$. The orbit of most interest for communications is a circular orbit where the satellite appears stationary in the sky. This will happen if its period is the same as the rotation period of the Earth and the plane of the orbit is the equatorial plane, Figure 7.12. The orbital period of the Earth is 23 h, 56 min and 4.1 s so inserting $T = 86164.1$ s in equation (7.15) gives $a = 42164.2$ km. This is called the *geostationary orbit*. The radius of the Earth, R_e is 6378.1 km so the altitude of the satellite directly under the satellite is $h = 35786.1$ km.

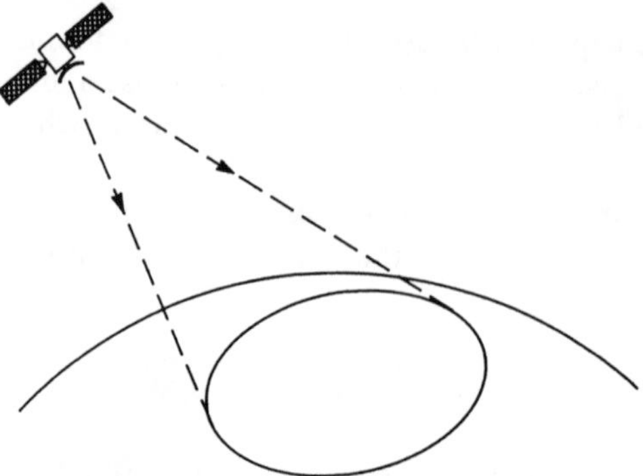

Figure 7.11. Communication satellite

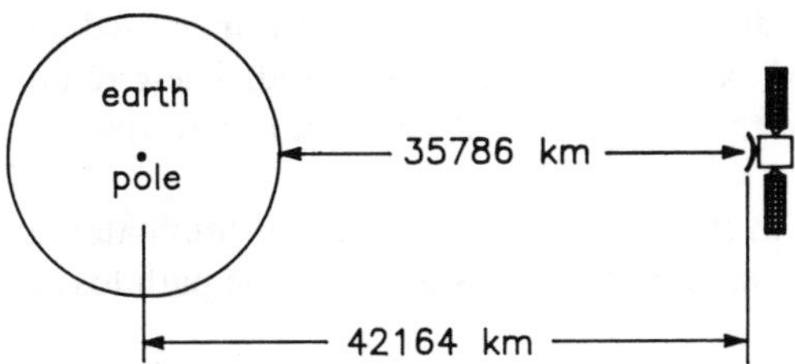

Figure 7.12. Geostationary orbit for satellites

This is the minimum range between an Earth station and the satellite. In practice the Earth station could be anywhere on the surface of the Earth that is visible to the geostationary satellite, so the actual range to the satellite will be greater than 35786.1 km. The distance from the Earth station to a geostationary satellite is found by spherical trigonometry as

$$R^2 = h^2 + 2aR_e[1 - \cos(\psi_e)\cos(\alpha_s - \alpha_e)] \quad \text{km} \tag{7.16}$$

where ψ_e and α_e are the latitude and longitude of the Earth station, and α_s is the longitude of the satellite above the equator. North and east directions, by convention, give positive angles.

The geostationary orbit is used by the majority of communication satellites because of the convenience of not having to move the antenna on the Earth station. There are some disadvantages. The range is long so the path loss is considerable. This also means that there is a noticeable time delay of about 0.25 s on a communication path to and from a satellite. The energy required to place a satellite in the high geostationary orbit (one tenth of the distance to the Moon), is considerable and this increases the costs of a launch. The geostationary orbit is particularly good for Earth stations in equatorial latitudes, but conversely is poor for parts of the Earth at high latitude, where the elevation angle will be low. For this reason the Soviet Union also uses communication satellites in a so-called Molniya orbit, which is a highly elliptical orbit giving improved coverage of the Soviet land mass.

The frequencies used by satellites are highly regulated and must be in one of a number of designated frequency bands. The lowest band is around 140 MHz and this is used for TT&C (Telemetry, Tracking and Command). The main bands used for transmitting voice, video and data are listed below. The actual frequency limits of these bands varies from region to region and country to country.

Band	Uplink	Downlink
C	6 GHz	4 GHz
Ku	14 GHz	11 GHz
Ka	30 GHz	20 GHz

The oldest band in use is C band which was used when satellites had relatively low power transmitters. Consequently large diameter Earth station antennas were needed, typically 30 m. As the demand for more bandwidth increased, and the power available on a satellite increased, Ku band started to be used. These require much smaller Earth station antennas, even though the signal decrease due to distance is greater than for C band.

The signal received at the end of a satellite link is dominated by the distance between the Earth station and the satellite. This is given by the path loss which can be calculated using equation (7.9).

Example 7.3 Path loss from a communication satellite in geostationary orbit

Calculate the path loss from an Earth station at latitude 45° North, longitude 5° West to a geostationary satellite at longitude 40° West for both a 6 GHz up link and a 14 GHz up link frequency.

Equation (7.16) is used to find the range with $\psi_e = 45°$, $\alpha_e = -5°$ and $\alpha_s = -40°$. This gives $R = 38\,819$ km.

The path loss is then given by equation (7.9) for $\lambda = 50$ mm (6 GHz) as

$$L_p = -20\log_{10}(\lambda/4\pi R) = -199.8\text{ dB}$$

For 14 GHz ($\lambda = 21.4$ mm) the path loss is -207.2 dB.

The path loss is counteracted by the power of the transmitter and the gains of the transmit and receive antennas. The signal received is given by the Friis transmission formulae, equation (7.8). Figure 7.13 shows the components which go to make up the losses for a satellite channel. The path loss is the main loss in a satellite system, but there are other losses, including atmospheric loss, polarisation mismatch loss, pointing errors and losses in the electronics which can be lumped together as receiver losses. These may amount to no more than a few decibels and it might seem at first sight that they could be ignored by comparison with the path loss. This is not the case because the costs of building and launching satellites are so large that the power budget must be optimised so that there is just enough signal at the receiver under the worst case scenario.

The amount of signal that is required at the input to the receiver is determined by two factors, the noise in the receiver and the carrier-to-noise ratio (C/N) required for

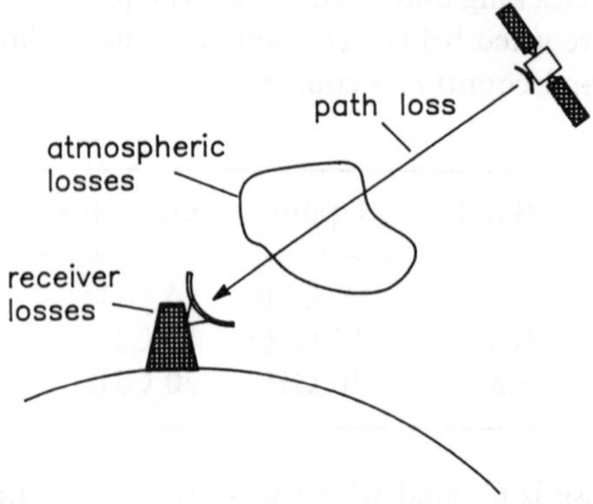

Figure 7.13. Components of power loss in a satellite communication link

satisfactory reception of the signals. The C/N depends on the information and modulation methods in use. For a general purpose communication satellite the C/N could be 18 dB above the noise level. For an FM TV channel the C/N could be 9 dB.

It is always easier to work out satellite power budgets in decibels so that the gains and losses can be added together. This means expressing the noise power, equation (7.12), $P = kTB$, in dB's as

$$P = (\text{dB}) = -228.6 + 10\log_{10}(T) + 10\log_{10}(B) \tag{7.17}$$

The power budget in decibels can be summarized as follows:

+ Transmitter power
+ Transmit antenna gain
− Path loss
− Atmospheric loss
+ Receive antenna gain
− Edge of coverage losses
− Reception losses
− Noise power
= Carrier-to-Noise ratio

All the factors have already been mentioned with the exception of the edge of coverage losses. These arise because the radiation from the antenna will not be uniform over all the portion of the Earth's surface. Allowance must be made for this variation and the power budget calculation must be done for an Earth station at the edge of the intended coverage zone. The calculation of a total power budget is best illustrated with an example. This uses values appropriate to a direct broadcast satellite, but the principles apply for any communication satellite.

Example 7.4 Power budget for a communication satellite

Calculate the down link power budget for a 12 GHz, direct broadcast satellite, with the parameters given below.

The power budget is as follows:

(a) Satellite		
200 W transmitter	23 dBW	
Transmit antenna gain (peak level)	37 dB	
= EIRP from satellite		60 dBW
(b) Path loss at 12.2 GHz over 39 000 km (equation (7.9))	−206 dB	
Atmospheric loss	−0.5 dB	
Total	−206.5 dB	
= Signal received at centre of coverage zone		−146.5 dBW

(c) Receiving antenna		
Diameter $(D) = 0.6$ m. Efficiency $(\eta) = 60\%$		
Gain $= \eta(\pi D/\lambda)^2$	35.5 dB	
= Signal at output from ideal antenna		− 111.0 dBW
(d) Reception losses		
Edge of coverage zone	− 3.0 dB	
Other losses (polarization, pointing, etc)	− 2.0 dB	
Total	− 5.0 dB	
(e) Signal at input to receiver		− 116.0 dBW
(f) Noise power in reciever (equation (7.17))		
Boltzmann's constant	− 228.6 dBW/°K/Hz	
Noise temperature of receiver (500 K)	27.0 dBK	
Channel bandwidth (27 MHz)	74.3 dBHz	
(g) Total noise power		− 127.3 dBW
(h) Worst case C/N (e)–(g) (equation (7.14))		11.3 dB

If the C/N necessary for acceptable TV reception in this example is 9 dB then there is a 2.3 dB margin available. If the receiving antenna is reduced in diameter to 0.45 m, its gain decreases from 35.5 to 33 dB. This eats up the margin and gives 0.2 dB too little signal. 3 dB has, however, been allowed for the receiving antenna being at the edge of the coverage zone, so for most users and the other assumptions above, there would be adequate signal with a 0.45 m diameter antenna.

7.7 MICROWAVE RADIO LINKS

Microwave ratio links are the terrestrial equivalent of satellite communications. information is transmitted between two fixed points on the Earth's surface. Microwave radio links are historically one of the first users of microwaves and were the first methods used by telecommunication authorities for high capacity communication over long distances. Antennas were mounted on towers to give individual path lengths of about 50 km which could be cascaded over as long a distance as required. Recently their use for long distance traffic has declined because optical fibre waveguide can offer higher capacity at a cheaper cost, even given the need to lay the waveguide in the ground. Microwave radio links are, however, still viable for distances of a few kilometres because the capital cost is small and a link can be established quickly. Planning of the early generation of radio links involved the need to cover the maximum distance commensurate with the curvature of the Earth and intervening hills. Usually the system consisted of reflector antennas mounted on towers, Figure 7.14, which used frequencies up to 12 GHz. The present generation of radio links often uses higher frequencies so that the reflector

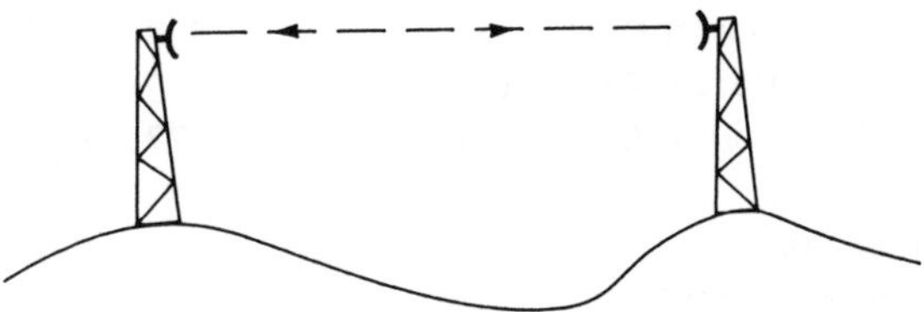

Figure 7.14. Microwave radio link

antennas can be made smaller and they are more likely to be mounted on buildings. Planning a clear line-of-sight may involve the avoidance of urban obstructions such as tall buildings.

The transmission path of a microwave radio link is very similar to that for satellite links. The power budget can be calculated in the same manner as in the last section, using the Friis transmission formulae. There are some extra losses which are not present in satellite links. Many microwave links use dual polarisation in order to double channel capacity. This means that two orthogonal polarised plane waves are transmitted simultaneously (see Section 5.12). There will be a small loss associated with this process. Fluctuations in the atmosphere due to rain and other weather conditions means that a terrestrial microwave link is prone to signal fading. A factor needs to be added to the power budget to allow for this fading. Usually the link is designed to remain operational for a certain percentage of the time, say 99.99% of the time. In order to reduce the need for a strong tower the high power transmitter and low noise receiver are at ground level and a cable or waveguide is used to feed the signals to and from the antennas. This will introduce some attenuation of the signal. Finally the antenna will be less efficient than a satellite Earth station antenna because it must be protected from the weather by a plastic radome and it must be designed to radiate a very low level of signal outside the main beam so as to avoid interference with other links which may be on the same tower.

Example 7.5 Microwave radio link

Calculate the received power on an 11.2 GHz microwave link which has a hop distance of 45 km, using antennas of gain 45 dB. The transmitter power is 10 W and the losses associated with the antenna and the waveguide feeder are 6 dB. The bandwidth is 60 MHz and the receiver has a noise figure of 8 dB.

At 11.2 GHz, the wavelength is 26.8 mm. The path loss over 45 km (equation (7.9)) is thus -146.5 dB. The noise figure (NF) is an alternative method of specifying the noise temperature, $T\ (\text{dB}) = T_0\ (\text{dB}) + \text{NF}$, where T_0 is the physical temperature of the receiver, assumed to be 290 K (or 24.6 dBK). The calculation is then as follows.

Transmitter power (10 W)	+ 10.0 dBW
Waveguide and antenna losses	− 6.0 dB
Transmitter antenna gain	+ 45.0 dB
Path loss	− 146.5 dB
Receiver antenna gain	+ 45.0 dB
Waveguide and antenna losses	− 6.0 dB
Receiver input power	− 58.5 dB

Boltzmann's constant	$-228.6\,\text{dBW/K/Hz}$
Noise temperature $(24.6+8\,\text{dB})$	$32.6\,\text{dBK}$
Channel bandwidth (60 MHz)	$77.8\,\text{dBHz}$
Total noise power	$-118.2\,\text{dBW}$
Carrier-to-noise level	$+59.7\,\text{dB}$

The system may require a C/N of about 20 dB for good quality signals. This leaves a difference of $59.7-20=39.7$ dB. This is necessary in order to allow for fading of the signal due to rain and other factors.

There is another type of microwave radio link which is useful for communication over distances of a few hundred kilometres. These are *tropospheric microwave links*, which use the troposphere to bounce signals from a transmitter to a receiver, Figure 7.15. The troposphere is the lowest region of the atmosphere. It has a base height of about 6 km at the poles and about 18 km at the equator. The troposphere has a refractive index which is slightly different from unity and is a function of temperature, air pressure and water vapour pressure. All these vary with altitude, h, and an empirical model gives

$$n=1+N_0 e^{-qh} \tag{7.18}$$

where N_0 and q are constants with typical values of $N_0=2.9\times10^{-4}$ and $q=0.135$. The important point to note is that the refractive index *decreases* as height increases. If the troposphere is broken into layers of constant refractive index and a microwave ray is traced through the layers using Snell's law of refraction in Chapter 6, the result is that the beam gradually turns over and is redirected back to Earth, Figure 7.16. Thus a radio link can be established between two points on the Earth's surface which are not in line-of-sight.

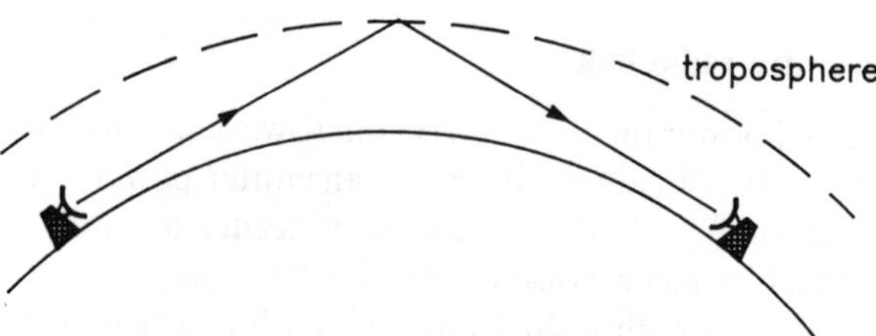

Figure 7.15. Tropospheric microwave link

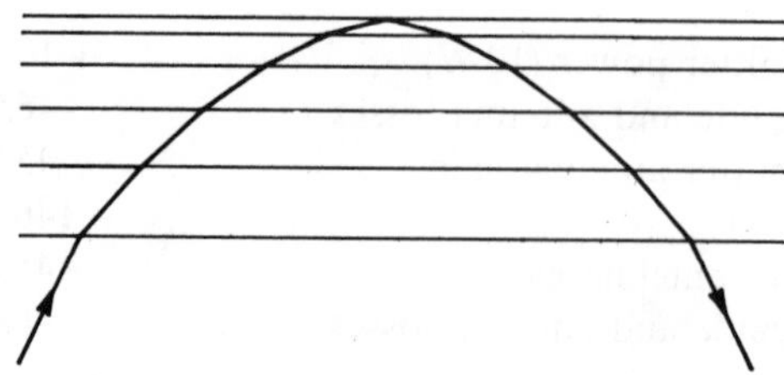

Figure 7.16. Propagation of a ray through a stratified troposphere

The signal calculations follow that for normal microwave links, but there is an important extra loss factor. The characteristics of the troposphere are continually changing which means that equation (7.18) is very approximate, and shows considerable short term fluctuations. These not only spread the beam out, but also modulate the signal with unwanted phase delays. The result is that communications are limited to relatively narrow bandwidths. To try to reduce the chances of losing the link altogether, two adjacent links are sometimes established using different parts of the troposphere.

7.8 BROADCASTING AND MOBILE RADIO

A first thought might be that broadcasting and mobile radio are different uses of radio waves. In fact from the point of view of the transmission path, they are, however, very similar. In contrast to satellite links and microwave radio links, there is one fixed base station and all the users of the signals are at variable points within the coverage area of the base station, Figure 7.17. In the case of broadcasting, the base station is a high power TV or sound transmitter mounted on a tall tower. The users have TV or radio receivers. In the case of mobile radio, the same principle applies but two way communications are necessary. The base station is a high power transmitter and a sensitive receiver. The user has a portable receiver and transmitter.

The discussions of the path loss in the previous sections has shown that this is the most significant factor in the power received by a user. The path loss is inversely proportional to the square of the distance and proportional to the square of the frequency, or expressing equation (7.9) in decibels

$$L_p(\text{dB}) = -32.4 - 20\log_{10}(f) - 20\log_{10}(R) \tag{7.19}$$

where f is in MHz and R is in km. For a fixed physical distance, this equation shows that the path loss can be reduced by using a lower frequency. For this reason terrestrial broadcasting and mobile radio prefer to use frequencies below the microwave part of the spectrum. TV and mobile radio mainly use UHF from 300 to 1000 MHz. The need for more channels and hence more frequency bandwidth means that there is a gradual move to microwave frequencies, especially for mobile radio. Unlike satellite services, there is little scope for improving the antennas gain, or reducing the noise in the receiver, so the range of each base station gets physically smaller if the path loss is to remain the same.

Broadcast stations are designed to cover a specific region, which will determine the height and power of the transmitter. Adjacent transmitters must use a different frequency so that a user does not pick up two radio waves from different transmitters. A receiver has no way of distinguishing one signal from the other and since the path lengths from

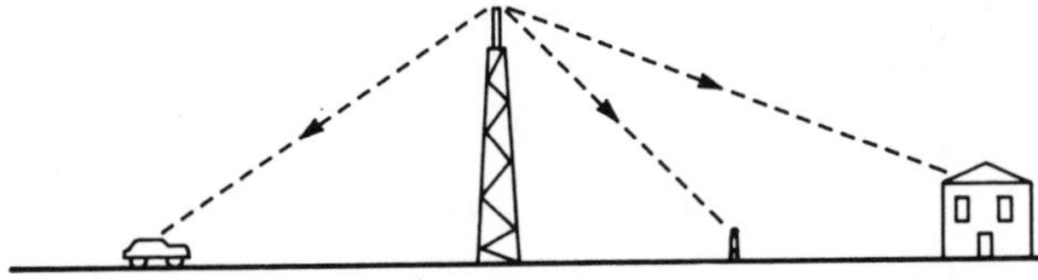

Figure 7.17. Broadcasting or mobile radio from a base station

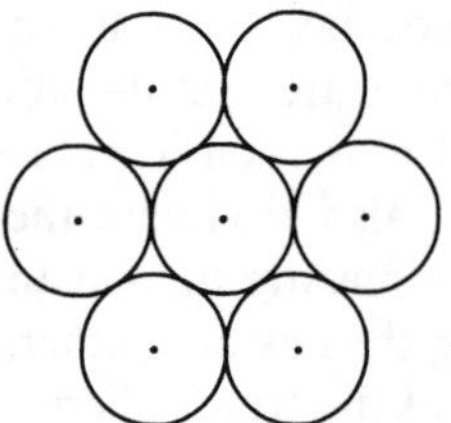

Figure 7.18. Cell arrangements of base stations for mobile radio

the transmitters will be different, the phase of the two plane waves will be different and cancellation or addition might occur. This must be avoided by careful planning. A process which is called *co-channel interference*. The same planning is needed with mobile radio. The transmitter areas are called *cells*, Figure 7.18, and are generally much smaller than equivalent broadcast areas because the low power mobile transmitter determines the maximum path distance. Adjacent cells can either use different frequencies, or, if digital transmission are used, different codes.

7.9 RADAR

The objective of *RADAR* (*RAdio Detection And Ranging*) is to detect the characteristics of objects in the beam of a microwave signal. A basic radar system, Figure 7.19, consists of a transmitter and antenna radiating a beam of plane waves. The plane wave is incident on any objects in the beam. The normal laws of reflection apply and some of the incident wave will be reflected from the surface of the object. In general the reflected waves will be scattered in all directions and will travel outwards from the reflecting object. Waves travelling in the direction of the receiver will be intercepted by the receiving antenna, detected and processed. Normally one antenna is used for both transmitting and receiving. This system is called a *monostatic* radar. If the transmitting and receiving antennas are separate, the system is called a *bistatic* radar.

Two microwave transmission paths are involved in radar systems, one from the transmitter to the object and the other in the reverse direction from the object to the transmitter. The object is called the radar *target* and its characteristics to be detected could be one or more of the following:

Position
Velocity
Polarising characteristics

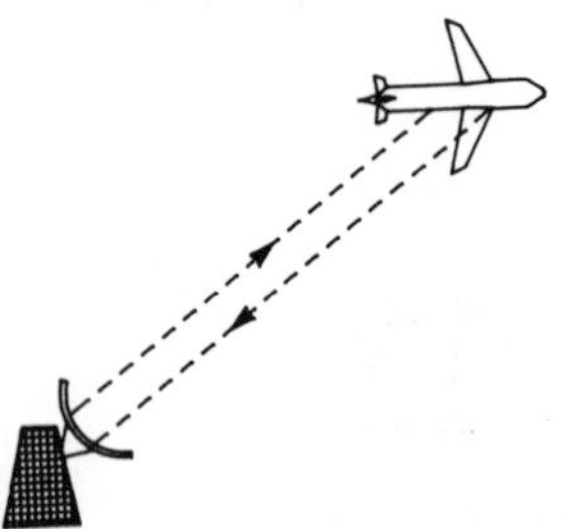

Figure 7.19. Monostatic aircraft detection radar

The position of the target is determined by the distance of the target from the radar, called the *range*, and the angular position of the radar beam. The velocity can either be found by processing the target position data over a known period of time, or by using the Doppler principle. The polarising characteristics are determined by transmitting a wave of known polarisation and comparing against the return signal. The study of reflection and refraction in the last chapter indicated how the characteristics of an object change the polarisation of the waves.

A description of the components of a radar system was given in Chapter 1, so will not be repeated. Here the electromagnetic principles of radar operation will be described. Radars use frequencies over a wide part of the spectrum, determined by the optimum transmission characteristics for the systems intention. Frequencies as low as a few MHz are used in *Over the Horizon* radar where the signals are bounced off the ionosphere so as to be able to detect objects which are not visible by direct line of sight. But most radars operate at microwave frequencies in defined bands throughout the range 1 to 18 GHz. There are also a growing number of radar applications in the millimetrewave part of the spectrum where the small size and high angular resolution are attractive. A band of frequencies around 94 GHz is used for precision military radars.

THE RADAR EQUATION

The radar equation describes the relationship between the range, the power transmitted, the power received and the gain of the antenna. The radar transmitter radiates waves which travel to the target at a range R. The power density incident on the target is given by equation (7.3). The target intercepts a portion of this power and scatters it in various directions depending on the shape of the target. The amount of power intercepted by the target and reradiated back in the direction of the radar is specified by the *radar cross-section*, σ, of the target. The radar cross-section is an important measure of the reflectivity of targets and will be discussed below. It is defined to have units of area. The scattered signal from the target also radiates in all directions so that the power density of the portion of the signal received back at the radar is

$$W = \frac{P_t G}{4\pi R^2}\frac{\sigma}{4\pi R^2} \tag{7.20}$$

where G is the gain of the radar antenna, which has an effective area given by equation (7.7). This area receives the power density given in equation (7.20), so the power received by the radar is obtained by combining equations (7.20), (7.5) and (7.7) as

$$P_r = \frac{P_t G^2 \lambda^2 \sigma}{(4\pi)^3 R^4} \tag{7.21}$$

This is the simplest form of the radar equation and shows that the received power decreases as the fourth power of the range so that the ability to detect a target falls off very rapidly with increasing range. Equation (7.21) is also proportional to the radar cross-section and the square of the gain of the antenna. Thus increasing the gain of the antenna increases the sensitivity and counteracts the range factor.

The minimum signal that can be received from a target will be determined by the noise level of the radar, as described in Section 7.5. If the minimum received power is

P_{min}, the radar equation, equation (7.21) can be reversed to give the maximum range at which a target can be detected,

$$R_{max} = \left(\frac{P_t G^2 \lambda^2 \sigma}{(4\pi)^3 P_{min}} \right)^{1/4} \tag{7.22}$$

This is another version of the radar equation.

Example 7.6 Radar equation

Find the maximum range for a 10.5 GHz air traffic control radar which radiates 10 kW through an antenna with a gain of 40 dB. The target has a minimum radar cross-section of 40 m^2. The receiver has a total system noise temperature of 2000 K and a bandwidth of 100 MHz.

The minimum power receivable is found from $P_{min} = kTB$ to be $P_{min} = 2.76 \times 10^{-12}$ W. Substituting all the values into equation (7.22) gives $R_{max} = 49.4$ km. If the receiver noise level was improved by 10 dB or ten times, the range would increase by a factor of $10^{0.25}$ or 1.78. Doubling the area of the antenna would, through equation (7.7), double the gain and increase the range by a factor of 1.41. This shows that increasing the maximum range is not easily achieved.

The maximum range given by equation (7.22) is optimistic in practice because there are losses in the radar equipment and along the transmission path due to the atmosphere which are not taken into account. In addition the process of detection is statistical in nature. The actual value of the radar cross-section of a target will have a statistical element and there will be a probability of return signals being lost in the noise level of the receiver. Quantifying these extra factors involves probability theory and is beyond the scope of this book. The cumulative effects of these extra factors is that the true maximum range can be as little as half the value predicted by equation (7.22).

RADAR CROSS-SECTION

The radar cross-section, or RCS, is the effective size of a target as *seen* by the radar. It is a characteristic of the size, shape and composition of the target at the angle of incidence. The RCS of targets is an important parameter in radar system design because it specifies how much of the incident signal is returned. It is particularly important in military radars because each target will have a unique RCS which can be used to classify it and to help to distinguish a target from background clutter. Conversely if the RCS of a target can be purposefully made to be small, the target will be difficult to detect.

The RCS of a target is defined as the area which scatters incident power equally in all directions so that the amount of power received at the radar is equal to that from the target. Assuming plane waves so that $P = E^2/Z_0$, this can be expressed as

$$\sigma = \lim_{R \to \infty} 4\pi R^2 \frac{|E_r|^2}{|E_i|^2} \tag{7.23}$$

where E_i is the electric field incident on the target and E_r is the electric field received at the radar. Equation (7.23) shows that σ is proportional to the power reflection coefficient of the target, since E_r is related to the electric field at the target through the path loss. Thus the RCS is effectively a three-dimensional representation of the reflection coefficient of an object.

The RCS of an object will usually be a function of the angle of the incoming plane waves. The object will scatter and diffract power in many directions and it may be that in one direction the RCS is high but in another direction it is very low. A typical polar RCS plot of an aircraft is sketched in Figure 7.20 to illustrate this feature. The plot is in the horizontal plane and indicates that reflections are high from the fuselage and from the leading and trailing edges of the wings, but low elsewhere. The plot is illustrative and does not necessarily represent a real aircraft.

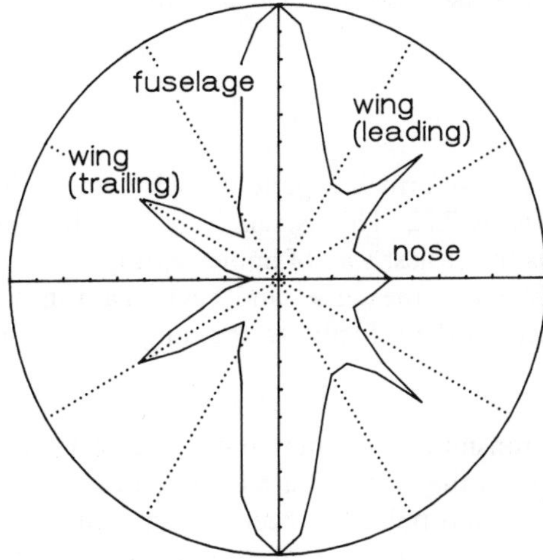

Figure 7.20. Radar cross-section in the horizontal plane of an aircraft, shown as a polar plot

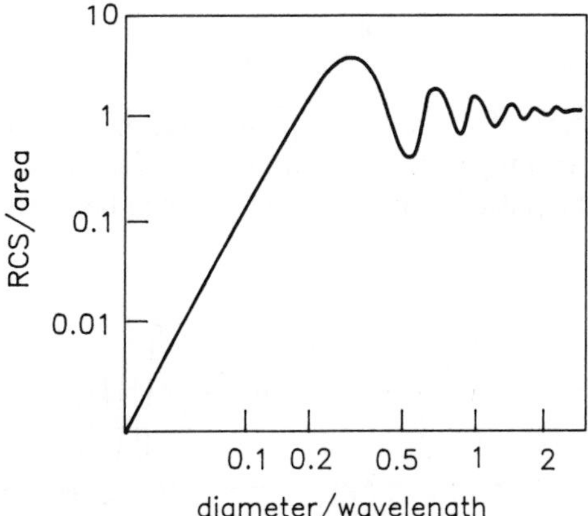

Figure 7.21. Radar cross-section of a sphere

RCSs are usually measured because it is not easy to calculate the RCS of real objects though it can be attempted by computer modelling. One object which can be calculated is a sphere which has the attribute that reflections are independent of direction. The RCS is sketched in Figure 7.21. It shows that for diameters under about 0.3λ, the RCS varies as f^4. This case is relevant to rain drops which approximate to spheres. For sphere diameters of between 0.3λ and 3λ, the RCS amplitude oscillates by as much as ten times. Above 3λ, the RCS is the same as that from a flat disc of the same diameter. The spherical curvature has no effect. This is called the optical region because it is the same as what the eye observes when it sees a sphere.

The peak RCS of practical targets is roughly proportional to the size of the object. A person has an RCS of about 1 m^2, whilst a bird has an RCS of about 0.01 m^2. A medium size boat or a medium size aircraft might have an RCS of about 20 m^2. The largest planes get up to about an RCS of 100 m^2. Perhaps surprisingly, automobiles can have as high an RCS, because in some directions they present a nearly plane surface to a radar.

PULSE RADAR

The most common radar waveform is a periodic pulse where the pulses are bursts of the carrier frequency, Figure 7.22. This is relatively easy to generate with very high peak power levels, and is the easiest way of implementing the radar equation discussed above. The pulse travels out to the target and back again in a time t_d. This is equal to twice the range divided by the velocity, so the range is given by

$$R = \tfrac{1}{2} c t_d \tag{7.24}$$

After a pulse has been transmitted, the next pulse cannot be sent until the first pulse has been received back from the target. If a return pulse from the first pulse did arrive at the receiver after the second pulse has been transmitted there would be no way of knowing whether it came from the first or second pulse, so the target could be at two possible ranges. This leads to the idea of a maximum unambiguous range for a given pulse repetition time. If this is expressed as a pulse repetition frequency, f_p, then equation (7.24) gives

$$R_{max} = \frac{c}{2f_p} \tag{7.25}$$

This range is not a function of the radar frequency but only of the pulse repetition rate. There is also a minimum range given by the width of a pulse. If a signal is received back before the end of the pulse has been sent then it will be mixed up with the outgoing signal.

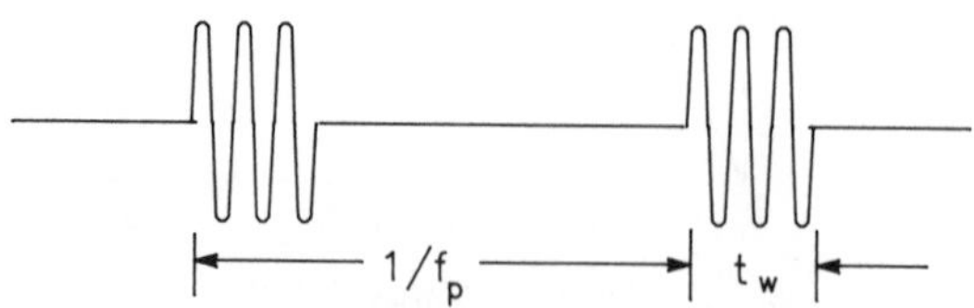

Figure 7.22.. Pulse radar waveform

Example 7.7 Pulse radar waveform

Find the pulse repetition frequency and the pulse width for a radar that is required to have a minimum range of 1 km and a maximum range of 100 km.

Equation (7.25) gives the maximum range in kilometres as $R = 1.5 \times 10^5/f_p$ km. Thus the required pulse repetition frequency is 1500 Hz, or a pulse repetition time of 666 μs. The pulse width is given by equation (7.24) as 6.6 μs. The ratio between the repetition rate and the pulse width is called the *duty cycle*. In this case it is 100:1 which is the ratio between the maximum and minimum ranges. This is relatively inefficient of signal usage because energy is only being radiated 1% of the time.

There are a very wide variety of pulse radars, ranging from long distance search radars to short distance tracking radars. They can be divided into *search radars* and *tracking radars*. A search radar has a radiated beam which continually searches over a chosen volume of the sky. It could be a complete hemisphere, or a much smaller volume. In either case, the radar must send out a pulse in one direction and then wait until any possible return signal has been received, that is the pulse repetition time, before moving to another position. In practice this is not a real restriction since the total travel time is relatively small by comparison with the time required to move the beam. This is even true for an electronically steered beam where the radar is physically stationary but the beam is steered electronically—a so-called *phased array radar*.

A tracking radar is required to continuously track a target. It is assumed that the target has been located by some means and the radar must then follow the target. The radar may itself be on a moving platform, say an aircraft, so both radar and target may be moving. The tracking has to be done by using the beam of the antenna to obtain position information. There are two basic techniques. One technique displaces the antenna beam by a small amount and compares the signal received from two or more positions, Figure 7.23(a). The difference in the signals received between the positions can be used to determine the angle and hence to track the target. The other technique is usually called *monopulse* radar and the antenna is designed to simultaneously transmit the main beam and a difference beam which has a null along the main beam direction, Figure 7.23(b). The return signal from the difference beam can be processed to give tracking data. The monopulse technique is better but requires a higher level of technology in the antenna.

DOPPLER RADAR

Doppler radars are used to detect the relative motion between a radar and a target by utilising the Doppler effect. In this case only a single carrier wave needs to be transmitted and received, Figure 7.24.

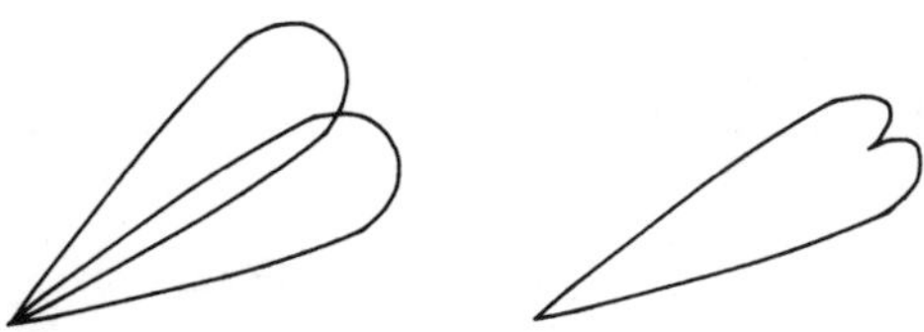

Figure 7.23. Tracking radar beams (a) Dual beams. (b) Monopulse beam

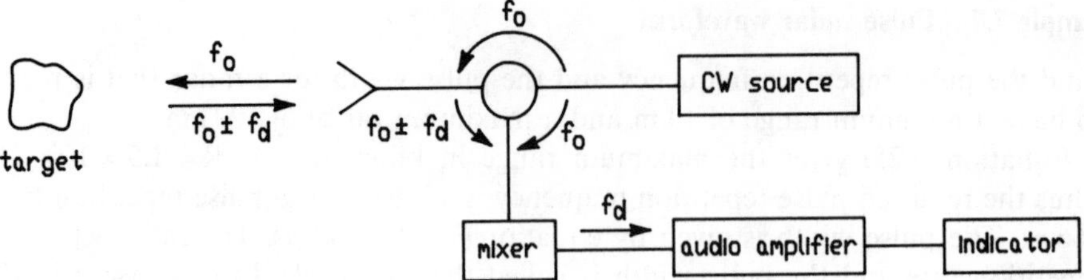

Figure 7.24. Doppler radar system

If the target is moving, the range, R, will be changing and this means that the phase of the waves, ϕ, reflected from the target will be changing with time. A change in phase with time is by definition equal to a frequency and this is called the Doppler frequency, ω_d

$$\omega = \frac{d\phi}{dt} \tag{7.26}$$

The distance travelled by the plane wave to the target and back is $2R$, so the total number of phase angles, ϕ, in the total distance will be $4\pi R/\lambda_0$ radians. Inserting into equation (7.26)

$$\omega_d = \frac{4\pi}{\lambda_0}\frac{dR}{dt} \tag{7.27}$$

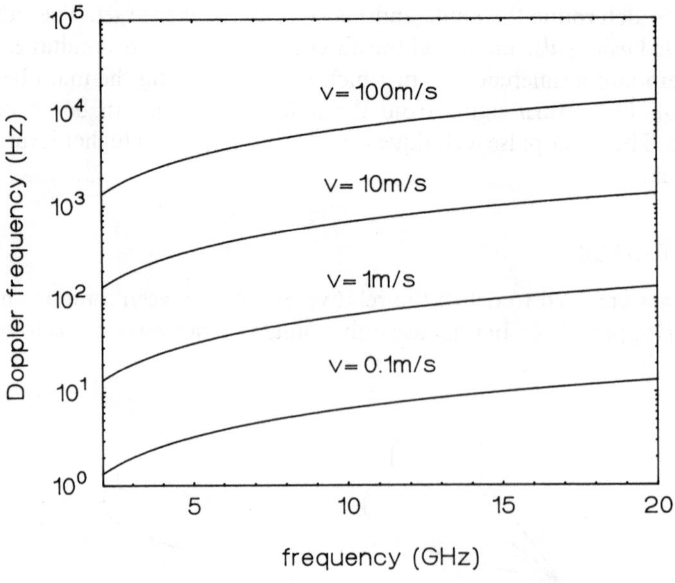

Figure 7.25. Doppler frequency against carrier frequency

dR/dt is the velocity of movement of the target, v, and $\omega_d = 2\pi f_d$, so

$$f_d = \frac{2f_0 v}{c} \tag{7.28}$$

where f_0 is the transmitted frequency. This is the basic equation for Doppler radar and is sketched in Figure 7.25 for a range of values.

The curves in Figure 7.25 show that the Doppler frequency is usually in the audio frequency range. This is very convenient and a Doppler radar can be implemented by mixing a portion of the transmitted signal with the radar return signal and detecting the audio output component. This can be done in a small piece of waveguide to form a compact hand held unit. Doppler radar has therefore found applications where a cheap and simple radar is required. Examples are intruder alarms and vehicle speed detectors.

FMCW RADAR

Frequency Modulated Carrier Wave, FMCW, radar operates by transmitting swept frequency signals. A carrier wave is swept over a band of frequencies. The target, which can be stationary or moving, reflects the signal and causes an effective frequency delay in the signal. At the receiver, the return signal is mixed with the transmitted signal to produce a beat frequency. Figure 7.26 shows a schematic of an FMCW radar.

The waveforms for an FMCW radar are shown in Figure 7.27. An equation will be derived relating the range and the output frequency assuming that the target is static so that there is no Doppler shift. A constant amplitude microwave signal is transmitted for a time t_s, which repeats itself in some form which could be a triangular, or sawtooth waveform. During the period t_s the frequency is swept linearly from frequency f_1 to frequency f_2. The return signal from a target at range R is mixed with the source signal to produce a beat frequency. f_d. This is equal to the difference in frequency $(f_2 - f_1)$ multiplied by the time delay, t_d, as a proportion of the length of the pulse t_s, or

$$f_d = (f_2 - f_1)\frac{t_d}{t_s} \quad \text{Hz} \tag{7.29}$$

The time delay is the time taken for the signal to travel from the radar to the target

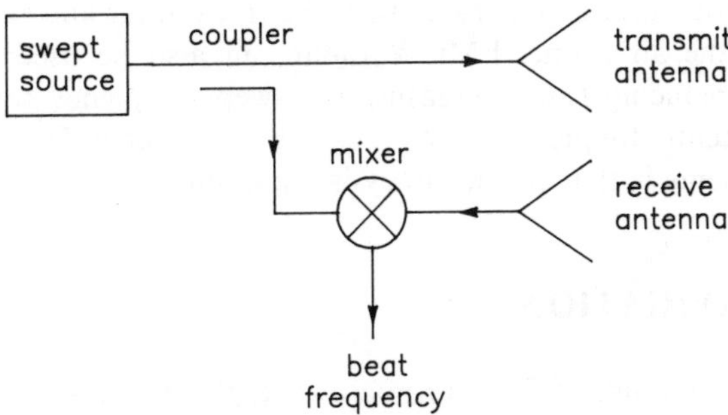

Figure 7.26. FMCW radar system

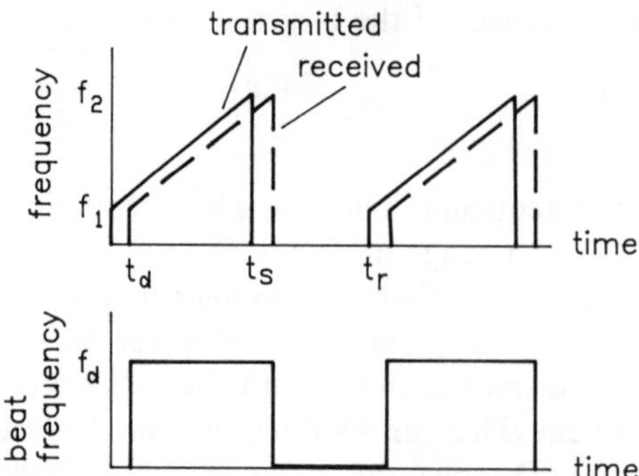

Figure 7.27. Waveforms for FMCW radar

and back again or

$$t_{\mathrm{d}} = \frac{2R}{c} \tag{7.30}$$

Combining equations (7.29) and (7.30) gives the FMCW equation for the range of a target as

$$R = \frac{c f_{\mathrm{d}} t_{\mathrm{s}}}{2(f_2 - f_1)} \tag{7.31}$$

This shows that the range is directly proportional to the beat frequency and the time period of the swept signal. It is inversely proportional to the difference in frequency. The frequency of the carrier does not appear in the equation.

The advantage of FMCW radar is the direct relationship between range and beat frequency and the fact that the beat frequency can be chosen to be any convenient frequency range by correct choice of the other parameters. It is used in radio altimeters to measure the height of an aircraft above the ground. The large cross-section of the ground from the aircraft means that low power sources can be used. One of the problems is to ensure that signal does not leak directly from the transmitter into the receiver. For this reason, a bistatic radar is preferable with the transmitter and receiver mounted on different parts of the air frame. FMCW radar can also be used for short distance measurements by reducing t_{s} or increasing the swept frequency range $(f_2 - f_1)$. This means choosing a centre frequency which is as high as possible. The advantage for short distance measurements is that the accuracy is maintained.

7.10 RADIO NAVIGATION

Radio navigation is the use of RF and microwave plane waves to determine position. A person, vehicle, ship or aircraft receives waves radiated from one or more sources and processes the information to derive position data. The principles are relatively simple and use techniques which have been covered in earlier parts of this text. Navigation techniques can be divided into four types:

Radar
Doppler radio
Hyperbolic systems
Measurement of time differences

Radar is used for marine and air navigation at short distances, for instance in ports and coastal waters, or at airports. Radar systems usually use pulse radar which is implemented as described in the last section.

The United States has had a satellite system using Doppler radio since 1969. The TRANSIT series of satellites are in polar orbits with periods of about 1.75 h. They transmit signals at 150 and 400 MHz which can be picked up by simple receivers. The satellite is moving at a known velocity and is at a known position so that the user can process the received signals to recover the Doppler data and use it to determine the position of the receiver with the theory described in the last section. The system enables position accuracies down to 200 m in two dimensions to be obtained but the limited number of satellites means that only one satellite per 90 minutes flies over any one observer and is only visible for about 10 minutes. The TRANSIT system has worked well but is gradually being phased out in favour of the much more accurate global positioning system described below.

Hyperbolic navigation systems were the first radio navigation systems to be implemented and are still widely used, particularly in local waters. There are two main systems, the Decca system and Loran C. Both use the principle of receiving plane waves from two or three fixed transmitters and comparing the phase of the received signals to obtain position data. If there are two transmitters, P_1 and P_2, Figure 7.28, a ship at position A will receive a signal from P_1 which has a phase determined by the distance $|AP_1|$ and a signal from P_2 which has a phase determined by the distance $|AP_2|$. The receiver will only detect the difference in phase of the two waves given by the difference in distance $|AP_1| - |AP_2|$. This difference in distance will have the same value when the ship moves to another position say B, because $|AP_1| - |AP_2| = |BP_1| - |BP_2|$. The curve joining the points A and B is, by definition, a hyperbola. A user will know that it is on a hyperbola but not the position along the hyperbola. To obtain this extra information, a third fixed transmitter is needed. This produces a second set of hyperbolas, shown dotted in Figure 7.28, which enables the position of the ship to be determined. In fact there is still a potential ambiguity because the signals received by a ship would also be received by a ship on the other side of the axis of the hyperbola. For unambiguous position information, a third set of hyperbolas are needed. Both the Decca system and Loran C use low frequency radio waves in order that the waves travel over long distances.

In the Decca system, the stations radiate CW signals with harmonics of $14\frac{1}{2}$ kHz. The master station in the English chain around the coast of the British Isles radiates at 85 kHz and the slave stations are at $85 \pm 14\frac{1}{2}$ kHz. All Loran stations radiate at 100 kHz but use pulses as the means of distinguishing stations. Loran-C stations exist throughout the Northern Hemisphere. The accuracy depends partly on the position of a ship relative to the base stations, and partly on the ionospheric propagation conditions.

The most accurate method of navigation is provided by measuring the differences in

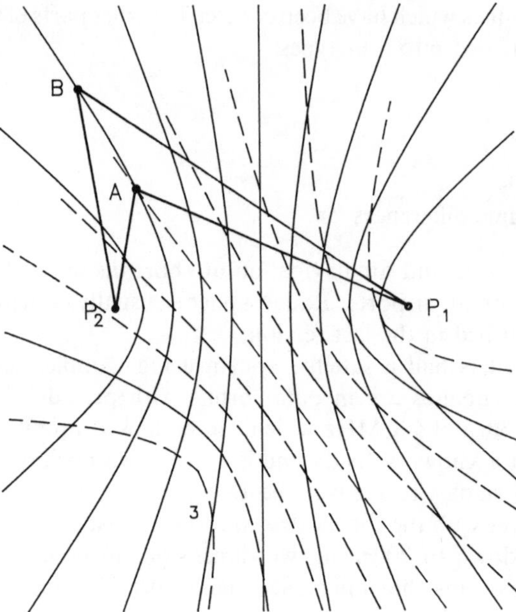

Figure 7.28. Hyperbolic radio navigation with transmitters at P_1, P_2 and P_3, and ships at A and B

the time of arrival of plane waves from a number of orbiting satellites. The high accuracy stems from the ability to make extremely accurate time clocks and to use digital circuits to measure time differences by counting frequency cycles very precisely. These principles have been implemented in the US Department of Defense Global Positioning System (GPS, also called NAVSTAR). This consists of up to 18 satellites with periods of 12 hours at an altitude of 20 200 km. The large number of satellites ensures that there are always four satellites visible to each user. GPS satellites carry very accurate clocks and transmit on frequencies of 1227.6 MHz and 1575.42 MHz. They use a spread spectrum method of modulation which means that the signal levels are very low and actually below the noise level at the receiver, but can be recovered by signal processing. The advantage of spread spectrum is the low power transmitters, immunity from interference, and the need for only a very simple antenna at the receiver.

The principle of GPS is to measure the distance from the satellite by measuring the time that the signals take to travel from the satellite, Figure 7.29. If a user is at P, the signal from two satellites will in principle enable the position of P to be determined. The satellite sends its position and the time that it transmits the signal. GPS is intended to give three dimensional data so a third satellite is needed. There is also an unknown time difference between the time generated by the clock on the satellite and the time generated by the users clock, called the time bias, t_b. To find the time bias, the signal from a fourth satellite is needed. If the satellite transmits a signal at time t_s (which is sent as a coded signal), and it is received at time t_u, the distance from the satellite is $c(t_u + t_b - t_s)$. If the satellite is at position (x_1, y_1, z_1) at the time t_s, and the user is at position (x_u, y_u, z_u), then

$$(x_1 - x_u)^2 + (y_1 - y_u)^2 + (z_1 - z_u)^2 = c^2(t_u + t_b - t_s)^2 \tag{7.32}$$

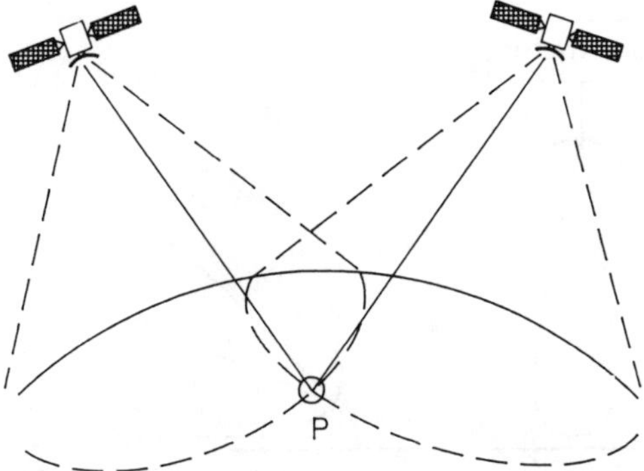

Figure 7.29. Signal received at point P from two satellites in GPS system

x_1, y_1, z_1, t_u and t_s are known but this leaves four unknowns in this equation, hence the need for four satellites. Four signals leads to four simultaneous equations so all the unknown can be determined.

The user needs to determine the travel time with very high accuracy. For instance an error of 0.1 μs in time corresponds to an error in the distance measurement of $ct = 3 \times 10^8 \times 10^{-7} = 30\,\text{m}$. The potential accuracy is, however, very high, particularly if the main requirement is the difference between the users position and another local user. This is referred to as differential GPS.

7.11 REMOTE SENSING AND RADIO ASTRONOMY

The last of the uses of RF and microwaves to be described in this chapter, remote sensing and radio astronomy, is different in character from the previous uses in that it is the electromagnetic waves emitted naturally by all objects which is being detected. There is no transmitter of waves and the system consists only of a sensitive receiver, Figure 7.30. The *source* of waves is nature. The principles of natural radiation were described in Section 7.5, where they were described as noise sources. In most microwave applications, the noise is a nuisance which must be overcome and limits the sensitivity of the system. In remote sensing, the noise is the signal to be received and the objective is to record and quantify the noise in terms of some parameter of the object emitting the natural radiation.

Remote sensing is the general term for the detection of any waves emitted by natural objects, but it is generally applied to detection of objects or volume regions on Earth such as vegetation, the sea, or clouds. The temperature of these regions will be around natural temperatures, and the aim will be to record the temperature as accurately as possible. Radio astronomy is remote sensing applied to natural sources outside the Earth. It has existed as a discipline since Karl Jansky discovered radio signals from space in 1931, but builds on optical astronomy which has a venerable history. The temperature of the stars and galaxies is very high by comparison with Earth-based sources and can exceed 1 million K. This helps to compensate for the long distances. In addition the stars are either point sources, or occupy a relatively compact angular region in space, so that high gain antennas can be used to increase the signal strength.

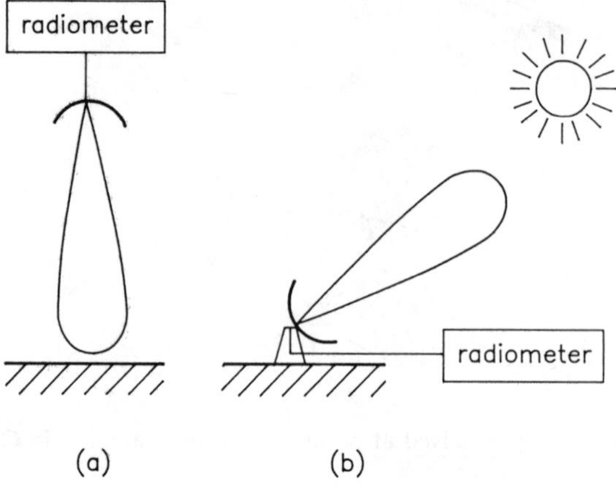

Figure 7.30. Remote sensing system. (a) Looking down at ground. (b) Looking towards extra terrestrial sources (radio astronomy)

In radio astronomy and remote sensing the strength of a source is measured by the *brightness, B*, which is the power density from 1 Hz over 1 steradian. Planck's law, equation (7.11) can be rewritten in terms of the brightness as

$$B = \frac{2hf^3}{c^2[\exp(hf/kT) - 1]} \qquad (\mathrm{W/m^2/Hz/rad^2}) \tag{7.33}$$

For the microwave part of the spectrum, this becomes

$$B = \frac{2kT}{\lambda^2} \tag{7.34}$$

The brightness is proportional to the temperature of the object and inversely proportional to the square of the wavelength. Hence the brightness increases as the observation frequency is increased.

Equations (7.33) and (7.34) give the brightness of a blackbody which is a perfect emitter or absorber of electromagnetic waves. The amount of noise power radiated by a natural objects on the Earth is always less than the amount which would come from a black body of the same volume. This means that the physical temperature, T_p, is less than the black body temperature. T. The ratio between the two is called the *emissivity*, e, or

$$e = \frac{T_p}{T} \tag{7.35}$$

The emissivity is used to characterize different objects and volume regions.

The crucial components in a radio astronomy or remote sensing system are a high gain antenna to receive as much signal as possible and a sensitive receiver. Remote sensing receivers are called *radiometers.* They must record the noise that is coming through the antenna, without it being influenced by the noise in the receiver. This is usually done in a radiometer by comparing the desired signal with an internal calibrated noise source which is cooled so that it has a low noise temperature, Figure 7.31. In one

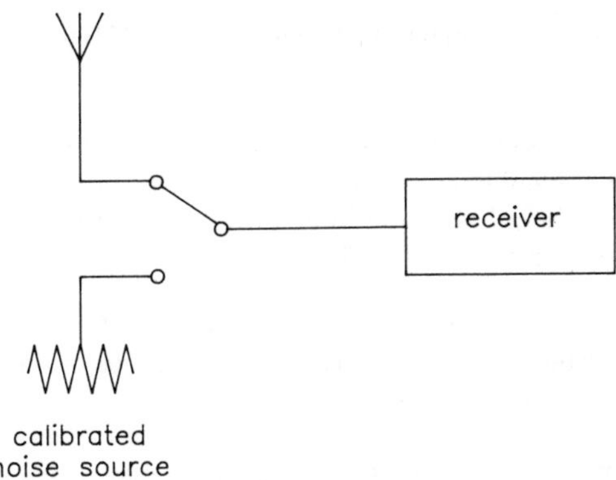

Figure 7.31. Schematic of Dicke radiometer

popular version, called the *Dicke* radiometer, the receiver is continually switched between the desired noise source and the calibrated source. Remote sensing and radio astronomy has one big advantage over other microwave users in that the noise source does not change rapidly with time and will normally ramain constant over a reasonable period of time. This fact can be used by collecting, or integrating, power over a period of time. This is analogous to increasing the bandwidth so that more noise power is detected.

PROBLEMS

1. A wave has an electric field of 1 V/m in free space. It is then incident on a dielectric with $\varepsilon_r = 2.0$. What is the incident power and the power in the dielectric, both in decibels?

2. The electric field of a 10 dBW microwave signal is reduced to 1/1000th of the original level before being amplified by 20 times. What is the decibel ratio of the output to input signal level?

3. A system with a gain of 10 dB has an input impedance of 50 Ω and an output impedance of 75 Ω. Calculate the output electric field amplitude if the input electric field is $10\,\mathrm{mV\,m^{-1}}$.

4. A receiving antenna has an area of $0.1\,\mathrm{m^2}$ and detects -50 dBm. Find the distance from an isotropic source which radiates $+5$ dBW. How much transmitter gain, in dBs, is needed to raise the detected signal level by 150 times?

5. A 400 MHz broadcast transmitter is to be mounted on a 200 m tall tower and is required to broadcast a minimum signal level of 1 mV/m to all receivers which can see the transmitter. What transmitter EIRP is needed? Assume that the Earth is perfectly smooth.

6. An antenna has an aperture efficiency of 65% and a circular diameter of 1 m. Find the gain in dBs at 12 GHz.

7. Find the path loss in dBs over 1 km for (a) an MF wave at 1 MHz, (b) a UHF wave at

400 MHz. (c) a microwave signal at 5 GHz, (d) an infrared wave at 10 μm, (e) a visible wave of blue light at 470 nm.

8. A base station for a mobile radio system operates at 2.5 GHz with a transmitter EIRP of 200 W. If the mobile units have a gain of 3 dB, find the received power 5 km from the base station. If the mobile units contain transmitters with a power of 1 W, what is the received power density at the base station?

9. A transmitter of 1 kW at 3 GHz is connected to a 30 m diameter antenna with an efficiency of 75%. Find the power density on the surface of the Moon. (Take the Earth to Moon distance as 380 000 km.)

10. Calculate the noise power per unit bandwidth produced by a black body radiator of 6000 K at (a) $\lambda = 400$ mm, (b) $\lambda = 400\ \mu$m, (c) $\lambda = 400$ nm.

11. Find the maximum frequency at which Planck's law can be approximated to within 10% by the Rayleigh–Jeans law for a source at room temperature.

12. The signal-to-noise ratio of a receiver with a bandwidth of 10 MHz is 20 dB. What is the minimum signal level if its noise temperature is 500 K?

13. Plot a graph of noise figure in dBs against noise temperature for $0\ \text{K} < T_n < 3000\ \text{K}$.

14. A satellite is placed at an altitude which assumes that the orbital period of the Earth is 24 h instead of the true value. By how much will the satellite move from the geostationary position after 24 h?

15. Prove that the range to a satellite is given by equation (7.16).

16. A 4 GHz geostationary communication satellite has an antenna which illuminates all the Earth. Find the difference in dBs between the path loss for an Earth station at the equator and an Earth station with an elevation look angle of 5° to the satellite. Both Earth stations are on the same longitude as the satellite.

17. Calculate the range and path loss for a 11 GHz geostationary satellite positioned at longitude 45°W for Earth stations at (a) London (0°W and 51°N), (b) Jamaica (77°W and 18°N).

18. What is the time delay on a satellite link from Moscow (38°E and 56°N) to Washington (76°W and 39°N) via a geostationary satellite at longitude 15°W?

19. Evaluate the up link power budget for a 6 GHz geostationary satellite at a range of 39 000 km. The earth station has an EIRP of 80 dBW. The receiving antenna on the satellite has a diameter of 2 m with an efficiency of 80%. Neglect other losses to work out the received power at the satellite.

20. Design the down link for a satellite system to operate at 20 GHz with a satellite EIRP of 50 dBW. The range is 38000 km and the total atmospheric and other losses are 3.5 dB. The receiving antenna diameter is to be chosen to yield a C/N of 20 dB with a bandwidth of 36 MHz and a receiver noise figure of 6 dB.

21. A 6 GHz microwave radio link has six equal spaced relay sections over a total distance of 300 km. All antennas are identical with gains of 40 dB and waveguide losses of 3 dB. The transmitters have a power of 10 W. Find the power available at the receiver.

22. A microwave radio link operates at 19 GHz with antennas of gain 30 dB. If the distance is 4 km, what transmitter power is required for a minimum received power level of − 60 dBm?

23. It is desired to design the radiation pattern of a UHF broadcast transmitting antenna so that the received signal is constant for all users. Find an equation for the shape of the antenna radiation pattern required to achieve this distribution in terms of the angle from the horizontal.

24. A bistatic radar has the receiver at a different location from the transmitter. Obtain a form of the radar equation for this case with distances R_T and R_R to the transmitter and receiver, respectively.

25. An 8 GHz radar has a minimum received power of − 80 dBm and a transmitter power of 10 kW. What antenna gain is required to give a range of 50 km and detect a target with a radar cross-section of 15 m^2?

26. Design a 74 GHz collision avoidance radar which has a minimum received power of − 50 dBm. It must detect a person on a cycle with an RCS of 1 m^2 at a distance of 20 m.

27. A pulse radar has a pulse width of 15 μs and a duty cycle of 200:1. Find the minimum and maximum unambiguous range to the radar.

28. A Doppler radar operating at 9.5 GHz is required to detect a range of velocities from 1 km/h to 150 km/h. What frequency band is required to accommodate this range?

29. An FMCW radar altimeter operates at 4 GHz and has a swept frequency range of 10 kHz which lasts for 10 μs. If it detects a frequency difference of 30 kHz, what is the height of the aircraft?

30. Calculate the brightness of a black body radiator at a temperature of 10 000 K and a frequency of (a) 30 GHz and (b) a wavelength of 500 nm.

8 Microwaves in lossy media

8.1 INTRODUCTION

The description and explanation of plane waves and microwaves in the preceding chapters assumed that the medium through which the waves were travelling introduced no attenuation to the waves. In other words the medium was loss-less. As the plane waves (microwaves or radio waves or optical waves) travel through a loss-less medium they suffer no attenuation and are represented as the sinusoidal wave in Figure 8.1(a) which has constant amplitude with distance. The medium through which the waves travelled was called either *free-space* or dielectric to indicate an idealised loss-less medium. The only case where this is actually true is for waves travelling in space. On Earth all media have a some loss including the atmosphere, although at many frequencies throughout the electromagnetic spectrum the atmosphere is nearly transparent. For other media such as dielectrics, wood and water the medium is lossy and this means that as a plane wave travels through the medium some of its energy is absorbed and turned into heat. The consequence for the plane wave is that the wave is attenuated and its amplitude decreases with distance as in Figure 8.1(b). This chapter starts by studying in detail what happens to any plane wave as it travels through a lossy medium and then describe a number of practical systems where microwave signals are influenced by the lossy medium. These divide into cases where the attenuation is undesirable, such as communications and radar, and cases where all the heat produced by the attenuation is used for heating and cooking. There is also the situation where the heat is undesirable and a maximum safety level is needed.

The amount of attenuation of a plane wave is determined by the conductivity of the medium. The higher the conductivity the greater is the attenuation of the plane wave. this can be seen qualitatively if it is recalled that the current density is equal to the electric field times the conductivity

$$\boldsymbol{J} = \sigma \boldsymbol{E} \tag{8.1}$$

where σ is the conductivity with units of Siemens per metre (S/m). The electric field of the incident plane wave $\boldsymbol{E}$ generates a current density $\boldsymbol{J}$. This is a conduction current and gives an attenuating part to the plane wave solution as will be demonstrated in the next section. The solutions will enable the characteristics of plane waves in any media to be determined.

Most practical media have finite conductivity. Dielectrics are good insulators, but even here there is a very small conductivity and consequently a very small loss. In previous chapters the concept of a perfect conductor was used which has infinite conductivity and is a perfect reflectors of electromagnetic waves. A perfect conductor does not exist in practice and all metals have high but finite conductivities. Some of these are listed in the Appendix. This means that when a plane wave is incident on the surface of a metal there is a small amplitude plane wave present in the conductor and

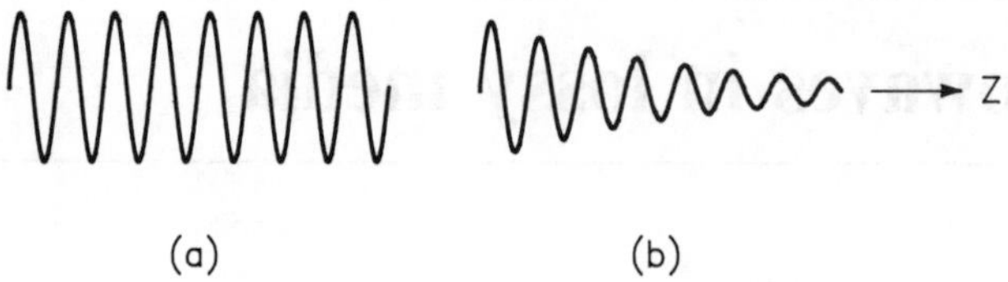

Figure 8.1. (a) Waves in loss less medium, (b) Waves in lossy medium

100% of the energy is not reflected. At most frequencies of interest to microwave and optical engineers only a very small amount is absorbed in the metal and it is a good approximation to assume infinite conductivity. The absorption, however, increases as frequency increases and so X-rays are substantially absorbed by metals.

8.2 CHARACTERISTICS OF PLANE WAVES IN A CONDUCTING MEDIUM

The way in which the characteristics of a plane wave are modified by the conducting or lossy medium can be demonstrated by reworking the plane wave analysis which was done in Chapter 5 for the case of a medium of finite conductivity. When free space was being considered one of the starting points was the second Maxwell equation under the assumption that $\boldsymbol{J} \equiv 0$. The full form of the second Maxwell equation is

$$\nabla \times \boldsymbol{H} = \boldsymbol{J} + \mathrm{j}\omega\, \varepsilon_0\, \varepsilon_r\, \boldsymbol{E} \tag{8.2}$$

Substituting equation (8.1) into this equation gives

$$\nabla \times \boldsymbol{H} = \sigma \boldsymbol{E} + \mathrm{j}\omega\varepsilon_0\, \varepsilon_r\, \boldsymbol{E} \tag{8.3}$$

or

$$\nabla \times \boldsymbol{H} = \mathrm{j}\omega\varepsilon_0\, \varepsilon_r \left(1 - \mathrm{j}\frac{\sigma}{\omega\varepsilon_0\varepsilon_r}\right)\boldsymbol{E} \tag{8.4}$$

comparison with the equivalent equation for a plane wave in free space shows that it will have the same form if a modified complex permittivity ε_d is defined as

$$\varepsilon_d = \varepsilon_r\left(1 - \mathrm{j}\frac{\sigma}{\omega\varepsilon_0\varepsilon_r}\right) \tag{8.5}$$

This can also be expressed as

$$\varepsilon_d = \varepsilon' + \mathrm{j}\varepsilon'' = \varepsilon_r - \mathrm{j}\frac{\sigma}{\omega\varepsilon_0} \tag{8.6}$$

where ε' is the real part of ε_d and is equal to ε_r and ε'' is the imaginary part of ε_d and is equal to $-\sigma/\omega\varepsilon_0$.

Equation (8.4) can be rewritten as

$$\nabla \times \boldsymbol{H} = \mathrm{j}\omega\varepsilon_0\, \varepsilon_d\, \boldsymbol{E} \tag{8.7}$$

This has exactly the same form as equation (8.2) with ε_d replacing ε_r and $\boldsymbol{J} = 0$. By analogy with the plane wave solution in free space, the solution for plane waves in a lossy medium will be

$$E_y = E_0 \exp[\mathrm{j}(\omega t - kz)] \tag{8.8}$$

where

$$k^2 = \omega^2 \mu_0 \varepsilon_0 \varepsilon_d \tag{8.9}$$

Equation (8.8) has the same form as the equation of a plane wave in free space except that β has been replaced by k. β was a real number which described the phase coefficient of a loss-less wave whereas k is a complex number with a real part β and an imaginary part α

$$k = \beta - \mathrm{j}\alpha \tag{8.10}$$

and is given by equations (8.9) and (8.5)

$$\beta - \mathrm{j}\alpha = \omega(\mu_0 \varepsilon_0 \varepsilon_r)^{1/2} \left(1 - \mathrm{j}\frac{\sigma}{\omega \varepsilon_0 \varepsilon_r}\right)^{1/2} \tag{8.11}$$

The real part, β, and the imaginary part, α, cannot be expressed simply in terms of the electrical parameters because the right-hand side of equation (8.10) involves a complex square root. The negative sign for α in equation (8.10) has been chosen in order that the wave is an attenuating wave as can be seen by substituting equation (8.10) into equation (8.8) to give

$$E_y = E_0 \exp[\mathrm{j}(\omega t - \beta z)]\mathrm{e}^{-\alpha z} \tag{8.12}$$

This equation can been seen to have a sinusoidal part which is similar to the plane wave solution for free space (although β is modified by the losses) and an exponential part whose magnitude decays as either α or z increases. This latter part describes the attenuation of the wave. The decaying plane wave is sketched in Figure 8.2. α is called the *attenuation constant* and depends on the frequency, the relative permittivity, the relative permeability and the conductivity.

Equation (8.12) describes the manner in which the electric field and magnetic field decay. The power density for a plane wave in free space is given by Poynting's vector

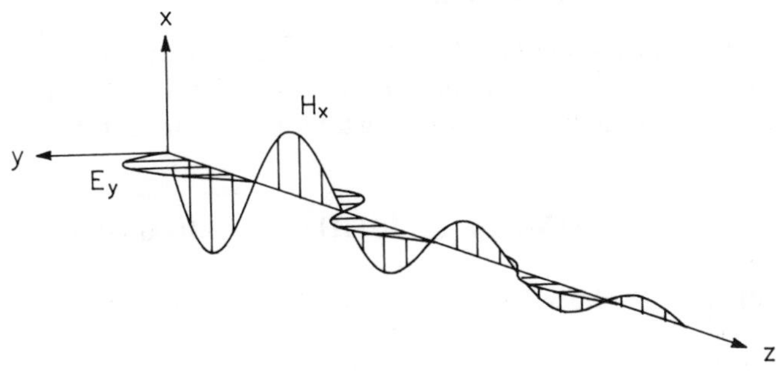

Figure 8.2. Decaying linear polarised plane wave

$\frac{1}{2}(\boldsymbol{E} \times \boldsymbol{H}^*)$. For a wave in a lossy medium this must be replaced by $\frac{1}{2}(\boldsymbol{E} \times \boldsymbol{J}^*)$. Referring back to Section 5.13 shows that the average power absorbed from a plane wave in a lossy medium is

$$P(z) = \tfrac{1}{2}\sigma E_0^2 \mathrm{e}^{-2\alpha z} \quad \mathrm{W\,m^{-3}} \tag{8.13}$$

In equation (8.13), E_0 is the *peak* amplitude of the electric field. If E_0 is the rms value then the equation becomes $P(z) = \sigma E_0^2 \mathrm{e}^{-2\alpha z}$. Substituting for σ from equation (8.6) into equation (8.13) gives an alternative form

$$P(z) = \tfrac{1}{2}\omega\varepsilon_0 \varepsilon'' E_0^2 \mathrm{e}^{-2\alpha z} \quad \mathrm{W\,m^{-3}} \tag{8.14}$$

This shows that the power absorbed per unit volume is proportional to the frequency of the waves, so for a fixed electric field, a higher power absorption can be obtained at a higher frequency.

The rate of attenuation is often expressed in decibels relative to the reference point $z = 0$, where $P(0) = \frac{1}{2}\sigma E_0^2$. Thus from equation (8.13)

$$\frac{P(z)}{P(0)} = 10\log_{10}(\mathrm{e}^{-2\alpha z}) = -8.686\,\alpha z \quad \mathrm{dB} \tag{8.15}$$

thus the rate of attenuation per metre is 8.686α.

8.3 LOSS TANGENT AND PENETRATION DEPTH

The complex permittivity in equation (8.6) contains all the information about the medium and the frequency of the plane wave. A parameter which is used to describe the amount of loss is the magnitude of the ratio of the imaginary and real parts of equation (8.6). This is called the *loss tangent*, $\tan\delta$, and

$$\tan\delta = \left|\frac{\varepsilon''}{\varepsilon'}\right| = \frac{\sigma}{\omega\varepsilon_0\varepsilon_r} \tag{8.16}$$

The loss tangent increases as the conductivity increases or the frequency decreases or the permittivity decreases. This indicates that the same materials can influence plane waves differently at different frequencies and can be demonstrated with examples.

Example 8.1 Loss tangent of different materials

Calculate the loss tangent for copper ($\sigma = 5.8 \times 10^7$, $\varepsilon_r = 1$), carbon ($\sigma = 3 \times 10^4$, $\varepsilon_r = 1$) and bakelite ($\sigma = 0.2$, $\varepsilon_r = 4.7$) at frequencies of 100 MHz, 10 GHz, and 1000 GHz.

Using equation (8.16) enables the following table of loss tangents to be drawn up:

	100 MHz	10 GHz	1000 GHz
Copper	1.0×10^{10}	1.0×10^{8}	1.0×10^{6}
Carbon	5.4×10^{6}	5.4×10^{4}	540
Bakelite	7.6	0.076	7.6×10^{-4}

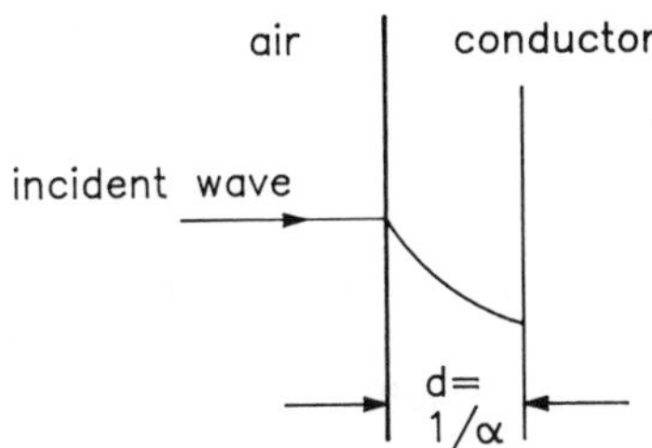

Figure 8.3. Penetration depth (or skin depth) in a conductor

The loss tangent is very high for copper throughout the microwave and millimetre wave band. Carbon is also high at 100 MHz. Bakelite has values around unity across the microwave spectrum.

As the plane waves travel into a conducting medium they are exponentially attenuated. It is useful to have a measure of the distance over which they decay by a standard amount. This is called the *penetration depth, d*, and is chosen to be the distance where the field amplitude is 1/e of the value at the surface.

At the penetration depth, equation (8.12) gives that $|E_y(d)/E_y(0)| = e^{-\alpha d} = e^{-1}$. This means that $d = 1/\alpha$. The penetration depth can be thought of as a layer of depth, d, within which lossy plane waves exist and beyond which the conductor is perfect, Figure 8.3. Since $1/e = (2.7183)^{-1}$ the penetration depth can be expressed in decibels as the depth at which the field has decayed to -8.686 dB. The penetration depth will be a very small distance for highly conducting materials and is like a skin over the conductor. For this reason it is also called the *skin depth.*

8.4 EVALUATION OF THE ATTENUATION CONSTANT

The evaluation of the attenuation constant α can be divided into three cases:

(a) waves in a general medium where $\tan\delta \approx 1$.

(b) waves in a low conductivity medium (or lossy dielectrics), where $\tan\delta \ll 1$.

(c) waves in highly conducting materials, where $\tan\delta \gg 1$.

In the last two cases it is possible to make approximations which give simple results. In the first case, the equation for the propagation coefficient, equation (8.11) must be evaluated directly, as will be illustrated with an example.

Example 8.2 Complex propagation coefficient

Find the complex propagation coefficient for a material with a conductivity of 5 S/m and relative permittivity of 20 at 5 GHz.

The loss tangent for this case is

$$\tan\delta = \frac{\sigma}{\omega\varepsilon_0\varepsilon_r} = \frac{5}{2\pi \times 5 \times 10^9 \times 8.854 \times 10^{-12} \times 20} = 0.9$$

This is within the regime where the exact expression in equation (8.11) must be evaluated. This involves finding the square root of a complex number. Some program languages, e.g. FORTRAN, can handle complex numbers directly. On a scientific calculator the complex square root is best found numerically by changing the complex number from rectangular to polar from ($Ae^{j\theta}$) then taking the square root of A and halving θ, then returning to rectangular form. Equation (8.11) gives $\beta - j\alpha = 468 \times (1 - j0.9)^{1/2}$. In polar form the complex number in the square root is $1.34e^{j(-41.9^\circ)}$ so its square root is $1.16e^{j(-20.95^\circ)}$. Converting this back to rectangular form gives $\beta - j\alpha = 507 - j194$. β is in rad/m and α in nepers per metre.

The two special cases of waves in a low conductivity medium and in a highly conducting medium will now be considered.

WAVES IN LOSSY DIELECTRICS OR A LOW CONDUCTIVITY MEDIUM

When $\tan\delta \ll 1$ the square root in equation (8.11) can be expanded as a binomial series. Taking the first term gives

$$k = \omega(\mu_0\varepsilon_0\varepsilon_r)^{1/2}\left(1 - j\frac{\sigma}{2\omega\varepsilon_0\varepsilon_r}\right) \tag{8.17}$$

or

$$\alpha = \frac{\sigma}{2}\left(\frac{\mu_0}{\varepsilon_0\varepsilon_r}\right)^{1/2}$$

The penetration depth becomes

$$d = \frac{2}{\sigma}\left(\frac{\varepsilon_0\varepsilon_r}{\mu_0}\right)^{1/2} \tag{8.18}$$

Both the attenuation coefficient and the penetration depth for low conductivity materials are independent of the frequency of the plane wave. It should, however, be remembered that $\tan\delta$ is inversely proportional to frequency and, therefore, the condition for independence of α and d on frequency only holds so long as $\tan\delta$ is much less than unity.

Example 8.3 Penetration depth in low conductivity material

An material with a low conductivity at radio frequencies is ice. At RF ice has a relative permittivity of about 3 and a conductivity of about 10^{-6} S/m. Find the penetration depth.

The loss tangent is

$$\frac{\sigma}{\omega\varepsilon_0\varepsilon_r} = \frac{6000}{f}$$

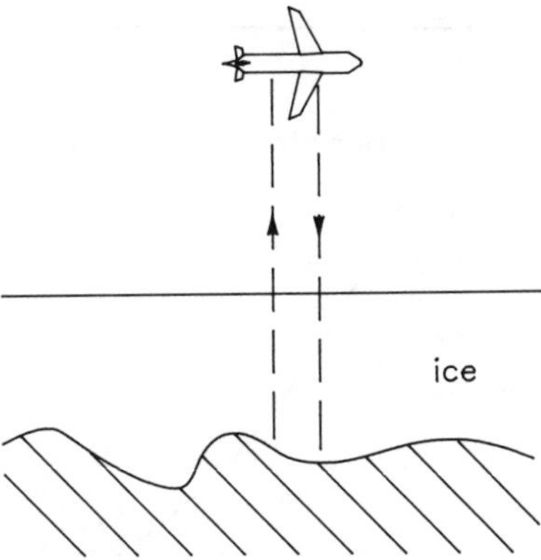

Figure 8.4. Probing of ice field using radar at radio frequencies

If f, the frequency, is in the MHz range then $\tan\delta \ll 1$. Inserting the values into equation (8.18) gives the penetration depth as 9.2 km. This means that RF radio waves travel very easily through ice. This fact has enabled large blocks of ice such as those covering Greenland or Antarctica to be probed using radar. Echoes are returned from the ground under the thick ice caps so that a map of the profile of the ground can be constructed, Figure 8.4.

WAVES IN A HIGHLY CONDUCTING MEDIUM

When $\tan\delta \gg 1$ the imaginary part of equation (8.11) is large and so the real part can be ignored. Hence

$$k \approx \omega(\mu_0\varepsilon_0\varepsilon_r)^{1/2}\left(-\mathrm{j}\frac{\sigma}{\omega\varepsilon_0\varepsilon_r}\right)^{1/2} = (\omega\mu_0\sigma/2)^{1/2}(1-\mathrm{j}) \tag{8.19}$$

The phase and attenuation coefficients have the same value. The penetration depth, or skin depth, is

$$d = \left(\frac{2}{\omega\mu_0\sigma}\right)^{1/2} \tag{8.20}$$

In this case most of the current is near the surface and then it is valid to think of the conducting medium as being made up of a layer of lossy material of thickness d and conductivity σ. This layer can be considered to behave as a resistive sheet and a *surface resistance* can be defined as

$$R_s = \frac{1}{\sigma d} = \left(\frac{\omega\mu_0}{2\sigma}\right)^{1/2} \quad \Omega \tag{8.21}$$

This is the effective resistance of the conductor. Notice that the penetration depth decreases as the square root of frequency, but the surface resistance increases as the square root of frequency.

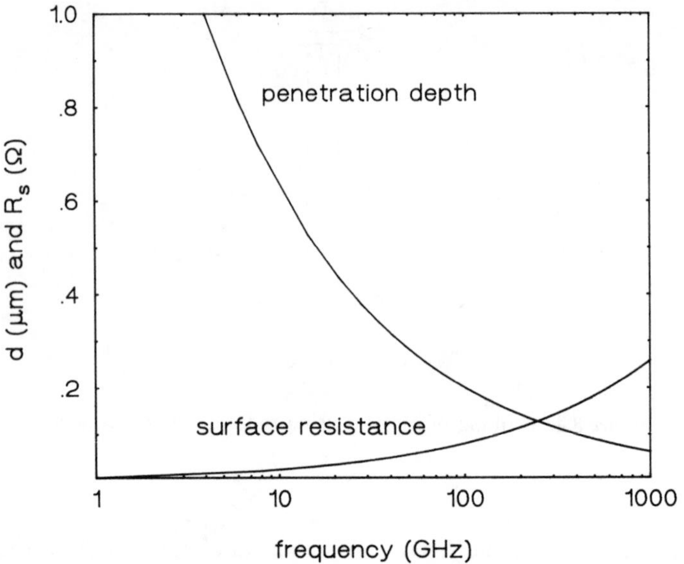

Figure 8.5. Penetration depth, d, and surface resistance, R_s, for copper

Example 8.4 Skin depth and surface resistance of copper

Good examples of highly conducting materials are metals. Plot the skin depth and surface resistance for copper over the frequency range 1 to 1000 GHz.

Copper has a conductivity of 5.8×10^7 S/m which gives a $\tan\delta$ of $10^{18}/f$, a penetration depth, or skin depth, of $0.07/f^{1/2}$ m and a surface resistance of $2.6 \times 10^{-7} f^{1/2}\,\Omega$. The values are plotted in Figure 8.5. The skin depth has values akin to optical wavelengths throughout this frequency range. For all frequencies below X-rays, copper is a good conductor. The surface resistance has a low value at 1 GHz, but is starting to be significant at 1000 GHz. It is partly for this reason that metals are not widely used in the submillimetrewave part of the spectrum. Another reason is that it is difficult to make the surface of the metal smooth so that surface roughness contributes to the attenuation.

8.5 MICROWAVE PROPERTIES OF WATER

One of the materials which is of most interest to microwave engineers is water. This occurs widely in its pure form and it also forms a large part of many other materials such as plants, animals and the ground. It is, therefore, of interest to know how microwaves are influenced by water based substances. Unfortunately water is a very complicated substance. It is known as a *polar* material because the hydrogen and oxygen bonds form polar molecules. When an electrical field is incident on the material, the molecules align themselves with the field. The electrical characteristics change radically depending on whether it is solid in the form of ice; partly frozen in the form of snow;

in normal liquid form; or in the form of water vapour. The liquid form is the most common and its relative permittivity and conductivity depend on three factors:

Frequency

Temperature

Proportion of other material absorbed

The characteristics of materials whose permittivity changes with frequency is called a *dispersive* material. The name comes from the fact that if a group of frequencies are sent through a material, the separate frequencies will be delayed by different amounts so that the energy will be dispersed at the receiver. The real and imaginary parts of the relative permittivity of pure water are sketched at room temperature in Figure 8.6 over the frequency range 1 to 30 GHz. The real part remains approximately constant for lower frequencies. Above 30 GHz its value oscillates due to molecular resonances but gradually decreases until it reaches a value of $\varepsilon_r = 1.78$ at visible wavelengths. The imaginary part peaks at about 15 GHz and then decreases. The peak is very temperature dependent. In the millimetrewave region of the spectrum, water has a number of resonances so that its attenuation characteristics oscillate as frequency changes. At optical frequencies the imaginary part of the permittivity has fallen to a low value.

The conductivity values of water can be obtained from equation (8.6) as $\sigma = \omega\varepsilon_0\varepsilon''$. This is apparently proportional to frequency but the frequency dependence of ε'' often dominates. It is well known that at low frequencies water conducts easily so that it is highly attenuating, whereas at optical frequencies it is possible to see through clear water which must mean that it is only slightly attenuating.

The typical variation of permittivity of water with temperature at 3 GHz is sketched in Figure 8.7. This shows that both the real part of the permittivity and the conductivity

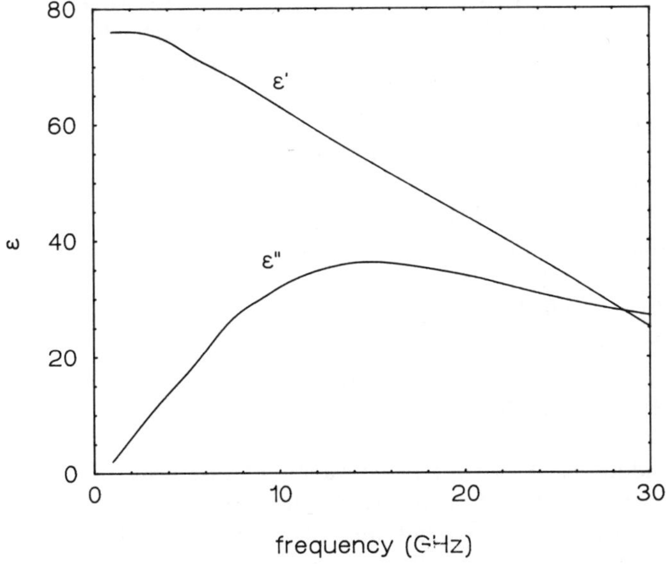

Figure 8.6. Permittivity of pure water in the microwave spectrum

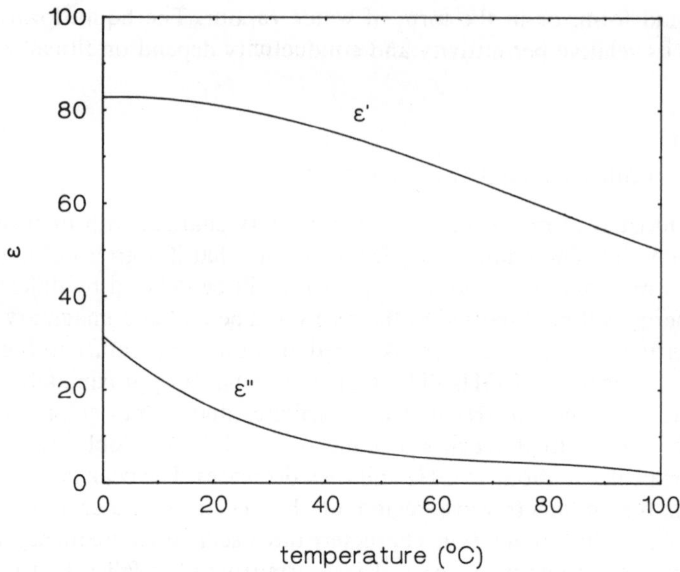

Figure 8.7. Variation of permittivity of water with temperature at 3 GHz

decrease with increasing temperature. This is a useful feature because equation (8.13) shows that the power absorbed as heat is proportional to the conductivity. As the temperature increases, the power absorbed decreases and the heating effect is thus partly self limiting.

The values given in Figures 8.6 and 8.7 are for pure water. The addition of any impurity, particularly salts, changes the properties considerably. The values of the real and imaginary parts of the permittivity change, and the relationship with frequency and temperature are altered, sometimes radically, so that it is very difficult to state any firm values for water when it is mixed with other substances. It is necessary to measure each case to find the properties.

At radio frequencies the attenuation of plane waves through water can be found from the general equation (8.11). An interesting case is the propagation of radio waves through sea water. At RF and low frequencies, sea water has typical properties, $\varepsilon_r = 80$ and $\sigma = 4\,\mathrm{S/m}$. At 10 MHz, equation (8.11) gives an attenuation of $\alpha = 12.5$ nepers/m or 108 dB/m. This value is very high and it can be concluded that radio frequency waves cannot be received under water. If the frequency is lowered to 100 Hz, then $\alpha = 0.34\,\mathrm{dB\,m^{-1}}$ which is a more reasonable value and shows that extremely low frequency (ELF) waves can be received under the sea. For this reason ELF is used as a method of communicating with submerged submarines. The low carrier frequency means, however, that the data rate is very small and only simple messages can be communicated.

There is an additional factor for plane waves entering water. The high value of the relative permittivity at RF and microwave frequencies means that the refections at the surface are very large so only a small amount of any wave incident on the surface of the sea actually enters the water. If $\varepsilon_r = 80$, then the voltage reflection coefficient for normal incidence is $|\rho| = 0.8$ so 64% of the incident power is reflected. By contrast at optical frequencies the relative permittivity is 1.78 so $|\rho| = 0.14$ and only 2% of the incident power is reflected. Light waves easily enter water and are then attenuated at a very low rate.

8.6 PLANE WAVES IN THE ATMOSPHERE

The description of microwave systems in Chapter 7 generally assumed that the atmosphere has no influence on the plane waves and is a good example of free space. This is a good approximation for RF waves and for lower microwave frequencies, but as the frequency increases the influence of the real atmosphere can no longer be ignored. There is also the additional factor that microwave signals may have to travel through rain where the atmosphere is loaded with water droplets. The losses incurred by the signals through the atmosphere can be accounted for by subtracting an additional factor, from the Friis transmission equation, equation (7.10). This will account for the losses caused by the atmosphere.

The atmosphere consists of a mixture of many gases. For electromagnetic waves the only two of significance are oxygen and water vapour. These have a number of resonances throughout the electromagnetic spectrum, particularly in the huge region between microwaves and light. The resonances cause energy to be absorbed. In all cases the atmosphere behaves as a low conductivity medium, so the attenuation is low. The distances travelled by signals are, however, usually long so that the total attenuation suffered by signals can be significant. Figure 8.8 shows the attenuation of horizontal signals at sea level through the atmosphere as a function of frequency over the microwave to optical parts of the spectrum. The solid curve shows the case for a standard dry atmosphere. The dotted curves show the *additional* attenuation due to fog and light rain at 0.25 mm/h and heavy rain at 25 mm/h.

For frequencies below about 10 GHz, the attenuation is very low, even in heavy rain, so radio waves and low microwave signals are not significantly attenuated by the atmosphere. The same is true for optical waves but only for clear weather conditions. If fog or clouds exist the attenuation is high—an observed fact! In the millimetrewave

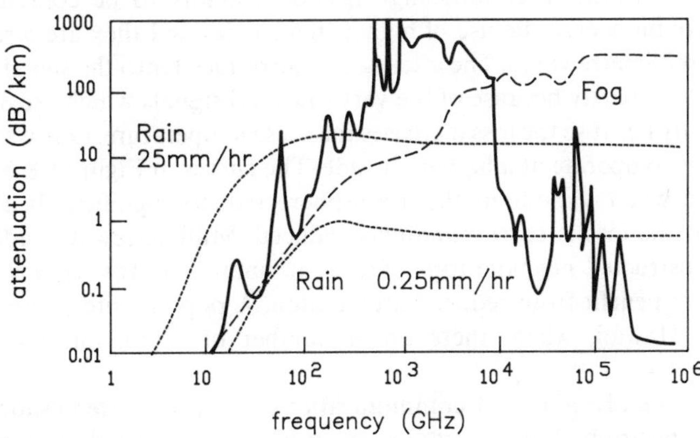

Figure 8.8. Attenuation in the atmosphere for horizontal propagation

and submillimetrewave region of the spectrum the situation is more complex. There are frequency windows where the attenuation is relatively low and other frequency bands where the attenuation is very high. In the millimetrewave part of the spectrum light rain and fog have relatively less influence so that these frequencies are useful for *seeing* in poor weather conditions.

The attenuation introduced into a microwave radio relay system or on a satellite link by the normal atmosphere is typically less than 1 dB and is probably within the overall uncertainty in the power budget. Of more importance for link design is the rain attenuation. This is a complicated subject at microwave frequencies. The rain causes multiple effects on signals. It attenuates the signals as discussed above, but it also causes depolarisation and scattering as the plane waves interact with the raindrops. The amount of interaction depends on the number of raindrops in the microwave beam and on the raindrop size and distribution. This problem gets more significant as frequency increases because the raindrops become larger in wavelengths. The significance of the rain on system design depends on the amount of rain and how often it occurs. Charts are published by the International Radio Consultative Committee (CCIR) showing rain rates which are exceeded for a certain percentage of time. Very heavy rain storms may cause a microwave or satellite link to black-out, whereas light rain may not cause more than a small increase in attenuation. A region of the earth which has infrequent but heavy rain storms is therefore more likely to cause signal loss in a microwave link. A region with frequent but light rain may have extremely few periods when the signal is lost. It is, thus, easier to design systems for northern Europe with light but continuous rain than for the tropics with heavy rainstorms.

The millimetrewave part of the spectrum is starting to be used for communications, radar and remote sensing. The advantages are the huge amount of frequency space available, the small size of components, the high angular resolution and high gain obtainable from compact antennas. This is because the size of natural objects is large relative to the wavelength. The disadvantages are the attenuation introduced by the atmospheric characteristics discussed above. These can be turned into an advantage because the distance over which signals propagate is limited by the attenuation. This means that the cell size of base stations for mobile communications is small. The lossy parts of the millimetrewave band can be used for short distance secure communications. The high angular resolution enables high accuracy radars to be constructed. Radio astronomers have pioneered the use of higher frequencies and they are already making extensive use of millimetrewaves. The attenuation introduced into the signal is significant for radio astronomy partly because of the very low level signals which are detected from the stars and partly because the loss increases the noise temperature. One way of reducing the attenuation is to operate at a higher altitude. The curves in Figure 8.8 were recorded at sea level. The water vapour in the atmosphere reduces significantly as the height increases so that the loss is correspondingly reduced. Millimetrewave radio telescopes are therefore constructed near the tops of high mountains in dry regions of the Earth to gain maximum benefit from reduced attenuation. A popular site is the mountain of Mauna Kea in Hawaii, where there are a number of radio, infrared and optical observatories.

Most applications of radar and communications use a low attenuation part of the millimetrewave spectrum, but the resonance peaks can be used in remote sensing applications. Figure 8.9 shows the attenuation across the millimetrewave part of the

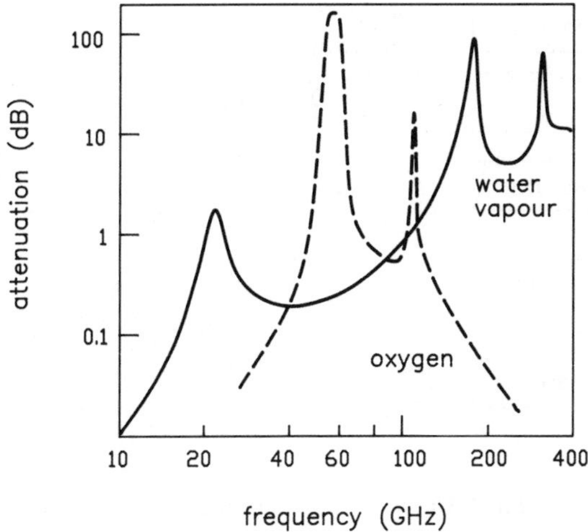

Figure 8.9. Vertical attenuation through the atmosphere over the millimetrewave spectrum

spectrum which clearly shows the resonances. Measurement of the amount of absorption of signals around the resonance peaks enables the amount of water vapour and oxygen to be deduced. The amount of oxygen can be related to the air temperature. This procedure is being developed for meteorological forecasting. A radiometer is mounted on an orbitting weather satellite which looks down onto the atmosphere and monitors the natural radiation emitted by the atmosphere. The water vapour resonance at 183 GHz can be used to monitor the water content in the atmosphere. Filters in the radiometer make it possible to monitor the emissions at a number of frequencies around the 183 GHz resonance peak in Figure 8.9. This means that the water content at different heights in the atmosphere can be deduced. This type of system can be extended to record the presence of other gases in the atmosphere.

8.7 GROUND PROBING RADAR

There is much interest in devising a method of detecting the presence under the ground of objects such as ammunitions, archaeological remains, human bodies, and structural foundations. If the objects are made of metal, an induction system can be successfully used which utilises the transformer principle. This will not, however, work if the object is non-metallic. Microwaves are an attractive possibility because metal and dielectric objects will be detected and the technology of radar can be used. The schematic of a system is sketched in Figure 8.10. A pulse, or chirp, of microwave signal is transmitted through a circulator which sends all the signal to an antenna. The signal is incident on the ground at an angle which means that the ground reflected wave is directed away from the antenna. The relatively high permittivity of the ground means that the wave is refracted towards the normal. A target in the ground will reflect a portion of the incident signal back to the antenna where the circulator directs it towards a detector, signal processor and display. The wavelength of the microwave signals needs to be a reasonable proportion of the size of the target in order that sufficient signal is reflected.

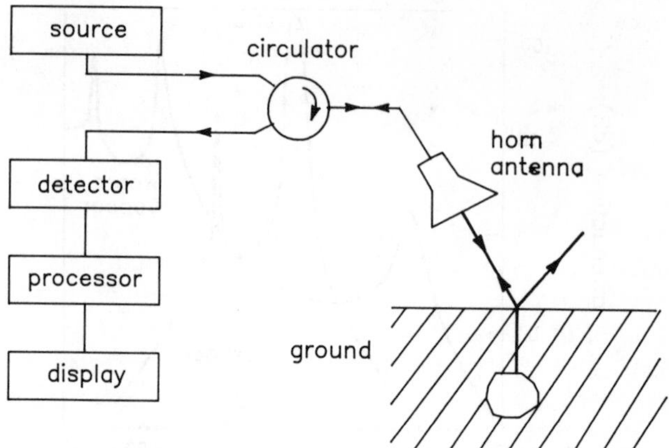

Figure 8.10. Microwave ground probing radar

This is called resolving the target and it limits the lowest frequency which can be used. The radar technique seems ideal but is limited by two practical considerations.

The first is the subject of this chapter, namely the losses due to the attenuation of plane waves in the ground. The ground consists of a mixture of sand, loam, rocks and water. The former have reasonably low loss but, as has been discussed above, the conductivity of the water at microwaves is very high. Typical attenuations against frequency are shown in Figure 8.11. If the water content of the ground is high, this severely limits the depth at which microwave signals can be used, particularly since the signal has to travel down through the ground and back up again. For instance the attenuation in wet clay at 10 GHz is nearly 100 dB/cm. This is an impossibly large value and rules out the use of 10 GHz. The solution is indicated in the figure and is confirmed by equation (8.14), namely to reduce the frequency of the radar signals. Reducing the

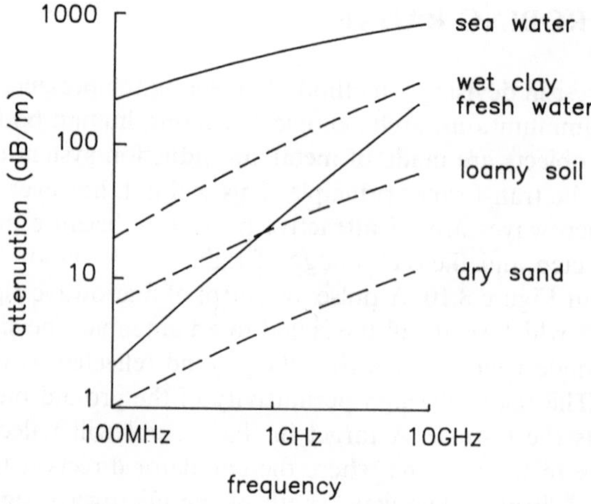

Figure 8.11. Attenuation of waves through sand, soil, clay and water

operating frequency to 1 GHz, lowers the attenuation by ten times in decibels to 10 dB/cm. This is still large and indicates that the operating frequency needs to be in the hundreds of MHz before a viable system for use in wet clay can be designed. On the other hand, if the ground consisted of dry clay, an operating frequency of 2 GHz would give an attenuation of about 5 dB/m. A radar system with a dynamic range of 60 dB could then detect targets down to $0.5 \times 60/5$ or 6 m. The 0.5 factor comes from the double traverse of the ground. This calculation assumes that all the signal would be reflected by the target which is unlikely to occur in practice.

The other constraint on the use of ground probing radar comes from the complexity of the composition of the ground. In general it consists of many different parts, each with different electrical characteristics. Since each interface reflects some signal the result is that the received signal is a complex mixture of the wanted signals from the target and the unwanted clutter signals. Processing of the radar returns using digital signal processing can be done in order to produce useful output displays which enable targets to be detected in the ground.

8.8 MICROWAVE HEATING

Microwave heating is the use of microwave power to heat or dry materials and it is used in industry for a variety of applications. It has a number of advantages over conventional heating techniques including, high efficiency of energy usage since energy is dissipated rapidly throughout the whole volume. This reduces drying times and can produce compact heating equipment. When a plane wave is absorbed by a lossy medium, the absorption occurs at all distances into the medium where there is finite energy. This is in contrast to conventional heating which relies on convection of heat from the surface to the inside of materials.

Microwave heating is a direct use of the theory developed in the early part of this chapter. Many materials contain a significant amount of water so the description of the interactions of water and microwaves in Section 8.5 is relevant to this application. If a high power plane wave is incident on the material to be heated, microwave power is lost as energy is absorbed in the material due to the finite conductivity. The power is turned into heat and as a consequence there is an increase in temperature of the material. The amount of power absorbed is given by equations (8.13) or (8.14) as

$$P = \sigma E_{\mathrm{rms}}^2 = \omega \varepsilon_0 \varepsilon_{\mathrm{r}} \tan \delta \, E_{\mathrm{rms}}^2 \qquad \mathrm{W/m^3} \tag{8.22}$$

where E_{rms} is the RMS amplitude of the electric field. The rate of power loss through a square metre of material as a function of distance is $\exp(-2\alpha z)$ W/m^2.

The rise in temperature can be calculated from the basic relationship between heat and energy: the heat required to raise 1 g of water by 1°C is 1 calorie or 4.18 J (W s). This is the specific heat of water. Thus if 4.18 kW is dissipated over 1 minute, 1 kg of water will be raised in temperature by 60 °C. The specific heat for other materials differs from water and is specified in units of kJ/kg, so that the specific heat of water is 4.18 kJ/kg. For many materials such as wood and common building materials it is around 1 kJ/kg.

The power required to raise a mass M, in kg, by a temperature ΔT in t seconds is

$$P = M s_{\mathrm{p}} \Delta T / t \tag{8.23}$$

where s_p is the specific heat. This expression together with equation (8.22) can be used for heat calculations.

Example 8.5 Heat transfer due to microwave heating

Find the time required to raise a block of wood which is uniformly illuminated by a 100 kW plane wave by 20 °C. The wood has a density of 700 kg/m^3, a specific heat of 2.8 kJ/kg, a conductivity of 0.003 S/m and a relative permittivity of 3.

The power in a plane wave is $E_{rms}^2 = PZ_0$ so for a 100 kW power, $E_{rms} = 6.14$ kV/m. Only a proportion of the incident power enters the wood because of the reflections at the surface. The proportion is given by the transmission coefficient for normal incidence at a dielectric interface is (see Section 6.2)

$$\frac{E_t}{E_i} = \frac{2}{(1+\varepsilon_r^{1/2})} \tag{8.24}$$

which equals 0.73 for $\varepsilon_r = 3$. The electric field entering the wood is thus $0.73 \times 6.14 = 4.48$ kV/m.

The rate of increase in temperature per second can be found by combining equations (8.22) and (8.23)

$$\frac{\Delta T}{t} = \frac{\sigma E_{rms}^2}{\rho s_p} = \frac{0.003 \times (4.48 \times 10^3)^2}{700 \times 2.8 \times 1000} \quad °\text{C/s} \tag{8.25}$$

Here, E_{rms} is the electric field transmitted into the wood and ρ is the density in g/m^3. Inserting the values specified into this equation gives $\Delta T/t = 0.031$ °C/s. The time required for a 20 °C rise in temperature is, therefore, 645 s or 10.8 min. Note that this example did not explicitly need the frequency of operation because this is incorporated in the value of the conductivity.

One problem that can arise with microwave heating of materials is *thermal runaway*. This occurs when there is a positive rate of change of conductivity or tanδ with temperature. In Section 8.5, it was noted that pure water has a negative rate of change of conductivity with temperature, but if too much salt or other impurities are introduced this can alter the situation. Some other materials have a natural positive rate of change with temperature. In this case, as the temperature increases, the conductivity also increases. By equation (8.22) more power is absorbed which increase the temperature, and so on. The result is an unstable situation and thermal runaway.

The practical implementation of microwave heating needs a high power source of microwave energy. This is usually generated by a magnetron connected to an antenna which is in close proximity to the material to be heated. The basic assumption behind plane wave analysis that the waves are a long distance from the source would appear to be violated, but there are so many approximations and roughly known parameters, that the foregoing analysis gives acceptable results. It is difficult to increase the accuracy because of the complexity of a practical microwave heating system.

A number of frequencies have been allocated internationally for industrial, scientific and medical purposes. The main ones for heating are 27.12 MHz, 896 MHz (915 MHz in North America), 2.45 GHz and 5.80 GHz. Of these 2.45 GHz is the most popular because the conductivity of materials at this frequency leads to reasonable heating times.

Successful applications of microwave heating have included the drying of paper, film, printing ink, adhesives, textiles and a wide range of foods. It is used for the curing of plastic foams, epoxy resins and synthetic rubber products. The vulcanisation of synthetic rubber has been particularly successful because rubber has ideal properties for microwave heating which reduce the time required for production considerably. It is used in the food industry for vacuum drying, pasteurisation, serialisation and thawing of a variety of foods.

8.9 MICROWAVE COOKING

Microwave cooking is the application of the principles of microwave heating to the cooking of food. It has become a very popular method of cooking and the *microwave* is a standard domestic item. For many people it is probably the only example that they have of the use of microwave energy. Microwave cookers use the heating frequency of 2.45 GHz.

The principles of microwave cooking have all been explained in previous sections. The bulk composition of food is mainly made up of water so microwave energy will be absorbed by the food and turned into heat. It follows from equations (8.22) and (8.23) that the amount of absorption and hence the rise in temperature and rate of cooking will depend on the amount of water in the food because the effective conductivity will be proportional to the water content. High water content foods will be heated more quickly than low water content foods. The other constituents in foods can influence the cooking, for instance sugar has a positive rate of change of conductivity with temperature so food with a high sugar content will burn easily unless the cooking times are short.

The knowledge of materials gained in earlier chapters leads to the conclusion that containers for the food in a microwave cooker should be made of a low loss, or very low conductivity, material such as plastic or foam. The container will then dissipate only very small amounts of power and its rise in temperature will be negligible, except for that due to conduction from the hot food. It is also follows from the discussion of the characteristics of waves incident on metals and conductors in earlier chapters that there should be no metal present around the food because the metal will reflect or absorb the energy. No energy would then be incident on the food. One of the boundary condition developed in Chapter 4 is that the tangential magnetic field equals the surface current flowing along the surface. This shows that a high incident field will generate a high surface current on any metal which can lead to arcing to the casing of the cooker. Hence this is an additional reason for not placing a metal container around the food.

Approximate calculations can be performed using the equations developed above to work out the cooking times and temperature rises. These are approximate because multiple reflections will take place within the food and the incident electric field in a real cooker might not be uniform.

Example 8.6 Microwave cooking

Find the penetration depth for a 0.5 kg piece of meat which has a relative permittivity of 40 and a conductivity of 1.6 S/m at 2.45 GHz. Find the approximate cooking time in a microwave cooker which is needed to raise the meat in temperature from 20 to 100°C. Assume that 350 W is absorbed by the food.

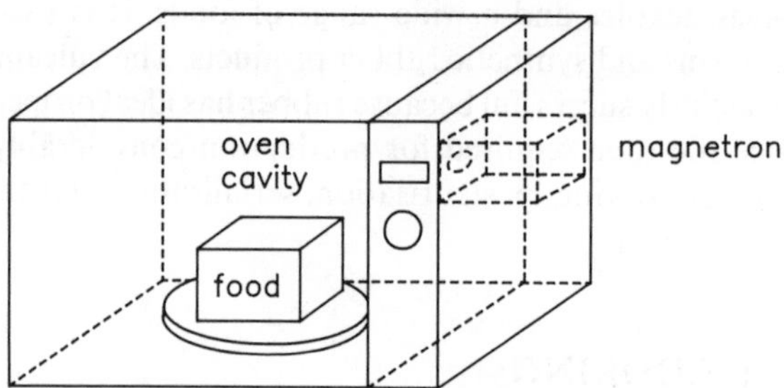

Figure 8.12. Typical layout of microwave cooker

The penetration depth must be found from the exact equation for the propagation constant, equation (8.11). Inserting the values gives $\beta - j\alpha = 325(1 - j0.3)^{1/2} \approx 325 - j49$. The penetration depth is $1/\alpha$ metres or 1/49 m or 2 cm. At this depth the power will have fallen to $1/e^2$ (14%) of the value at the surface. The energy required to raise 500 g by 80 °C in temperature is found from equation (8.23) to be $500 \times 4.18 \times 80\,\text{J} = 167\,\text{kJ}$. If 350 W is dissipated in the food the cooking time will be therefore be 447 s or 8.0 min.

A microwave cooker, Figure 8.12, consists of a magnetron which produces pulses of 2.45 GHz energy (commonly rated at 650 W mean) feeding into a metal box which is the cooker. This forms a resonant cavity to contain the energy. A typical cooker has dimensions of about $300 \times 300 \times 250\,\text{mm}^3$. Since the wavelength in free space of 2.45 GHz is 122.4 mm, there will be many resonances in the cavity. Prediction of these is complicated by the presence of the food which changes the distribution of electric field, otherwise energy would not be dissipated in the food. The natural distribution of electromagnetic field inside the cavity is not ideal and so some form of crude device is usually provided in order to break up the natural resonant field pattern. This is either a turntable to rotate the food or a mechanical stirrer which is basically a rotating piece of metal at the output of the magnetron. The cooker door is an important component because it must be made so that there is no leakage of microwave energy when the magnetron is operating. This is achieved by a quarter wavelength deep gap or *choke* around the door which provides a very effective electrical seal. The explanation of chokes and quarter wavelength sections is dealt with in Chapter 11.

8.10 MICROWAVE SAFETY LEVELS

Human beings and animals consist of a mixture of bone, muscle, fat and skin, all of which are composed partly of water. If electromagnetic waves are incident on the body absorption of power will take place just as in the case of materials or food. The power absorbed will be dissipated as heat and the body will increase in temperature. It is obviously necessary to know how much energy will be absorbed and what the biological effects on humans will be of any absorption. Safety levels then need to be set at which it is hoped that the interaction between the radiation and the body is negligible. This can be a controversial subject because the long term influence on the body of external electromagnetic fields always carries an element of uncertainty. Considerable research has been, and is being, undertaken into the biological effects of electromagnetic fields. This divides into non-ionising radiation and ionising radiation. Non-ionising radiation causes energy to be lost as heat. It is reasonably well understood, the effects are quantifiable and it is the subject of this section. Any effects from ionising radiation on, for instance, nerves, is not understood. There is conflicting evidence but few reliable results. Further research is being conducted to be able to try to quantify the interactions between electromagnetic fields and humans or animals.

Electromagnetic waves of all frequencies can in principle cause absorption by the body tissues but the most significant effects come at microwave frequencies due to the fact that the penetration depths are comparable with the dimensions of the body. This can be seen by considering the electrical permittivity of tissue. Measurements of the electrical characteristics of typical tissues have been able to divide them into two types:

Fat and bone have a low water content with a relative permittivity of about 5.5 and a conductivity of about 0.03 S/m at 1 GHz. This gives a $\tan\delta$, equation (8.16), of 0.1 and using the formula for low conductivity materials, equation (8.18), a penetration depth of 41 cm for frequencies above about 1 GHz.

Muscles and skin have a high water content with a much higher relative permittivity of about 50 and a conductivity of about 1.4 S/m at 1 GHz. This gives a $\tan\delta$ of 0.5 and using the exact expression to calculate the attenuation, equation (8.11), the penetration depth is then 27 mm at 1 GHz. The measured conductivity increases with frequency so the penetration depth decreases with increasing frequency.

These penetration depths are compatible with the sort of distances occurring within the body so exposure to high levels of microwave signals will cause heating of the tissues. The need is to set power levels which will not cause heating.

The interaction of RF and microwave radiation with the body is complex and the amount and localization of energy absorption will vary with the size of the human body. The orientation of the electric field will also be important. The quantity used to measure the amount of energy absorption is the *specific energy absorption rate*, or SAR, which is the rate at which energy is absorbed by 1 kg of material. For sinusoidal waves it is

$$\mathrm{SAR} = \frac{\sigma E_{\mathrm{rms}}^2}{\rho} \qquad \mathrm{W/kg} \tag{8.26}$$

where ρ is the density in $\mathrm{kg/m^3}$. Safety levels have been recommended by setting SAR values that do not cause perceivable heating on the body. Research has shown that behavioural disturbances in exposed animals and humans which have been continuously exposed to microwave radiation is just perceivable with an SAR of between 4 and 8 W/kg. A safety factor of ten is added to give the safety level recommended by the International Non-Ionizing Radiation Committee (INIRC) of a SAR = 0.4 W/kg. This is for people working near electromagnetic sources and is for exposure over any 6 minute period. To obtain a value for the maximum permissible amount of incident power density an average human body is assumed. A SAR of 0.4 W/kg is caused by an incident plane wave power density of 50 $\mathrm{W/m^2}$ on the average human body. The power density limit for the general public is set at one fifth of the above amounts, or 10 $\mathrm{W/m^2}$.

These INIRC recommended safety levels for the general public are plotted in

Figure 8.13. Below 2 GHz the human body is only a proportion of a wavelength so the permissible limits are scaled with frequency down to 400 MHz. Between 10 and 400 MHz a constant level is again recommended. Below 10 MHz, the electromagnetic fields are essentially static and the levels are set in terms of separate electric and magnetic field levels. The levels in Figure 8.13 are only recommended levels. Individual countries are free to adopted different levels if they wish, but most follow the guidelines with small variations.

The sources of possible exposure of the general public to RF and microwave signals is potentially considerable because of the large range of communication and radar systems in use. Very few systems are, however, likely to radiate fields at levels near to the recommended safety levels. This is partly because the safety levels have been built into the design, partly because antennas have been designed to send signals only in the direction of the main beam so that spurious radiation is reduced, and partly because it is inefficient of energy usage to send energy in unwanted directions. The most common continuous exposure that the general public has to RF signals is the low level fields from radio and TV transmitters operating in the VHF and UHF bands. The estimated SAR when 300 m from the base of a high power UHF TV transmitter with an EIRP of 1000 kW is, however, less than 0.001. Higher exposures for a short time arise in close proximity to mobile radio antennas, or in the main beam of a tracking radar. The latter is unlikely, but a person in the main beam of a 300 W 10 GHz tracking radar and 100 m from the radar is subject to a SAR of about 0.04, which is ten times less than the permitted maximum level. The same SAR is predicted at 0.3 m in front of a microwave cooker with the maximum permitted leakage of power. Since the power decays as the square of the distance, this level implies the power immediately in front of the cooker will be ten times larger which means that it will be at the safety limit. Microwave cookers are designed to radiate considerably lower levels than this limit. Even though all the above power levels are small, considerable effort is devoted by RF and microwave designers to reducing them even further.

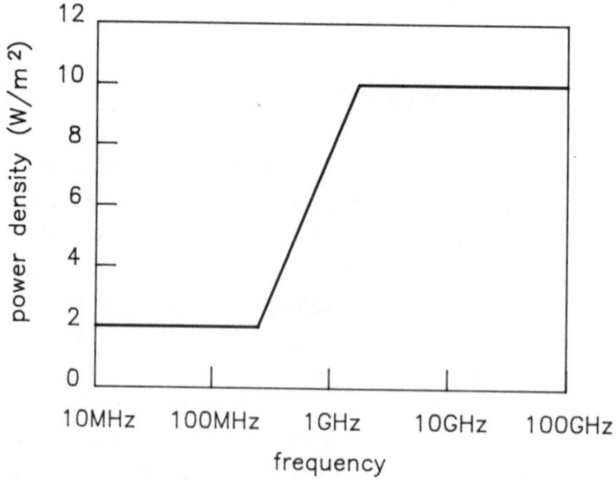

Figure 8.13. INIRC recommended safety levels for continuous exposure of the general public to microwave signals over a period of six minutes

8.11 MICROWAVES AND MEDICINE

The previous section was concerned with the safe levels of exposure to microwaves, implying that no intrusion of waves into the body can be allowed. There are, however, some applications in medicine where microwaves can be used beneficially. The main use to date is controlled microwave heating for cancer treatment. This is called microwave hyperthemia and is used as an ancillary to radiation therapy and chemotherapy. The aim in microwave hyperthermia is to selectively heat tumors to 43 °C, at which temperature the cancerous cells start to die. A focussed beam of microwave power is directed onto the tumour. This is done with an applicator which is a small antenna placed on the skin above the cancer. The applicator is designed to focus the microwave power directly onto the tumour and not to allow power to spread to other parts of the body. The high relative permittivity of body tissues helps in this respect because energy entering at a finite angle of incidence is refracted towards the normal and travels directly down below the surface. Deep seated tumors can be heated by a small coaxial probe inserted into a patient. Hyperthermia has been found to be effective in some cases as an additional method of cancer treatment. Other areas of surgery and medicine may benefit from carefully controlled heating and are being researched.

Passive microwave remote sensing can be used to monitor the temperature of deep parts of the body without the need for incisions. The system is the same as the remote sensing described in Section 7.11, except that the 'remote' area is very local, although still hidden from view. The temperature is measured with a radiometer.

Short distance, very low power Doppler radar is being investigated for non-invasive monitoring of the motion of the heart or arteries. The amount of movement of the muscles is small, but significant as a proportion of the wavelength in the body tissues. In all cases of the use of microwaves in medicine, the objective is to see if, and how, microwaves can aid medicine by using some of the unique properties of plane waves in dielectric tissues.

8.12 ELECTROMAGNETIC COMPATIBILITY

A subject of increasing interest to microwave engineers is *electromagnetic compatibility*, or EMC. This refers to the suppression of spurious RF and microwave radiation. Items of equipment need to be electromagnetically compatible with each other so that they are not adversely influenced by spurious radiation.

Unwanted electromagnetic radiation comes from equipment which is badly shielded. Computers and digital equipment are some of the worst offenders because the fast digital pulses have many harmonics which can radiate as a sort of unintentional noise source. The amount of spurious RF and microwave radiation has increased in recent years due to the large number of new equipment and systems which are being used. The spurious radiation is not usually at a level which imposes any health hazard on humans, but it can cause interference with other RF and microwave equipment.

EMC has two aspects. For the source of interference it is the limitation of the generation of electromagnetic disturbances so as to allow equipment to operate as intended. For the receiver of interference it is the provision of adequate levels of immunity so that equipment can operate as intended. Manufacturers of all types of electronic equipment have to design for good electromagnetic compatibility and then to measure the electromagnetic signals emitted by the equipment to ensure that it complies with legislation. Legal limits are being introduced which specify the level of interference which is acceptable for equipment.

Interfering signals can come out of a piece of equipment either by conduction along wires and cables or by radiation. Conducted interference is mostly at low frequencies of a few kHz, so it is radiated interference which is mainly of concern in the microwave band.

The study of EMC requires a general knowledge of electromagnetic waves and radiation mechanisms described in this book. Specific theory is, however, limited. This is partly because each item of equipment is different and partly because the physical design of equipment is so complicated that few situations can be analysed or modelled. Radiated interference must be suppressed by careful design to reduce, or preferably, eliminate radiation. Most of the discussion in Chapter 7 and this chapter assumed that radiated energy will always come through the antenna. Any circuit carrying RF and microwave signals will, however, potentially radiate from bad contacts, abrupt junctions and mismatched components. The layout of microwave circuits needs to be designed to suppress unwanted radiation. The techniques for suppressing this unwanted radiation include providing a common grounding system and carefully bonding all parts to the ground. Shielding of the circuits or equipment by enclosing them in an electromagnetically sealed box is important. This does not mean that the equipmen has to be physically sealed in an closed conducting volume. The examples in Chapter 4 showed that a cage or grounded shield can be effective, as long as the conducting elements are close enough together. If the current induced in a conductor by an interfering field has its flow path interrupted by an air gap, then radiation occurs. The effectiveness of shielding depends on the avoidance of radiating gaps. Any system which suppresses radiated interference from leaking out of a piece of equipment will also stop outside interference from being picked up by the equipment. It will not, however, stop any equipment with a receiving antenna from detecting and receiving interference through the antenna.

PROBLEMS

1. Calculate the imaginary part of the permittivity and the loss tangent at 1 MHz, 1 GHz and 1 THz for the following materials: silver ($\sigma = 6.17 \times 10^7$ S/m, $\varepsilon_r = 1$); carbon ($\sigma = 3 \times 10^4$ S/m, $\varepsilon_r = 1$); polystyrene ($\sigma = 10^{-16}$ S/m, $\varepsilon_r = 2.5$).

2. Find the power absorbed over a distance of 10 mm for a plane wave in carbon. The wave has an RMS amplitude of the electric field of 100 mV m^{-1} and a frequency of 1 MHz.

3. Evaluate the propagation and attenuation coefficients at 2 GHz for a plasma with $\varepsilon_d = 0.5 + j1.5$.

4. Calculate the penetration depth at the frequencies and for the materials specified in Problem 1.

5. Find the frequency range over which dry sandy soil with a conductivity of 10^{-3} S/m and a relative permittivity of 4 can be treated as a low conductivity medium.

6. Calculate the skin depth and the surface resistance for mercury ($\sigma = 10^6$ S/m and $\varepsilon_r = 1$) at frequencies of 10 GHz and 100 THz.

7. A piece of germanium semiconductor has a conductivity of 2 S/m and $\varepsilon_r = 16$. Find the attenuation of plane waves in germanium for frequencies of 1 MHz and 100 MHz.

8. A 10 MHz pulse radar is used to probe the depth of an ice field ($\varepsilon_r = 3, \sigma = 10^{-6}$ S/m). The pulse delay is recorded as 0.5 μs. Find the depth of the ice and the attenuation due to loss in the ice. (The signal would also suffer the normal path loss.)

9. A sheet of aluminium foil has a thickness of 1/1000 inch and is used to shield a radio receiver at 100 MHz. By how much will the signals be attenuated? ($\sigma = 3.5 \times 10^7$ S/m for aluminium.)

10. An aircraft with a radar altimeter is flying above ground which is covered by snow to depth of 3 m. Snow has a relative permittivity of 1.5 and $\tan\delta = 9 \times 10^{-4}$. Find the error caused in the altimeter reading if the aircraft is at an altitude of 100 m.

11. Use the values for the permittivity of water given in Figure 8.6 to obtain the attenuation in dB/mm of waves at 2 GHz, 10 GHz and 20 GHz.

12. A microwave beam is incident at right angles on the surface of pure distilled water with $\varepsilon_r = 80$ and $\sigma = 0.0001$ S/m. Find the reflection and transmission amplitudes and the attenuation in the water at 10 GHz.

13. Repeat Problem 12 for red light at $\lambda = 700$ nm assuming that the properties of the water are $\eta_r = 1.33$ and the conductivity is the same as above.

14. A submarine is submerged at a depth of 10 m below the surface of the sea. It can receive ELF signals at 10 kHz. A 100 kW signal is radiated isotropically by a transmitter 2000 km away. Find the power received by the submarine. (Take sea-water to have electrical constants of $\varepsilon_r = 80$ and $\sigma = 4$ S/m.)

15. An experimental communication link with a power of 1 W operates over 1 km at frequencies of 100 GHz, 10 THz and 1000 THz. Use the data in Figure 8.8 to estimate the received power level in (a) clear weather and (b) fog.

16. A microcellular mobile radio system could use the 60 GHz oxygen absorption band to naturally limit the range of signals. This would reduce interference from adjacent cells. The attenuation at 60 GHz is 60 dB/m. Assuming that the base stations radiate 10 W and have a gain of 40 dB find the maximum cell size if the receivers have a gain of 10 dB and a sensitivity of -80 dBm.

17. A ground probing FMCW radar has a 100 mW source which sweeps 1.5 ± 0.5 GHz in 10 ms. It is used to detect an object buried in the ground in soil ($\varepsilon_r = 4$) with an attenuation of 10 dB/m. If the received difference frequency is 2 kHz, what is the depth of the object? If all the signal enters the ground and 10% is reflected by the target back in the source direction, what is the detected power level?

18. Find the power absorbed in water at 3 GHz for temperatures of 20 °C, 60 °C and 100 °C for an incident field of $E_{rms} = 200$ V/m. Use the graph in Figure 8.7 to obtain ε''.

19. Microwave heating is used to dry paper with a density of $500\,kg/m^3$, a specific heat of 3.5 kJ/kg, a conductivity of 0.003 S/m and a relative permittivity of 3. If 10 kW of power illuminates the paper for 5 min, what is the rise in temperature?

20. Find the approximate time required to boil 0.5 l of water from room temperature in a 650 W microwave cooker in which 80% of the power is effectively dissipated in the water. Find the penetration depth if the water has $\varepsilon' = 70$ and $\varepsilon'' = 10$, for all temperatures.

21. Find the specific energy absorption rate for a 10 W wave incident on 2 l of distilled water with $\sigma = 10^{-4}\,S/m$.

22. A 10 GHz open-ended waveguide has a cross-section area of $200\,m^2$. Use the graph of safe levels in Figure 8.13 to estimate the maximum power level for a person who is 300 mm in front of the waveguide.

9 Ray optical systems

9.1 INTRODUCTION

Optical waves are radiated naturally from the Sun, and artificially from lamps, LEDs and lasers. There are consequently many applications of optical plane waves ranging from human vision to optical data processing. There are traditional optical instruments such as spectacles or telescopes and modern optical systems such as compact audio disc players. All these use beams of optical plane waves that are guided through the air or space. This chapter will apply the knowledge of plane waves developed in Chapters 5 and 6 to free space optical beams. The guidance is done by lenses, mirrors and other optical components, and the principles of these components will be described.

Part of the subject matter in this chapter forms an introduction to a subject that is classically described as *optics.* Optics is, however, only one part of modern *optical engineering* which is developing rapidly into a broad subject with a wide range of applications. *Optical fibre communications, optoelectronics, integrated optics* and *optical computing* are just some of the new subjects which make use of optical waves. Guided optical waves will be described in Chapter 13. In this chapter the interest is in optical systems which use power produced by either natural sources or artificial sources and in which the energy forms beams of plane waves travelling in free space. Even this is, however a broad subject and there are many aspects which are beyond the scope of this book. In particular, the topics of diffraction and interference will not be treated.

Optics is one of the oldest branches of physics because it is concerned with understanding natural visible light. The basic principles of optical propagation have been known for centuries, but classical optics was exclusively concerned with natural radiation. This meant that only waves in the visible part of the spectrum were of interest and light was treated as an incoherent bundle of energy. Modern optics is concerned with all waves which have wavelengths that are very small by comparison with objects. This certainly means waves with frequencies which fall anywhere in the optical and infrared parts of the spectrum. The exact borders are diffuse but cover frequencies from 3000 GHz (3 THz) to 3000 THz, or wavelengths from 100 μm to 100 nm. Modern optics is also concerned with waves produced by artificial sources which produce beams of *coherent* energy. In this context, coherent means that all the parts of a wavefront are in phase, so all the waves studied so far in this book are coherent waves. *Incoherence* means that beams are composed of a random jumble of waves of mixed frequencies, amplitudes, phases and polarisations. All light from the Sun is incoherent. Any incoherent bundle of waves can, however, be analysed as a series of coherent waves with appropriate frequencies, amplitudes, phases and polarisations. There is, thus, no fundamental difference between incoherent light and coherent light and the theory developed in this book applies to incoherent light.

This chapter will use the traditional notation used in optics of primarily describing phenomena by their wavelength and not by their frequency. This is different from that used in microwave systems, where frequency is used in preference to wavelength. Also the electrical characteristics of dielectrics are specified by their refractive index, n, and not their relative permittivity, ε_r, where $n = \varepsilon_r^{1/2}$.

9.2 GEOMETRIC OPTICS

In Chapter 5, it was stated that electromagnetic energy could be considered as either waves or rays. This is generally adequate for the purposes of this book but it is worth noting that for optical systems there are actually three ways of dealing with optical energy:

(1) Optical energy can be treated as consisting of mass less particles called *photons.* These are small quanta of energy which have zero electrical charge and zero mass and which travel at the velocity of light. This method of dealing with light comes about because the wavelength of optical waves approaches that of the basic properties of matter. It is, however, only of relevance in optical systems when the radiant energy is very small. For low power levels the *noise* is caused by photons, and for very low levels, the photons can be counted.

(2) Optical energy can be treated as *waves.* This is the electromagnetic approach which is the main theme of this book. It enables all the properties of optical systems to be rigorously explained, except at the quantum level. The rigour brings with it the problem of solving the wave equations and associated boundary conditions. This can be done for plane waves and guided waves in rectangular, cylindrical and spherical coordinates. When waves pass through apertures they diffract and scatter energy. Waves in free space have already been studied and guide waves will be treated in later chapters. Scattering and diffraction will not be studied and it becomes increasingly difficult to solve problems where waves are incident on arbitrary shaped objects. Advanced electromagnetic wave theory treats these problems. When the wave theory is applied to larger objects, it is sometimes called *physical optics.*

(3) Optical energy can be treated as *rays.* The energy is assumed to be wave like, but to travel *in straight lines in a homogeneous medium.* It assumes that there is no diffraction or scattering of energy. The ray approach is valid as long as the object is large in wavelengths, which is almost always true at optical wavelengths. Treating the waves as rays of energy which travel in straight lines from point to point means assuming that the waves are everywhere plane waves and solving problems by drawing straight lines and applying geometry. For this reason it is called *geometric optics.* The method has already been used in Chapter 6 to explain reflection and refraction. The basic principles will now be stated more formally, and then used to explain optical behaviour with prisms, mirrors and lenses.

Geometric optics is an approximation which is valid as the wavelength approaches zero ($\lambda \rightarrow 0$). If it was really necessary to obey this limit, then geometric optics would have limited appeal. Fortunately experience has shown that it can be used in a wide variety of situations where the wavelength is not physically small, but it is small in comparison with the size of the objects being illuminated by waves. The laws of reflection and refraction were thus derived in Chapter 6 with the intention that they apply to microwaves as well as optical waves. The derivation in Chapter 6 used geometric optics without explicitly stating the name. Snell's law of refraction was derived from considering the time taken for a ray to travel through a dielectric surface. That approach was based on a principle formulated by the French mathematician Pierre de Fermat (1601–1665).

Fermat's principle was based on an idea originally propounded by Hero of Alexandria in the second century B.C. who said that light always took the shortest path between two points. This is true for reflection, but does not work for refraction. Fermat reformulated it into the statement that *light always takes the path which involves least time.* This principle can be used to rederive Snell's law in a more elegant fashion and then to generalise it for multiple media.

Example 9.1 Snell's law from Fermat's principle

Derive Snell's law using Fermat's principle.

Consider the geometry of a ray refracted at a surface between two dielectrics, with $n_2 > n_1$ so that the ray is refracted towards the normal, Figure 9.1. In order to find the minimum time for a ray to travel between A and B, the total travel time $t(x)$ is minimized with respect to the distance x along the surface as this is proportional to the distance travelled. The time t is

$$t = \frac{\mathrm{AO}}{v_1} + \frac{\mathrm{OB}}{v_2} = \frac{(a^2 + x^2)^{1/2}}{v_1} + \frac{[b^2 + (d - x)^2]^{1/2}}{v_2} \tag{9.1}$$

where $v_1 = c/n_1$ and $v_2 = c/n_2$ are the velocities in the two dielectrics. Differentiating with respect to x and setting the result to zero to find the minimum value of x gives

$$\frac{\mathrm{d}t}{\mathrm{d}x} = \frac{x}{v_1(a^2 + x^2)^{1/2}} + \frac{-(d - x)}{v_2[b^2 + (d - x)^2]^{1/2}} = 0$$

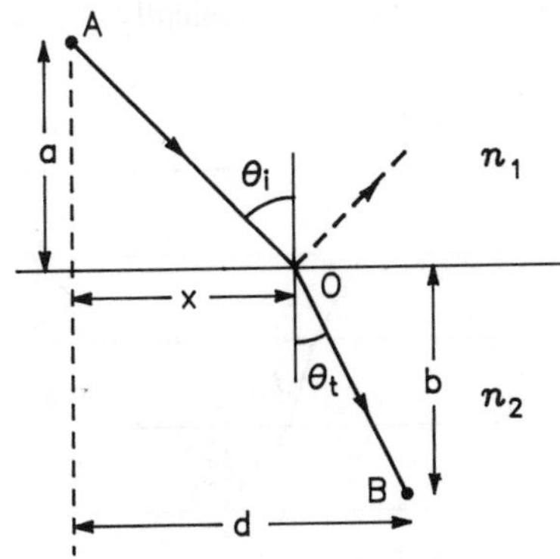

Figure 9.1. Geometry for Snell's law derived using Fermat's principle

The geometry in Figure 9.1 shows that this can be expressed as

$$\frac{\sin\theta_{\mathrm{i}}}{v_1} = \frac{\sin\theta_{\mathrm{t}}}{v_2} \tag{9.2}$$

or

$$\frac{\sin\theta_{\mathrm{t}}}{\sin\theta_{\mathrm{i}}} = \frac{n_1}{n_2} \tag{9.3}$$

This is Snell's law.

If there are a number of layers with different refractive indices, Figure 9.2, then the total travel time will be a generalisation of equation (9.1)

$$t = \frac{l_1}{v_1} + \frac{l_2}{v_2} + \frac{l_3}{v_3} + \cdots + \frac{l_m}{v_m} = \sum_{i=1}^{m} \frac{l_i}{v_i} \tag{9.4}$$

where l_i and v_i are the lengths and velocities, respectively, for a ray in the ith layer. Replacing v_i by c/n_i gives

$$t = \frac{1}{c} \sum_{i=1}^{m} n_i l_i \tag{9.5}$$

The multiplication of the refractive index and the path length, $n_i l_i$, is called the *optical path length*, OPL, because it is the plane wave path length in the layer. Equation (9.5) can be generalised further for an inhomogeneous medium where the refractive index is a function of position, $n(l)$, by replacing the summation with an integral

$$t = \frac{1}{c} \int n(l)\,\mathrm{d}l = \frac{1}{c}\,\mathrm{OPL} \tag{9.6}$$

where the integration is taken over the complete medium. This equation can be used to restate Fermat's principle as *light always takes the smallest optical path length.* An example of an inhomogeneous medium is the atmosphere, which is more dense at the bottom than at the top, so the refractive index increases as the altitude decreases. Rays entering at the top of the atmosphere take the least path length to reach the ground which means that they are bent towards the Earth. This is apparent from the well known phenomena that the Sun can be seen when it is physically below the horizon, Figure 9.3.

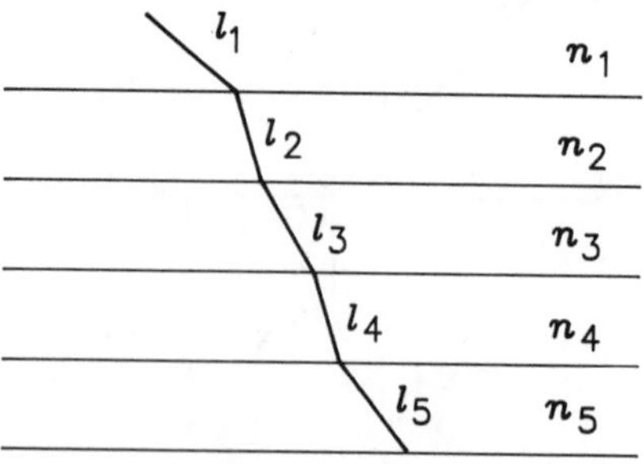

Figure 9.2. Fermat's principle applied to multiple dielectric layers with refractive indices n_1 to n_5 and optical path lengths nl_1 to nl_5

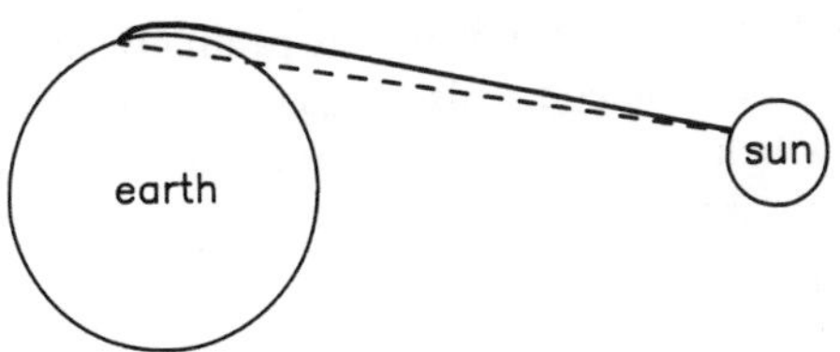

Figure 9.3. Demonstration of Fermat's principle for sunlight refracted in the atmosphere

Fermat's principle as expressed above is not quite complete because it does not state what happens when a ray in Figure 9.1 is just to one side of the ray drawn on the figure. Intuitively it would not be expected to make much difference to the result, and this is confirmed by replacing θ_i in Snell's law, equation (9.3), by $\theta_i + \delta\theta$. This is because a plot of dt/dx for Snell's law produces a broad curve like a U. There will be a region around the base of the curve where $dt/dx \approx 0$. This is called the stationary point of inflexion. Fermat's principle in its most general form is this that *light always takes the optical path length which is stationary with respect to the variations of the path.* This helps to explain why geometric optics can be widely applied. Waves do not have to only follow the exact optical path. Adjacent rays will also obey the rules.

The statement of Fermat's principle describes the path that rays will travel. It makes no mention of the direction of travel and it is fairly obvious that a ray travelling from A to B will follow the same path as one travelling in the reverse direction from B to A. This is called the *principle of reversibility*. It only applies if the medium is isotropic which means that the refractive index must be a scalar which is independent of direction. An anisotropic medium has a vector refractive index n and this can lead to non-reciprocal paths.

When geometric optics is applied to analyse mirrors and lenses, rays are traced through the mirror or lens using Fermat's principle to find the ray direction after reflection or refraction, Figure 9.4. This produces equations which involve trigonometric functions that are solveable but quite complicated. A simplification which is always used in optics is to assume that the angle, α, between the optical axis of the surface and the ray direction is small. In this case sines and cosines can be approximated by the first terms of their series expansions. This gives

$$\begin{aligned} \sin\alpha &\approx \alpha \\ \cos\alpha &\approx 1 \end{aligned} \tag{9.7}$$

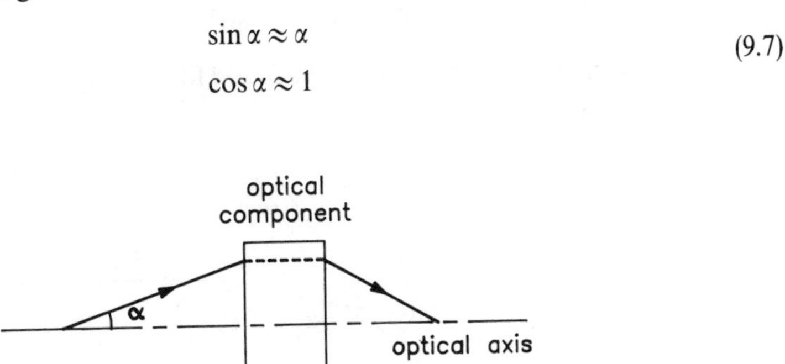

Figure 9.4. Geometric optical rays through a general optical component

Applying these approximations invariably leads to much simpler equations. It was first formulated by Gauss in 1841 and so the use of these approximations is sometimes called *Gaussian optics.* Rays which satisfy Gaussian optics and arrive at shallow angles with respect to the optical axis of the systems are called *paraxial rays* and the region in which they exist is called the *paraxial region.*

9.3 PRISMS

Prisms are one of the simplest optical components that are widely used in systems. They can be used to measure wavelengths, to redirect or reorientate rays, to split rays, and to create polarised light. The principles can be deduced from Snell's law, total internal reflection and the geometry of the prism. There are two broad classes of prisms, those in which the rays goes straight through the prism without reflection, and those where internal reflection takes place. The two classes will be described in turn.

If a ray passes through a prism, the requirement is to relate the direction of the output ray to the direction of the incident ray. This is measured by the angular deviation, δ, Figure 9.5, through the prism. Using the same notation as was used in Capter 6, θ_i represents the angle of incidence and θ_t the angle of transmission. The angular deviation δ is

$$\delta = (\theta_{i1} - \theta_{t1}) + (\theta_{i2} - \theta_{t2}) \tag{9.8}$$

From the geometry of the prism, the angle, $\alpha = \theta_{t1} + \theta_{i2}$, so equation (9.8) is

$$\delta = \theta_{i1} + \theta_{t2} - \alpha \tag{9.9}$$

This relates the angular deviation to the angle of the prism and the incident and transmitted angles. In order to eliminate θ_{t2} from the equation, Snell's law is used to give a second relation for θ_{t2}

$$\sin\theta_{t2} = n\sin(\alpha - \theta_{t1}) \tag{9.10}$$

where n is the refractive index of the prism and it is assumed that the incident medium is air. Expanding the right-hand side and inverting the left-hand side so that it can be substituted into equation (9.9) gives

$$\delta = \theta_{i1} + \sin^{-1}[(\sin\alpha)(n^2 - \sin^2\theta_{i1})^{1/2} - \sin\theta_{i1}\cos\alpha] - \alpha \tag{9.11}$$

This equation enables the deviation to be found for a given θ_{i1}, α and n. It is a reasonably

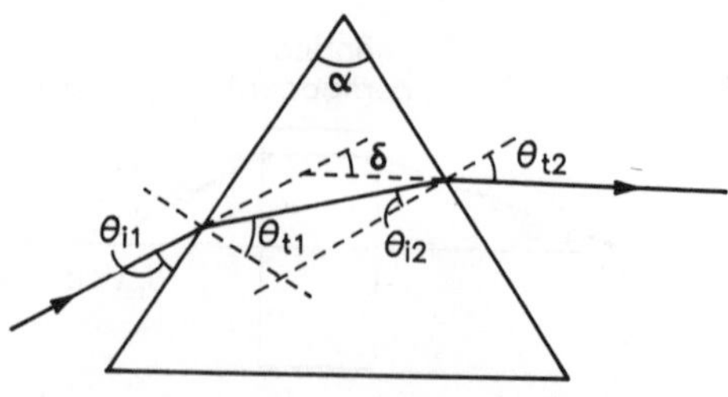

Figure 9.5. Geometry of rays through a refracting prism

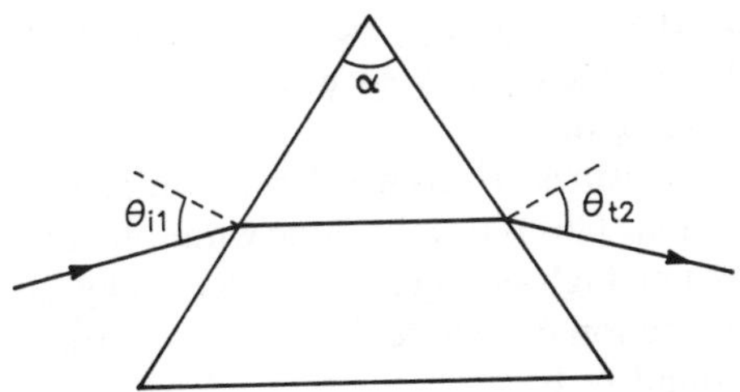

Figure 9.6. Minimum deviation of a ray through a dispersive prism

complicated equation, but the important point to note is that the deviation δ is a function of the refractive index n. This means that in principle, the equation can be used to measure n. There is, however, a simplification which is easier to use. This comes from the case where there is minimum deviation through the prism. Simple consideration of the geometry indicates that this must occur when the incident angle is equal to the transmitted angle, $\theta_{i1} = \theta_{t2}$. The internal ray is now parallel to the base, Figure 9.6. Then, $\theta_{t1} = \alpha/2$ and equation (9.9) gives, $\theta_{i1} = (\delta + \alpha)/2$ so that Snell's law at the incident boundary gives

$$n = \frac{\sin[(\delta + \alpha)/2]}{\sin(\alpha/2)} \tag{9.12}$$

Equation (9.12) can be used to find the value of the refractive index of any prism. It is easiest to use with visible light because the rays can be seen, but it will work at any optical or infrared wavelength. The angle of the prism and the angular deviation can both be measured accurately, so that this method provides a simple and accurate measurement technique.

It has been assumed so far that the dielectric has a constant refractive index as wavelength changes. If the dielectric is glass at visible wavelengths, this is not true. Glass shows a small but significant variation of its refractive index over the visible band. The general name for material which are wavelength dependent is *dispersive* materials and the prisms analysed above are called *dispersive prisms* because the deviation depends on wavelength.

Example 9.2 Dispersion in a glass prism

Find the minimum deviation through a 60° prism made of glass with $n = 1.733$ for standard blue light at $\lambda = 486.1$ nm to $n = 1.708$ for standard red light at $\lambda = 656.3$ nm.

The result is obtained by inserting the values into equation (9.12). This gives that δ changes from 60.09° to 57.25°, which is a appreciable change over the visible band. The change of refractive index with wavelength varies considerable from one type of glass to another type, though all of them follow the trend of this example and they have a refractive index which gets less as wavelength increases.

A dispersive prism provides a good means of displaying the colours of the spectrum by shining white light (which consists of light of all wavelengths) onto the prism. The different colours will deviate by different amounts and the output rays will form a coloured beam with red light showing least deviation and violet light showing maximum deviation, Figure 9.7. This principle is used in *spectrometers* as a means of measuring the wavelength of incident light.

The other broad class of prism is prisms which rely completely or partially on reflection to change the beam direction. This is usually done with total internal reflection, although if the range of angles is not high enough, the surface may be silvered. Total internal reflection was studied in Section 6.6 where the critical internal angle of incidence which must be exceeded was found to be

$$\sin \theta_c = 1/n \tag{9.13}$$

The deviation that occurs in a reflecting prism is only a function of the incident angle and the prism angle, so no dispersion takes place. A simple example is the right angle prism shown in Figure 9.8(a). This turns the ray through 90°. It is important to note the polarisation direction through the prism. If the ray is a single plane wave, then the

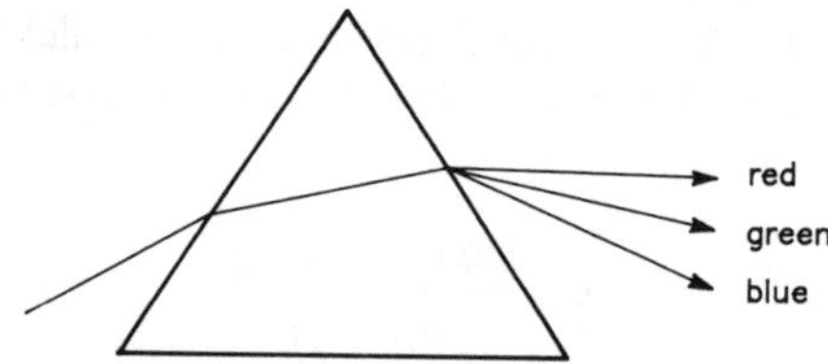

Figure 9.7. Deviation of path lengths of coloured light through a dispersive prism

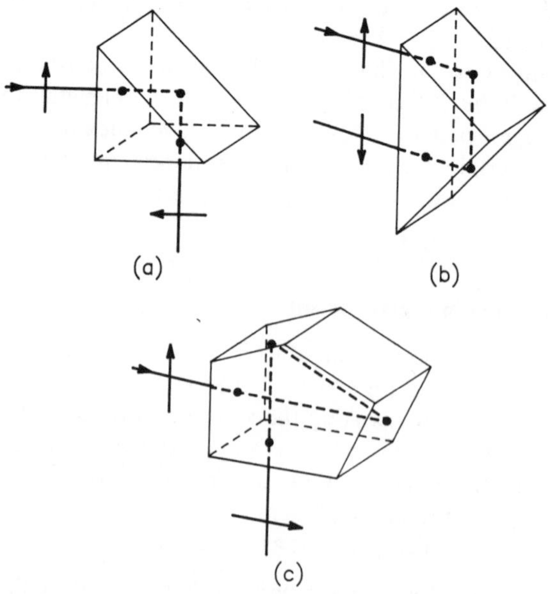

Figure 9.8. Reflecting prisms. Arrows show direction of electric field vector. (a) Right angle prism. (b) Porro prism. (c) Penta prism

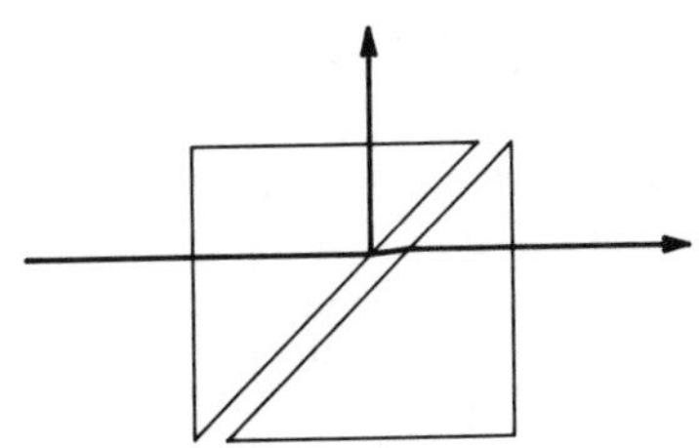

Figure 9.9. Beam splitter using two prisms and frustrated internal reflection

arrow represents the electric field vector. If the ray represents an optical image, the arrow shows the direction of the image.

A wide range of combinations of angles and shapes can be used make prisms which turn images through desired angles. Figure 9.8(b) and 9.8(c) show two of these. The *Porro* prism turns the beam back in the direction from which it came and works over a wide range of angles of incidence. A three dimensional version can be made by slicing a cube through a cross diagonal. Then light entering the prism from any direction will be reflected back in the same direction. The *penta* prism is designed to turn the beam through 90° without altering the orientation of an image. It is used in single lens reflex (SLR) cameras in conjunction with a plane mirror to view the image at the focal plane.

There are many other types of prisms, of which just two more will be mentioned. It was explained in Section 6.6 that total internal reflection was not perfect and that a field exists outside the surface which decays away very rapidly from the surface. In optics this is called *frustrated internal reflection.* It can be used to split a beam by placing a second prism close to the first prism in order to capture some of the power, Figure 9.9. By adjusting the gap, any amount of coupling of power can in principle be obtained.

There are a class of materials called birefringent crystals which are anisotropic at optical wavelengths and have different refractive indices for orthogonal polarisations. These will have different critical angles for the two polarisations, say θ_{c1} and θ_{c2}. By using rays with incident angles between θ_{c1} and θ_{c2}, the orthogonal polarisations can be separated since one polarisation will be transmitted and the other reflected.

9.4 IMAGING IN OPTICAL SYSTEMS

Mirrors and lenses are optical components which change the shape of wavefronts. Optics is concerned with what happens as optical rays from extended sources pass through optical systems. Examples of an extended source are the view of a scene as seen through a camera, or a physical object as seen through a microscope. This is different from the coherent applications described in most of the book, where the source is effectively at a single point and energy is received at a single point. The extended source needs a notation to deal with the way the object is perceived to change due to the action of the optical system. The source is called the *object* and the result after the action of the optical system is called the *image*, Figure 9.10. If the medium surrounding an optical system is isotropic and homogeneous then the optical rays from the object will always be spherical wavefronts which *diverge* from the object in the real object space. Each of the parts of the wavefronts are rays which obey the laws of geometric optics. The optical

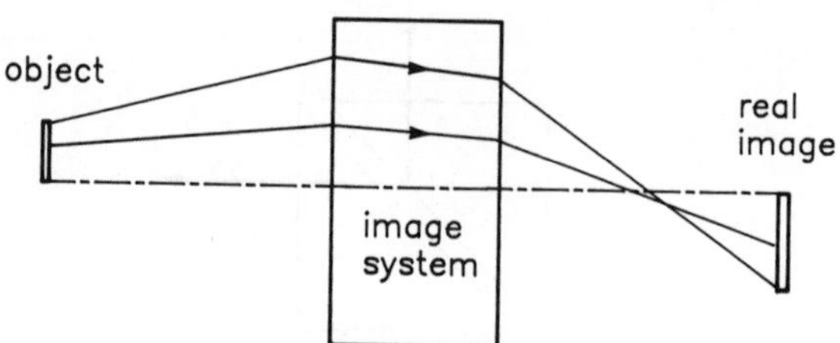

Figure 9.10. Object and real image with an optical imaging system

system will redirect the rays and ideally will eventually bring the rays back to a point. This is the image point and the region in which this happens is called the real image space. Fermat's principle will apply to every ray path through the optical system so the process is reversible and the object and the image are interchangeable. If the object has finite dimensions, then the image will also have finite dimensions but will be modified by the optical system. This is measured with the *magnification* of the system, which could be less than one. The process of determine the shape and orientation of the image is called *imaging*. This is important in optical design.

Sometimes the optical system will create a *virtual image*, as in Figure 9.11. In this case the reflecting or refracting surface causes the rays from the object to diverge even more than originally. They then *appear* to emanate from a virtual image point which is behind the surface. The action of a plane mirror is the obvious example of this effect. The region behind the surface is called the virtual image space.

The position of the object is specified with the *object distance*, s_0, which is measured along the optical axis of the system. The position of the image is similarly specified by the *image distance*, s_i. One of the most common situations is to image an object which is at infinity, as in viewing a distant scene. In this case $s_0 = \infty$.

The description of optical systems in this chapter assumes that the components are capable of producing a perfect image from a point object. In practice this is never completely true, although optical systems manage to get very close to perfection because of the short wavelengths and relatively large sizes of components. Distorted images can be caused by three different mechanisms.

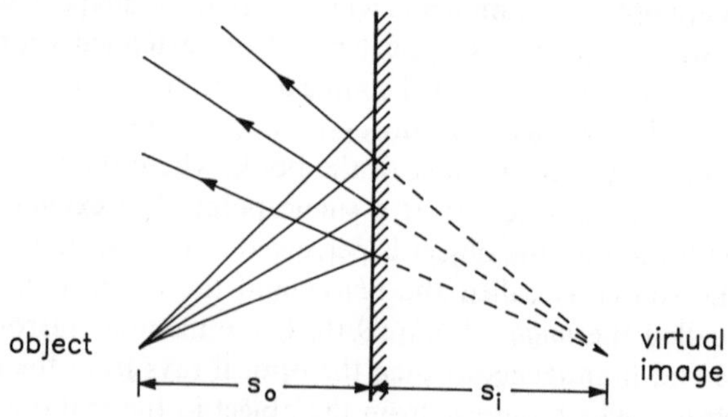

Figure 9.11. Reflections off a plane mirror and its virtual image

(a) The surfaces of the mirrors and lenses can be slightly imperfect so that some of the incident light is scattered by the rough surface. This inevitably happens to some extent at optical wavelengths because of the very fine mechanical tolerances required.

(b) The shape of the surfaces of components can be slightly incorrect, perhaps due to the need to make a surface mechanically. For instance, a spherical profile might be used where a parabolic profile would be ideal. This causes *aberrations* which distort the images.

(c) The optical components might not be large enough to avoid the effects of rays hitting the edge of the component. The edge rays are *diffracted* in a cone of angles centred around the incident ray so that energy is spread in many directions. Diffraction is of major importance at microwave frequencies, but can often be ignored at optical wavelengths. There are optical components which are specifically designed to use diffraction effects, but these will not be treated in this chapter.

9.5 MIRRORS

Mirrors are designed to reflect all the incident optical rays in a defined way. They have wide applicability and are used at all frequencies from microwaves to X-rays. At microwaves they are called *reflector antennas*, but the principles are the same as optical mirrors. Practical mirrors and reflectors have the great advantage that they are very broad band components. They operate over wide ranges of wavelength because the principles of reflection do not involve any wavelength dependent factors. This is not true for refracting systems using lenses. The material doing the refraction often has some dispersion, as discussed in Section 9.3.

Reflector antennas are made from metal, and optical mirrors can also be made from polished metal surfaces. It is, however, preferable to make optical mirrors by coating a shaped piece of glass. Glass is very stable and can be made to have a high quality surface with only small amounts of roughness. The coating is either a thin layer of silver or vacuum evaporated aluminium. High quality mirrors are front coated so that the metal is on the front of the glass, but lower quality mirrors for domestic use are rear coated to give more protection to the reflecting surface.

The plane mirror, Figure 9.11, is designed to redirect the rays from an object. The image is inverted in a plane mirror, so that left becomes right etc. The actual image has the same shape and size as the object which means that the magnification is one. This is to be expected since the plane mirror could be replaced by an identical object at the virtual image.

Mirrors which produce perfect point images from point objects are called *conic surfaces*. These are paraboloids, ellipsoids and hyperboloids. The paraboloid, Figure 9.12, has the property that an object at infinity, which means that the rays are parallel to the axis, is imaged at a point. This is called the *focal point* of the paraboloid. The process is reciprocal so that an object at the focal point produces an image at infinity. The paraboloid turns a spherical wave centred on the focal point into a plane wave. The optical path lengths for all rays from the focus to the corresponding point on a

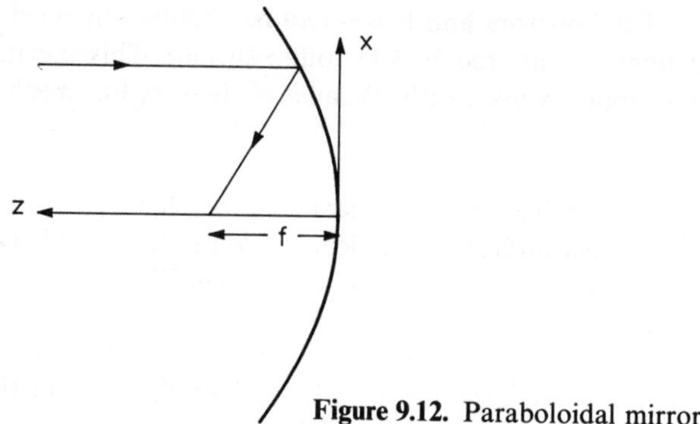

Figure 9.12. Paraboloidal mirror

plane wavefront must be equal. This gives the equation of the surface of a parabola as, Figure 9.12,

$$x^2 = 4fz \tag{9.14}$$

where f is the focal length from the focal point to the vertex of the parabola. The parabola is used as a means of creating plane waves. It is also the basis of most microwave reflector antennas where it is used as a means of magnifying the power from a source at the focal point.

Ellipsoids and hyperboloids can be concave or convex, Figure 9.13. Both ellipsoids and hyperboloids have two focal points. If one is the object point, a perfect image (real or virtual) is produced at the other focal point. Ellipsoidal mirrors are used in a Gregorian telescope or dual reflector antenna to increase the magnification. Hyperboloids are similarly used in the Cassegrain telescope or dual reflector antenna.

Conic surfaces are ideal for imaging, but not ideal to make. The surface needs to have

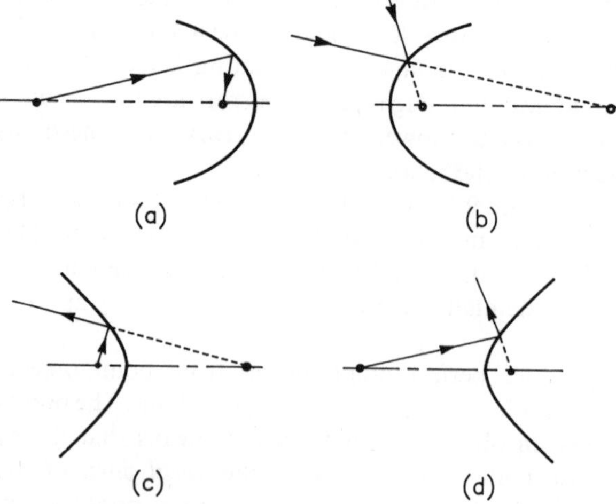

Figure 9.13. Ellipsoidal and hyperboloidal mirrors. (a) Concave ellipsoidal. (b) Convex ellipsoidal. (c) Concave hyperboloidal. (d) Convex hyperboloidal

a precise profile in order to produce a perfect image, and this is difficult to manufacture accurately. On the other hand spherical mirrors can be manufactured accurately and for this reason most optical mirrors are spherical in shape. Spherical surfaces have the unique property that if a concave and a convex spherical surface are brought into contact, they will fit together irrespective of the orientation. This means that a spherical mirror can be made by taking two roughly spherical glass surfaces, one concave and the other convex and grinding them together. Irregularities will be ground away and eventually a pair of perfect spherical surface will exist.

Although spherical mirrors can be made accurately, they appear not to be good imaging devices, because an object at infinity will be imaged as a curve and not a point, Figure 9.14. Fortunately, it turns out that in the paraxial region with rays near the axis, the sphere approximates to a parabola. This is shown by expanding the equation of the circular cross-section of a sphere. If the origin of the sphere is at the vertex,

$$x^2 + (z - R)^2 = R^2 \tag{9.15}$$

This can be solved for z as

$$z = R \pm R\left(1 - \frac{x^2}{R^2}\right)^{1/2} \tag{9.16}$$

Expanding the square root binomially and using values of $x < R$ gives

$$z = \frac{x^2}{2R} + \frac{x^4}{8R^3} + \cdots \tag{9.17}$$

The paraxial region is when x is small, so only the first term of equation (9.17) will be significant. Comparing this to the equation of a parabola, equation (9.14), shows that it is identical if $R = 2f$. The spherical mirror thus approximates to a parabola in the paraxial region and the focal length is half the radius of curvature of the sphere.

Mirrors are used to image from one point to another point so an equation is needed to relate the object distance, s_0 to the image distance, s_i, Figure 9.15. This can be done by considering the angles that the rays make with the axis. From Figure 9.15,

$$\theta = \alpha_0 + \phi \qquad \text{and} \qquad 2\theta = \alpha_0 + \alpha_1$$

combining these two to eliminate θ gives

$$\alpha_0 + \alpha_i = -2\phi \tag{9.18}$$

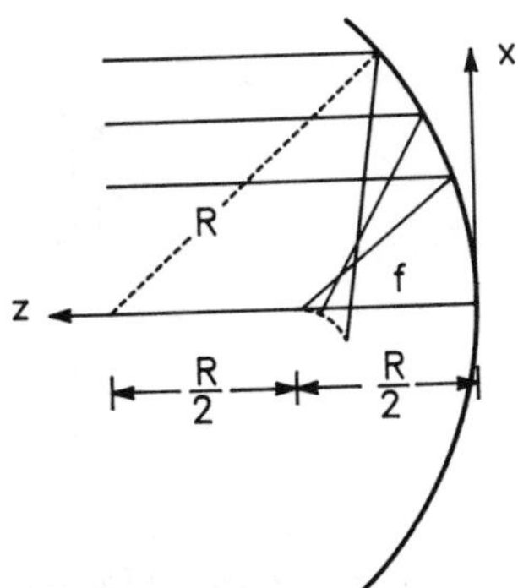

Figure 9.14. Spherical mirror showing focal curve

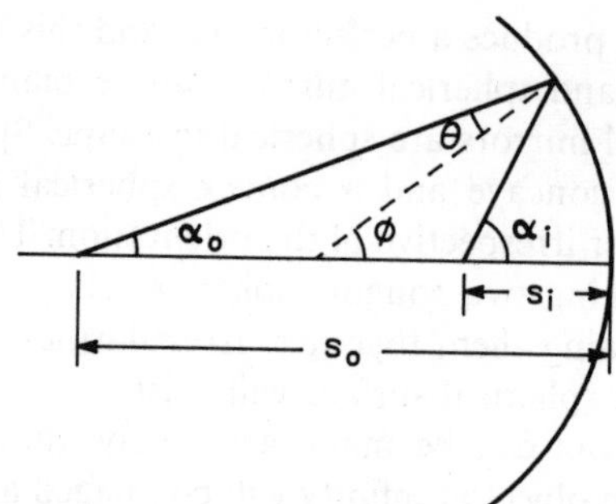

Figure 9.15. Geometry for reflections at a spherical mirror

Now $\tan \alpha_0 = x/s_0$. For small angles in the paraxial region, $\alpha_0 \approx x/s_0$. Inserting this approximation and a similar one for α_i into equation (9.18) gives

$$\frac{1}{s_0} + \frac{1}{s_i} = -\frac{2}{R} \tag{9.19}$$

This can also be expressed in terms of the focal length, f, as

$$\frac{1}{s_0} + \frac{1}{s_i} = \frac{1}{f} \tag{9.20}$$

in which $f = -R/2$. The negative sign is present because by convention the radius R of concave mirrors is negative. Equations (9.19) and (9.20) are sometimes called the mirror equations. They can be used to find the radius of a mirror for a particular object and image distance.

If the object has a lateral height, h_0, Figure 9.16, and the image has a lateral height h_i, then the magnification, M, is defined as the ratio of the image height to the object height. Simple geometry in Figure 9.16 shows that

$$M = \frac{h_i}{h_0} = \frac{s_i}{s_0} \tag{9.21}$$

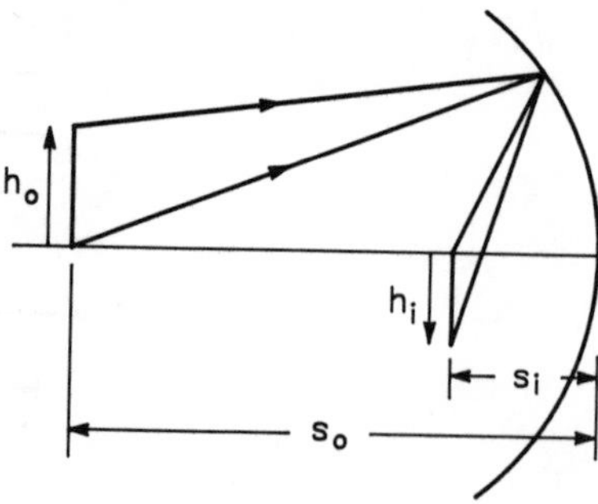

Figure 9.16. Image height through a spherical mirror

The magnification is just the ratio of the image distance to the object distance. If the object is at infinity, then the magnification is zero, since the image is a point, irrespective of the height of the object. Alternatively, if the image is at infinity, the magnification is infinite, because a point source at the focal point appears spread out as a plane wave.

9.6 LENSES

Lenses change the shape of a wavefront by refraction at two dielectric surfaces. Optical lenses are usually made of glass and are very widely used. Glass is stable and can be accurately ground to a spherical shape. The existence of two refracting surfaces means that a wide range of imaging devices can be made since either surface can be convex, plane or concave. Dielectric lenses can also be used at wavelengths longer than optical or infrared and they are used in millimetrewave systems and occasionally at microwaves, though the weight tends to be a deterrent.

The discussion on the ideal shape of a mirror in the last section applies also to refracting surfaces. Perfect images are formed with conic sections. It is, however, difficult to make paraboloidal, ellipsoidal or hyperboloidal shaped lenses so they are not often used. They can be replaced by a spherical shaped lens as long as the paraxial region is used. Thus most lenses have spherical surfaces. There is an additional constraint which is imposed in the theoretical analysis in order to obtain simple design formulae. This is the *thin lens* approximation which assumes that the thickness of the lens can be neglected by comparison with the object and image distances. The approximations mean that the real image in a practical lens will be slightly distorted due to aberrations away from the perfect image.

A lens which does not satisfy the thin lens approximation is called a *thick* lens. The surfaces of a lens can be *concave, planar* or *convex*, so the combinations of shapes shown in Figure 9.17 are possible. Convex lenses are also called *converging* or *positive* lenses because they cause rays from an object at infinity to converge. Concave lenses are called *diverging* or *negative* lenses because they cause rays from an object at infinity to diverge.

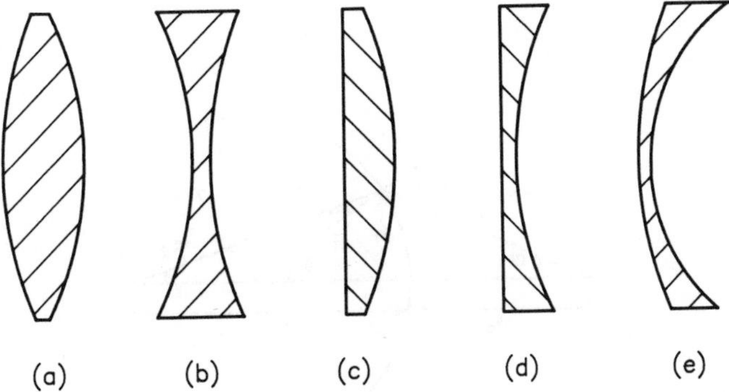

Figure 9.17 Lenses. (a) Double convex. (b) Double concave. (c) Planar convex. (d) Planar concave. (e) Meniscus convex or concave

REFRACTION AT A SINGLE SPHERICAL SURFACE

The basis of any lens is the refraction that takes place at a spherical surface. The geometry is shown in Figure 9.18. The object is at a distance s_0 in a medium of refractive index n_1 and the image is at a distance s_i in a medium of refractive index n_2. It is desired to find the relationship between the refractive indices, object and image distances and the radius R of the spherical surface. This can be found either by minimising the optical path lengths using Fermat's principle, or by using Snell's law. Here the paraxial approximation to Snell's law at the refracting surface is used so that the $\sin\theta$ functions in Snell's law are replaced by θ, leading to

$$n_1\theta_i = n_2\theta_t \tag{9.22}$$

θ_i and θ_t are now expressed in terms of the axial angles α_1, α_2 and ϕ, Figure 9.18. This gives $\theta_i = \phi + \alpha_1$ and $\theta_t = \phi - \alpha_2$ and equation (9.22) becomes

$$n_1(\phi + \alpha_1) = n_2(\phi - \alpha_2) \tag{9.23}$$

Now the paraxial approximation allows the tangent of each of these angles to be approximated by just the angle, so $\phi = h/R$, $\alpha_1 = h/s_0$ and $\alpha_2 = h/s_i$, where h is the lateral distance of the refraction point. Inserting these into equation (9.23) and re-ordering gives

$$\frac{n_1}{s_0} + \frac{n_2}{s_i} = \frac{n_2 - n_1}{R} \tag{9.24}$$

This gives a relationship between the parameters of a lens. The sign convention for R is the same as for mirrors. Concave surfaces have $R < 0$ and convex surfaces have $R > 0$ with respect to the incident beam direction. The surface in Figure 9.18 thus has R positive. If s_i is at infinity, then s_0 is called the object focal length, f_0 and equation (9.24) gives

$$f_0 = \left(\frac{n_1}{n_2 - n_1}\right)R \tag{9.25}$$

If $n_1 = 1$, then the focal length decreases as the refractive index of the second medium increases.

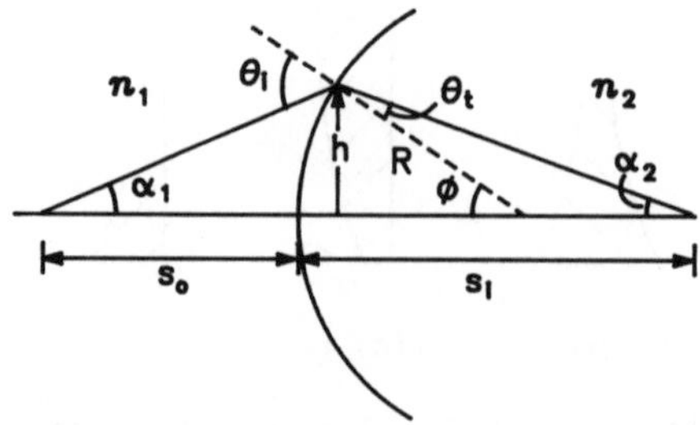

Figure 9.18. Refraction at a single spherical surface

THIN LENSES

The *thin lens*, Figure 9.19, consists of two spherical surfaces, back to back. It is shown as a double convex lens, but any combination of concave, plane or convex is implied. Equation (9.24) can be used with the notation shown in Figure 9.19, to find the relationship between object and image distances. Assuming that the lens is in air so $n_1 = 1$ and $n_2 = n$, equation (9.24) gives for the first surface

$$\frac{1}{s_0} + \frac{n}{s_{i1}} = \frac{n-1}{R_1} \tag{9.26}$$

where s_{i1} is the image due to the first surface. Similarly for the second surface

$$\frac{n}{s_{01}} + \frac{1}{s_i} = \frac{1-n}{R_2} \tag{9.27}$$

where s_{01} is the second object distance, which is related to the first image distance by

$$s_{01} = t - s_{i1} \tag{9.28}$$

where t is the thickness of the lens. The optical thin lens approximation states that it is possible to ignore the thickness of the lens, t, so that $s_{01} = -s_{i1}$. Substituting this into equation (9.27) and combining with equation (9.22) to eliminate s_{i1} gives

$$\frac{1}{s_0} + \frac{1}{s_i} = (n-1)\left(\frac{1}{R_1} - \frac{1}{R_2}\right) \tag{9.29}$$

As with the single spherical surface, if s_i is at infinity, then the object distance becomes the focal length, f of the lens, and from equation (9.29)

$$\frac{1}{f} = (n-1)\left(\frac{1}{R_1} - \frac{1}{R_2}\right) \tag{9.30}$$

Equation (9.29) and (9.30) are called the *lens maker's equations* because either one is the basic design equation for lenses. Combining them together gives the same relationship between object distance, image distance and focal length as for mirrors, equation (9.20)

$$\frac{1}{s_0} + \frac{1}{s_i} = \frac{1}{f} \tag{9.31}$$

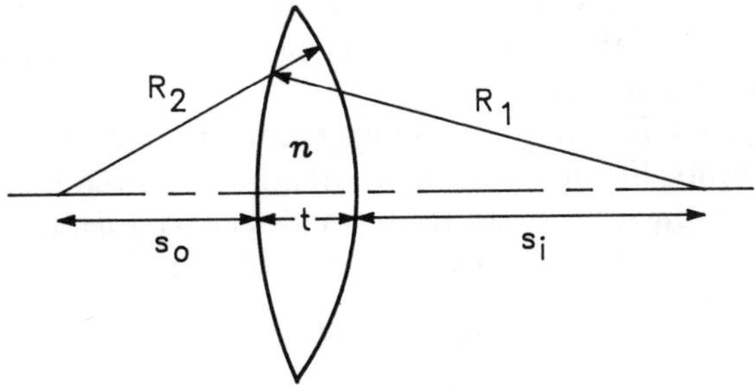

Figure 9.19. Geometry for thin lens

This is called the *Gaussian lens equation.* It is important to remember when using the lens maker equations that the convention on the sign of the radii of curvature are with respect to the incident direction. Thus in Figure 9.19, R_1 is a convex surface so is positive. R_2 is a concave surface for the incident rays so is negative.

The lateral magnification through a thin lens is also the same as for mirrors, as can be seen from the geometry, so $M = s_i/s_0$ as in equation (9.21).

Example 9.3 Thin lens design

Design a planar convex lens to have a focal length of 100 mm with glass of refractive index 1.5, and then find the image distances and magnifications for object distances of infinity, 500 mm, 200 mm, 100 mm, 75 mm and 50 mm.

Equation (9.31) shows that it makes no difference which side of the lens is the object side or the image side. The focal length is the same whichever way round the lens is used. For a planar surface, $R_1 = \infty$ and the radius of curvature R_2 can be obtained from equation (9.30) with $R_1 = \infty$ and $f = 100$ mm. This gives $R_2 = -50$ mm, which is the radius of curvature of the convex surface. If $R_2 = \infty$ then $R_2 = 50$ mm. This has the same value but the change in sign indicates that the rays are incident from the opposite direction. The image distances are found from equation (9.31) to be 100 mm, 125 mm, 200 mm, ∞, -300 mm and -100 mm, respectively. This demonstrates the reversibility of the lens. For an object 200 mm away, the lens does nothing. For an object inside the focal distance, the image rays diverge so there is a virtual image. The corresponding magnifications are found from equation (9.21) to be 0, 0.25, 1, ∞, -3 and -2, respectively. The negative magnifications indicate a virtual image. Thus only reduction in real size occurs with a planar convex lens.

9.7 MULTIPLE OPTICAL COMPONENTS AND RAY MATRICES

An optical system usually consists of a number of optical components. This might be a series of lenses or a series of mirrors, lenses and prisms. The equations developed in the last sections enable the results of a series of components to be found in sequence. This was done in the last section to find the lens makers equation from refractions at two spherical surfaces. The mathematics, however, rapidly gets tedious. A more elegant and simpler approach is to use *ray matrices* which relate the output lateral distance and slope of the ray to the input lateral distance and slope of the ray. For mirrors or lenses, the input is the object and the output is the image, but this method is more general and applies to any component or medium.

The simplest situation is a ray in a homogeneous medium which does not pass through any components, Figure 9.20. The relationships between the input and output angles and lateral coordinates is that $\alpha_1 = \alpha_2$ and $x_2 = x_1 + d \tan \alpha_1$. In the paraxial region, $\tan \alpha_1 = \alpha_1$, so x_2 and α_2 are related to x_1 and α_1 by

$$x_2 = x_1 + d\alpha_1$$

$$\alpha_2 = \alpha_1$$

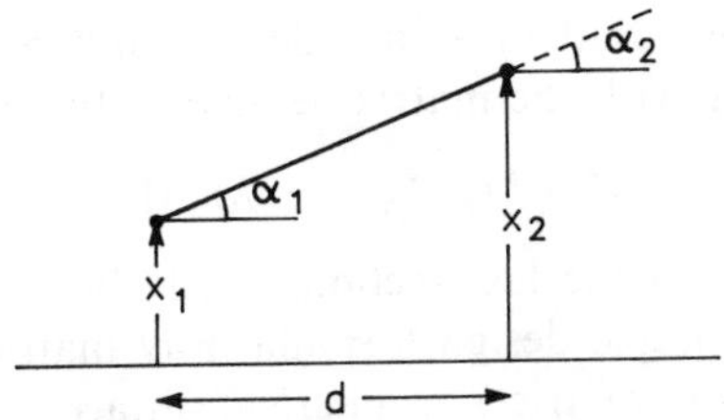

Figure 9.20. Geometry for ray in a homogeneous medium

These two equations can be expressed more conveniently in matrix form as

$$\begin{pmatrix} x_2 \\ \alpha_2 \end{pmatrix} = \begin{pmatrix} 1 & d \\ 0 & 1 \end{pmatrix} \begin{pmatrix} x_1 \\ \alpha_1 \end{pmatrix} \tag{9.32}$$

The square matrix is essentially a translation matrix between two points. This is a trivial case but the general form of equation (9.32) is

$$\begin{pmatrix} x_2 \\ \alpha_2 \end{pmatrix} = \begin{pmatrix} A & B \\ C & D \end{pmatrix} \begin{pmatrix} x_1 \\ \alpha_1 \end{pmatrix} \tag{9.33}$$

Expressions for the *ABCD* matrix can be found for all optical components by rewriting the appropriate equations in terms of x_1, x_2, α_1 and α_2.

Snell's law for the paraxial region, equation (9.22), becomes

$$\begin{pmatrix} x_2 \\ \alpha_2 \end{pmatrix} = \begin{pmatrix} 1 & 0 \\ 0 & \dfrac{n_1}{n_2} \end{pmatrix} \begin{pmatrix} x_1 \\ \alpha_1 \end{pmatrix} \tag{9.34}$$

The refraction at a spherical interface, equations (9.23) and (9.24) may be recast as

$$\begin{pmatrix} x_2 \\ \alpha_2 \end{pmatrix} = \begin{pmatrix} 1 & 0 \\ \dfrac{1}{R}\left(\dfrac{n_1}{n_2} - 1\right) & \dfrac{n_1}{n_2} \end{pmatrix} \begin{pmatrix} x_1 \\ \alpha_1 \end{pmatrix} \tag{9.35}$$

Similarly the thin lens maker's equations, (9.29) to (9.31), become

$$\begin{pmatrix} x_2 \\ \alpha_2 \end{pmatrix} = \begin{pmatrix} 1 & 1 \\ -\dfrac{1}{f} & 0 \end{pmatrix} \begin{pmatrix} x_1 \\ \alpha_1 \end{pmatrix} \tag{9.36}$$

The advantage of ray matrices comes when components are in series. If two components with ray matrices having suffices 1 and 2 are in series then the overall result is found by cascading the matrices:

$$\begin{pmatrix} x_3 \\ \alpha_3 \end{pmatrix} = \begin{pmatrix} A_2 & B_2 \\ C_2 & D_2 \end{pmatrix} \begin{pmatrix} A_1 & B_1 \\ C_1 & D_1 \end{pmatrix} \begin{pmatrix} x_1 \\ \alpha_1 \end{pmatrix} \tag{9.37}$$

Cascading can be done for any number of matrices and is a powerful method of dealing with multiple parts of an optical system. If the square matrices are designated M_1,

$M_2, \ldots, M_{N-1}, M_N$, then the result of a cascade is found by multiplying together the individual matrices, starting with the matrix nearest to the output, M_N,

$$M = M_N M_{N-1} \cdots M_2 M_1 \tag{9.38}$$

The thin lens was analysed in the last section, partly because the thick lens was too complicated to produce a simple design formula. Ray matrices enable thick lenses to be analysed by cascading three matrices: a spherical surface matrix, a translation matrix, and a second spherical surface matrix. Computers make the manipulation of matrices relatively straightforward so there is no need to perform laborious mathematics.

Example 9.4 Two thin lenses in cascade

Use ray matrices to work out the combined focal length of two thin lenses which are in contact.

If the focal lengths of the two lenses are f_1 and f_2, then using equations (9.36) and (9.37) gives

$$\begin{pmatrix} x_3 \\ \alpha_3 \end{pmatrix} = \begin{pmatrix} 1 & 1 \\ -\dfrac{1}{f_2} & 0 \end{pmatrix} \begin{pmatrix} 1 & 1 \\ -\dfrac{1}{f_1} & 0 \end{pmatrix} \begin{pmatrix} x_1 \\ \alpha_1 \end{pmatrix} \tag{9.39}$$

Multiplying the two square matrices together gives

$$\begin{pmatrix} x_3 \\ \alpha_3 \end{pmatrix} = \begin{pmatrix} 1 & 1 \\ -\dfrac{1}{f_2} - \dfrac{1}{f_1} & 0 \end{pmatrix} \begin{pmatrix} x_1 \\ \alpha_1 \end{pmatrix} \tag{9.40}$$

Comparing this equation with equation (9.36) shows that the combined focal length, f_c, is

$$\frac{1}{f_c} = \frac{1}{f_1} + \frac{1}{f_2} \tag{9.41}$$

The combined focal length is the reciprocal sum of the separate focal lengths.

9.8 OPTICAL SOURCES AND DETECTORS

Every optical system needs a source of energy and a method of detecting signals. This is true for all electromagnetic systems, but at optical wavelengths there is a much wider range of sources and detectors than at other wavelengths. Most optical sources are distinguished by being incoherent sources of energy which radiate over a broad band of wavelengths. This is acceptable at optical wavelengths where there is a large amount of frequency space available. At microwaves frequency space is too precious so incoherent sources are generally not acceptable. At visible wavelengths optical sources are unique in having available the strong natural source of energy from the Sun and the natural detector of the human eye. A brief review of optical sources and detectors is now presented. The eye is treated separately in the next section because of its importance.

OPTICAL SOURCES

Optical sources can be divided into generators of coherent energy and generators of incoherent energy, and subdivided as follows:

- Coherent sources
  - Lasers
- Incoherent sources
  - Semiconductor light emitting diodes (LEDs)
  - Discharge lamps (fluorescent lamps, arc lamps, flash lights)
  - Incandescent lamps
  - Sunlight

The only coherent optical source is the *LASER*, or Light Amplification by Stimulated Emission of Radiation. The discovery of lasing action in the late 1950s was one of the great breakthroughs in physics and has led to the development of many different types of lasers. These range from ruby lasers capable of generating gigawatts (10^9 W) of pulsed light to semiconductor lasers generating milliwatts of CW energy. In a laser, energetic photons are absorbed by atoms which raises, or *amplifies*, the atoms to an excited state. The atoms are then *stimulated* into *emission* of energy by the presence of electromagnetic waves. The photons and the stimulating electromagnetic waves have identical phases and polarisations because they are resonant waves. This is done in a resonant cavity. Physically a laser consists of an external source of power (electrical or chemical or optical) feeding a resonant cavity. One end of the resonant cavity has a mirror which is totally reflecting, the other end has a partially transparent mirror so that some of the coherent waves can radiate. The mirrors are usually spherical mirrors in gas lasers and plane mirrors in solid state lasers.

The principles of resonant plane waves were described in Chapter 6 where it was found that standing waves are supported in a cavity which has a length of $N\lambda/2$, where N is an integer and λ is the wavelength of the waves in the medium inside the cavity. Looked at as a source, the length therefore determines the wavelength of the radiation. In lasers, the integer N is very large and many lasers are multi-mode, meaning that resonance occurs at a number of values of N, each giving a different frequency of radiation.

There are a wide variety of types of laser. The main types break down into *solid state lasers*, *gas lasers* and *semiconductor lasers*, although there are also liquid laser and chemical lasers. The first lasers were solid state lasers and used ruby crystals. The highest power lasers used in plasma and nuclear research are solid state lasers capable of generating very high pulse power and CW powers of a few hundreds of watts. The efficiency of conversion of external power into optical waves is very low, usually less than 1%. Gas lasers generate powers from milliwatts to tens of watts of CW or pulsed waves. They are versatile and relatively cheap to produce. The most common is the helium–neon laser resonating of 632.8 nm. This is widely used for experiments requiring coherent light and forms a very convenient source of plane waves in the laboratory. The semiconductor laser is attractive because of its small size and integratability with other optical components. They are the main source of power in optical fibre communications and optoelectronics. Most use gallium arsenide as the base material with the addition of other elements. They are generally quite complex geometries. A schematic is sketched in Figure 9.21. The active region is a layer of pure GaAs, surrounded by doped GaAlAs layers and metal conductors on the top and bottom. Semiconductor lasers for communications are required to generate wavelengths which have longer wavelengths than visible light as the lowest attenuation is achieved around 1.4 μm.

Incoherent optical sources exist in many forms. This partly reflects the need for many sources of visible light. The most compact, but low power devices, are p–n junctions of GaAs semiconductor. Optical energy is produced by the recombination of electrons and holes around the junction. These LEDs can produce outputs in both the visible and infrared parts of the spectrum. They emit over a band of wavelengths, Figure 9.22, which is broad by comparison with a laser, but much narrower than other incoherent sources.

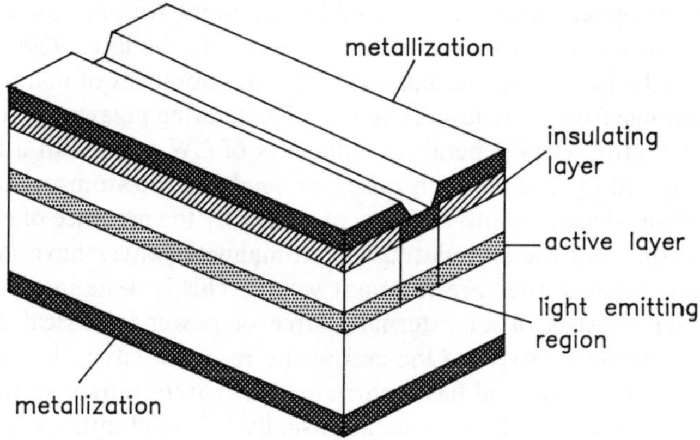

Figure 9.21. Cross-section of a semiconductor laser

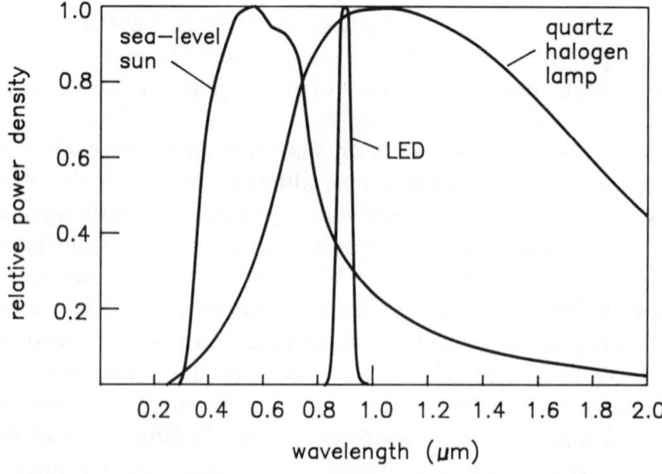

Figure 9.22. Relative spectral response of the Sun, a quartz halogen lamp and a light emitting diode (LED). The peak power of all sources is normalised to unity

Discharge lamps depend on an electrical discharge in gases. A current is passed through an ionized gas between two electrodes which causes a discharge of energy in the visible range. There are a wide variety of types, including sodium arc lamps, photoflash tubes and fluorescent lamps. The spectral emission depends on the gas and is characterised by many peaks at individual frequencies. This means that discharge lamps are not very suitable for optical instruments as the spectrum of radiation is too complex. Incandescent lamps are the most common artificial source of light produced by heating a material until it is incandescent. Tungsten filament lamps are popular for simple and cheap sources of light which can be run at a wide range of supply voltages. The output from a tungsten filament lamp changes considerably with its age due to evaporation. The addition of quartz–halogen or tungsten–halogen gases helps to slow down the evaporation and gives a more stable source which radiates over the visible nd infrared parts of the spectrum.

The last source of incoherent light is the most important for life on Earth, namely sunlight and skylight. Direct light from the Sun is called sunlight, whereas reflected light in the atmosphere is called skylight. The Sun behaves as a blackbody with a temperature of about 6000 K at its centre and 5000 K at the edge. At the mean Earth radius from the Sun, the peak incident power density is about 2.2 kW/m^2 at a wavelength of 500 nm which is in the middle of the visible spectrum. The atmosphere absorbs about 36% of the energy from the Sun so that the power density at the Earth's surface peaks at 1.4 kW/m^2. The relative spectral response is shown in Figure 9.22 in comparison with a quartz halogen lamp and in LED. The peak energy from the Sun comes in the visible region, but in absolute terms there is a considerable amount of energy in the infrared and longer wavelength parts of the spectrum.

OPTICAL DETECTORS

The most common optical detector is the eye, which provides a subjective response of incident visible energy. The eye is actually a complete optical system of lenses and detectors and is worth separate consideration in the next section. Artificial detectors aim to give an output which is a quantitative measure of the amount of light incident on them. The absolute measure of optical power is done with thermal detectors where the incident radiation causes a rise in temperature of a metal or semiconductor. The consequent change in resistance can be accurately measured. Detectors using metal are called *bolometers*, detectors using semiconductor are called *thermistors*. A change in resistance can be measured with a sensitive current bridge.

Changes in temperature are relatively slow to occur, so thermal detectors are not good for detecting quick changes in optical power levels or for conveying signals. Detectors which respond to the rate of incidence of photons are called *quantum* detectors. The photons interact with electrons in the detector and cause a change in the electrical or chemical state. *Photocells* were an early example of quantum detectors. Popular modern detectors are semiconductors which rely on either *photo conductivity* or *photo voltaic effects*. In photo conducting detectors, the photons produce additional electron–hole pairs in the semiconductor. This changes the electrical conductivity. An example of photo conducting detectors are arrays of charge coupled devices (CCDs) used in solid state video cameras.

Photodiodes are voltaic detectors and are the most common type of optical detector.

They consist of a highly doped, reverse biased, p–n junction at which the incident photons are absorbed. This increases the number of free electron–hole pairs and causes a change in voltage which can be measured by an external electrical circuit. The spectral response can be chosen according to the types of semiconductor and doping element. Photodiodes are very widely used because they are small and rugged, have low noise and high sensitivity, have a fast response rate and can be integrated easily into electronic circuits. They are the main choice for many communication and signalling applications.

An optical detector which relies on chemical changes, rather than electrical changes to record incident optical energy is *photographic film*. This consists of an emulsion of silver halide crystals. The halide ions receive energy from incident photons. These then combine with silver ions to produce a stable silver atom. The light falling all over the film causes a latent image of the intensity of the incident light to be stored in the silver atoms. The action of a chemical developer is used to effectively amplify the stored image and make a permanent record. Unlike other detectors, photographic film stores a record of how much optical energy was incident on it. This can be used to detect very weak sources, by allowing the film to integrate the incident power over a period of time.

9.9 OPTICS OF THE EYE

The eye forms a dynamic optical system which senses visual light of different strengths and wavelengths. The output is passed to the brain for processing and interpretation. The strength of the incident light is interpreted as the amount of lightness or darkness and the different wavelengths are interpreted as colours. The actual process of interpretation involves physiological and psychological aspects which mean that absolute and quantitative measures cannot be applied. Nevertheless the optics of the eye implements the basic principles of optical waves in a remarkable manner. It is of interest not only because most humans and animals rely on vision for interpretations of the surrounding environment, but also because it indicates the relative simplicity of the artificial optical devices which have been constructed and are described in the rest of this chapter. The eye has the dynamic ability to adjust to a very wide range of light levels, it can view images at a wide range of distances from a few centimetres to infinity, it can do this over a broad field of view, it can simulteneously see the whole field of view at one time so that objects can be localised in three-dimensional space, finally it can distinguish subtle shades of colour. On the negative side, the eye, being natural, often does not work correctly and so needs correction. It also only detects visual light which is a very small part of the electromagnetic spectrum, though, as described in the last section, it happens to be the frequency range where the Sun emits most of its energy.

A simplified drawing of the human eye is shown in Figure 9.23. It is almost spherical with a diameter of about 22 mm and is composed of a jelly-like mass surrounded by a tough skin which is opaque except at the front. The front surface bulges away from the main sphere and forms the *cornea*. The convex surface of the cornea is the first and strongest lens in the optical system. The cornea has a refractive index which varies slightly over its thickness but mainly has a value of 1.38. The air–cornea interface causes most of the bending which occurs to incident rays of light. It is interesting to note that the refractive index of the cornea is very similar to that of water at optical wavelengths

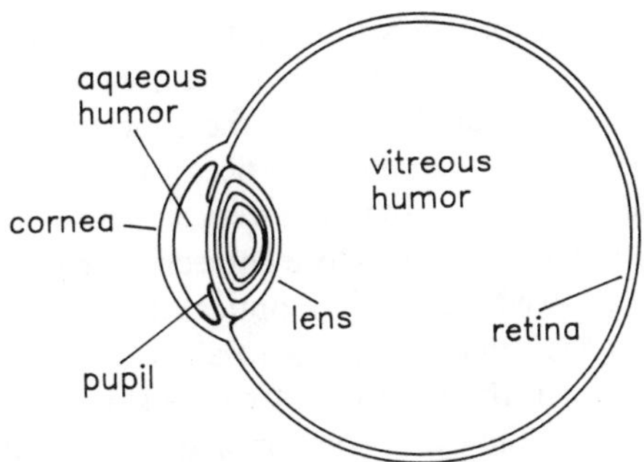

Figure 9.23. Cross-section of the human eye

($n \approx 1.33$). This has the consequence that it is not easy to see under water because very little refraction takes place.

Behind the cornea is a clear watery fluid called the *aqueous humor*, which has a refractive index of 1.34. Immersed in the aqueous humor is a variable *iris* which controls the diameter of the aperture of the eye. The aperture through which the light passes is called the *pupil*. The iris adjusts itself to changing light levels to let the correct amount of light through to the *crystalline lens* which is immediately behind the iris. The lens is a complex elastic membrane about 9 mm in diameter and 4 mm in thickness. The refractive index varies from 1.41 in the centre to 1.39 at the outer parts. The crystalline lens changes its shape by means of the muscles which support it. This provides a fine focusing mechanism, called *accommodation*, so that an external object forms a sharp image on the retina. When the muscles are relaxed, the lens is flattish and little bending of rays occurs. In this state, distant objects are in focus. When the muscles supporting the lens are in tension, the lens becomes more curved and objects close to the eye are in focus. This is the state of the eye for close vision and reading. The lens provides the second and final refraction for incident light rays, thus the eye consists of a double lens system.

The crystalline lens is followed by a relatively large chamber filled with *vitreous humor* with a refractive index of 1.34. After traversing this chamber the light falls on the *retina*, which is made up of an array of photoreceptor cells called *rods* and *cones*. These absorb photons and are linked to nerve cells which transmit the visual images to the *optic nerve*. The photons are absorbed in the rods and cones by electrochemical reactions. Behind the retina is a non-reflecting coating which absorbs any photons not absorbed by the rods and cones. This is the case in the human eye. In animals which hunt at night, such as the cat, the coating is reflective so that light has a second chance of being detected by the rods and cones. This effectively gives cats double the sensitivity to low light levels, by comparison with humans.

The shape of the rods and cones are described by their name. The rods are long and thin and there are over 100 million distributed unevenly across the retina, with more towards the outer edges. The rods respond only to the intensity of the incident light and cannot distinguish the wavelength of the light. They provide the eye's sensitive to dim light. There are about 10 million cones clustered near to the centre of the retina, they are thicker and shorter than the rods. They are not very sensitive to the intensity of light but can distinguish between different wavelengths so that the cones are responsible for the perception of colours. A sketch of a cone is shown in Figure 9.24. There is a spherical focussing lens at the entrance followed by a tapered cone. The refractive index varies from 1.36 at the start of the cone to 1.39 at the end of the cone. Thus the difference between the refractive indices of the cone and the surrounding vitreous humor is very small. In this respect it is very similar to an optical fibre waveguide (which will be described in Chapter 13). More correctly the optical waveguide is similar to the eye, since the eye evolved long before optical fibres.

The cones do not have the same sensitivity to all wavelengths. They have a spectral response curve which peaks in the yellow part and is shown in Figure 9.25. This is the average standard response of the eye as determined by the International Commission on Illumination. It is an important curve because it relates physical intensity levels of light to those perceived by the eye. The absolute measure of light is done by *radiometry*, which is the same as the measure of radiation in any part of the electromagnetic spectrum. The perceived response of the eye to light is measured by *photometry*. Figure 9.25 provides the link between radio metric quantities and photometric quantities. The cones in the retina do not respond to low light levels, so the ability to distinguish colours decreases

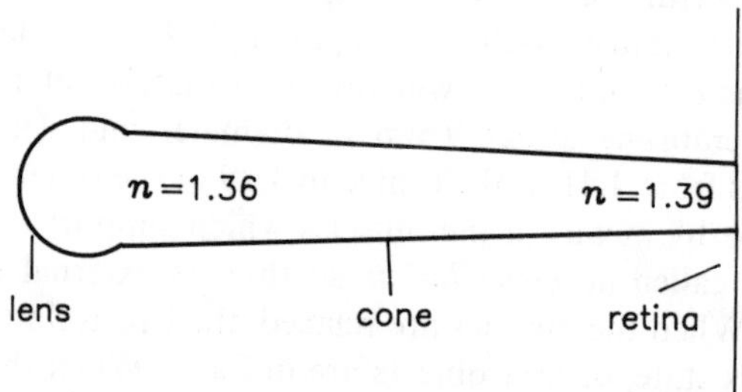

Figure 9.24. Cross-section of a cone in the eye

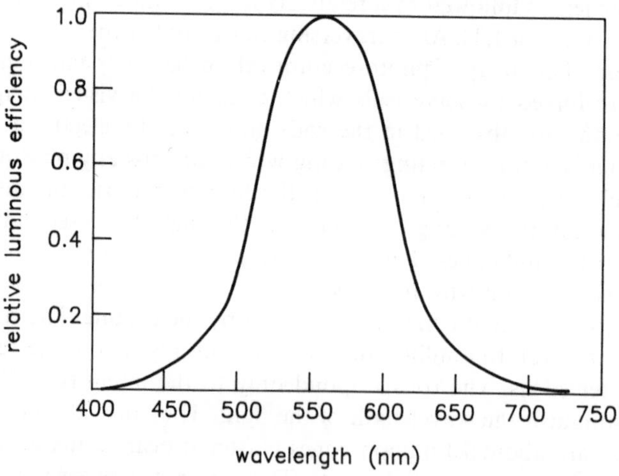

Figure 9.25. Standard spectral response of the eye

as the incident light level decreases. Rods also have a spectral response curve, even though they cannot distinguish colour. The curve for rods has the same shape as the curve for cones, but moves slightly to a shorter wavelength and peaks at 500 nm instead of 555 nm.

As an optical system, the eye works extremely well, but the cornea lens is not perfect and does not always produce sharp images on the retina. There are three main errors which lead to the defects of short-sightedness, long-sightedness and astigmatism.

Myopia is the correct name for near-sightedness or short-sightedness. It occurs when parallel rays entering the eye are brought to a focus in front of the retina. Only objects which are near to the eye will be focused correctly. Objects further away will appear blurred. Myopia can be corrected with a negative lens which diverges the rays by a small amount, Figure 9.26. The object will then be in focus for all incoming rays. The accommodation of the lens in the eye will keep objects close to the eye in focus. Opticians measure the power of lenses by the *dioptric power, D*, of a lens. This is the reciprocal of the focal length in metres, so a focal length of 0.5 m has $D = 2$. This is convenient because the lens makers equations, equation (9.30) contains the reciprocal of the focal length. In terms of dioptric power it becomes

$$D = (n-1)\left(\frac{1}{R_1} + \frac{1}{R_2}\right) \tag{9.42}$$

Similarly the Gaussian lens equation, equation (9.31) becomes

$$D = \frac{1}{z_1} + \frac{1}{z_2} \tag{9.43}$$

where z_1 and z_2 are the object and image distances. These can be used to design a correcting lens for myopia. The focal point at which the image is in focus when the eye

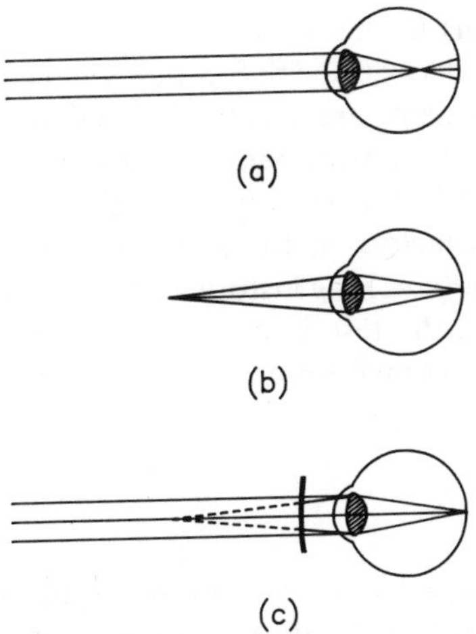

Figure 9.26. Myopia and its correction. (a) Image at infinity is focused before the retina. (b) Near point where image is focused on the retina. (c) Correcting lens so that image at infinity is focused on the retina

is relaxed will be $-z_2$ and it is required to refocus this to $z_1 = \infty$. For instance if $z_2 = -2\,\text{m}$, then equation (9.43) gives $D = -0.5$ diopters. This simple calculation assumes that the correcting lens is immediately in front of the cornea, so applies to contact lenses. For spectacles, a small correction needs to be introduced to allow for the separation of the lens and the cornea.

Hyperopia is the opposite of myopia and is far-sightedness or long-sightedness. The cornea causes parallel rays to focus behind the retina. Clearly hyperopia needs to be corrected with a positive lens that causes the rays to converge.

Astigmatism is caused by the cornea having an uneven curvature to its surface, so that rays in different planes have different focal points. If the incoming light was linearly polarised, then it would be equivalent to the horizontal and vertical polarisations having different focal points. Regular astigmatism is when the two planes are perpendicular, as in the polarisation analogy. This can be corrected by a lens which has a small cylindrical shape to its surface. If the astigmatism is not in two orthogonal planes, then it is more difficult to correct.

9.10 INCOHERENT OPTICAL SYSTEMS—CAMERAS, TELESCOPES AND MICROSCOPES

There are a large number of practical ray optical systems, meaning collections of optical components which are positioned in series in order to perform some useful function. It is possible to divide the applications into those using incoherent beams of light and those using coherent optical waves. The latter are dependent on the laser as the source of optical plane waves and will be described in Section 9.12. This section is devoted to the applications using incoherent bundles of light, and brief descriptions of the optical transmission system through cameras, telescopes and microscopes will be given. Spectacles to correct defects in the eye have already been described in the last section. The following section will then discuss applications using LEDs as the incoherent light source.

Incoherent optical systems in this section have a number of general characteristics. The main source of optical energy is the Sun, although, the stars and artificial incoherent source are also used. The systems are usually designed to image an extended source in other words to look at some physical object. This indicates that the imaging is going to be performed by the eye which is usually the final detector in the system. There may be intermediate artificial detectors in the system to aid imaging, for instance video cameras. The optical systems have long histories, although recent advances in technology have considerably improved the resolution and sensitivity. For instance the telescope was invented centuries ago, yet modern optical glasses and lens designs, have improved the ability to identify distant objects.

CAMERAS

Camera used to mean *photographic cameras*, in which the image of a subject is recorded on photographic film. Camera now also means *video cameras* in which the image of a subject is recorded by either a phototube or, more commonly, by an array of solid state photo conductive cells. The optical action of either a photographic camera or a video camera is the same. In both cases a combination of lenses is designed to focus an image over a rectangular focal plane. The principles of a camera are very simple and can be seen from a pinhole camera, Figure 9.27. This consists of a very small hole through which rays of light pass. The pinhole means that every point on the image plane is illuminated by rays which come from approximately the equivalent point of the object. As Figure 9.27 shows, the image is inverted and this has remained the practice on all cameras. The pinhole camera is very simple and has the advantage that all objects at any distance are in focus on the image plane because there are no optical focussing elements. In camera terms this means that the *depth of field* is unlimited. The disadvantage, which makes it impractical, is that the pinhole allows very little light energy to pass through it. Thus it only works in high light levels or with very long exposure times.

The use of a larger aperture and a convex focusing lens solves the disadvantages and gives a bright, sharp, image from an object. The distance between the lens and the image is now fixed for a given object distance. A focusing mechanism must thus be provided to change the lens–image distance so that objects at different distances can be sharply imaged. An important quantity is a measure of the amount of light incident on the photographic plate. The power density will depend on two factors; (1) it will be proportional to the area of the aperture at the lens and (2) it will be inversely proportional to the area of the image plane. The larger the image area, the more light will be required to provide the same power density for a fixed aperture. The photographic plate is always by convention rectangular, but the lens aperture is circular, so the image area should be measured by the circular cross-section, which is proportional to the square of the diameter. The power density received is thus

$$P \propto \frac{\text{area of aperture}}{\text{area of image}} \propto \frac{D^2}{d^2} \quad \text{W m}^{-2} \tag{9.44}$$

where D is the diameter of the aperture and d is the diameter of the image, which is proportional to the focal length, f, of the lens, hence

$$P \propto \left(\frac{D}{f}\right)^2 \tag{9.45}$$

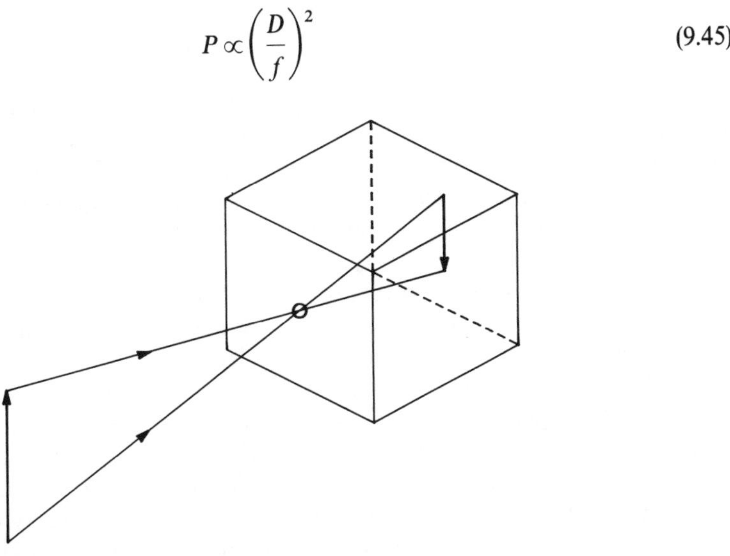

Figure 9.27. Principle of a pinhole camera

The quantity D/f thus measures the amount of light incident on a photographic plate or solid state array. The inverse of D/f is the relative aperture size of the lens. It is called the *f-number*

$$f\text{-number} = \frac{f}{D} \tag{9.46}$$

Unlike most other technical units, the f-number is expressed with the symbols $f/$ in front of the relative aperture size. Thus an aperture which has a focal length to diameter ratio of 16 is written as $f/16$. Cameras are designed with variable irises so that the aperture size and hence f-number can be varied. The larger the aperture diameter, the smaller the f-number. The iris is designed to change in steps which double, or half, the power density on the image for each step. Equation (9.45) shows that the f-number must therefore change in $\sqrt{2}$ steps. Increasing the power levels by factors of two from unity gives f-number of $f/1.4$, $f/2$, $f/2.8$, $f/4$, $f/5.6$, $f/8$, $f/11$, $f/16$. The weakest light levels will be recorded with the largest aperture size, so camera lenses are specified by their smallest f-number. Solid state video cameras are more sensitive than photographic films, so the lens can be smaller in size and more compact.

A camera with a single lens will only produce a sharp image if the aperture is small. If it is not small, the aberrations in the lens will produce an image which is not sharp over the whole area. In addition the sharpness must occur over a wide angular field of view so that scenes near or distant can be photographed. The greater the flexibility of the camera, the more complex the optical task which the lens has to perform. For this reason, modern cameras usually have multiple lenses to provide a wide range of operating conditions. The design of a modern camera lens is a sophisticated task and computer analysis with ray tracing is used to optimise designs. The most complicated type of camera is the single lens reflex (SLR) with a zoom lens that can remain automatically in focus over a wide range of object distances. An SLR incorporates a plane mirror and a penta prism so that the image to be recorded can be viewed though the lens.

TELESCOPES AND BINOCULARS

Telescopes are designed to produce magnified images on the retina of the eye of an object which is a long distance away. In astronomical telescopes, the objects are the planets or the stars. In terrestrial telescopes, the object distance are anything from a few metres to the horizon. There are two types of telescopes. Those using lenses which are called *refracting telescopes* and those using mirrors which are called *reflecting telescopes.*

Refracting telescopes consist of two sets of lenses, Figure 9.28. Each of the lenses might actually be composed of more than one lens. The light from a distant object arrives as nearly parallel rays at the *objective* lens which has a large aperture to collect as much power as possible. The objective is a positive convex lens which forms a real image at its focal plane. This is viewed by the second lens which is called an *eyepiece* lens and has the function of magnifying the image and producing nearly parallel light rays. The eye is placed near to the eyepiece, which has about the same aperture diameter as the pupil of the eye. The eye appears to see an image at infinity but magnified. The magnification, M, is given by the ratio of the incoming angle α_0 and the angle of the rays at the eyepiece, α_e. The geometry of the telescope shows that this ratio is inversely

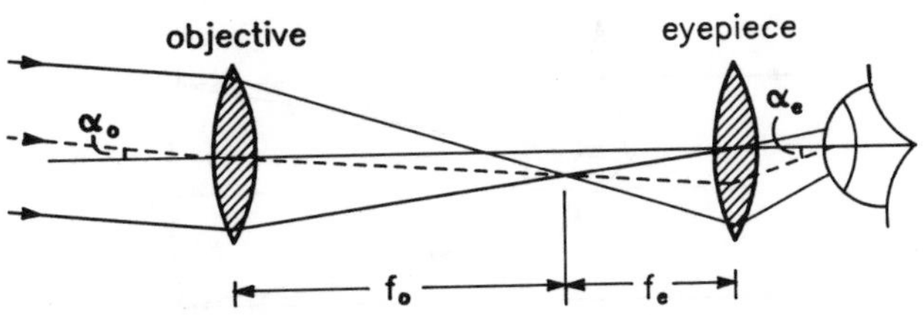

Figure 9.28. Lens system in a refracting telescope.

proportional to the focal lengths, so

$$M = \frac{\alpha_e}{\alpha_0} = -\frac{f_0}{f_e} \tag{9.47}$$

The magnification can be increased by increasing the focal length of the objective lens. This will clearly increase the length of the telescope.

The telescope shown in Figure 9.28 produces an inverted image at the eye. This does not matter much for an astronomical telescope, but is unacceptable for a terrestrial telescope. The image can be corrected by inserting an inverting lens in between the objective lens and the eyepiece. In *binoculars*, the inverting is done with a pair of Porro prisms (see Section 9.3). Binoculars are simply folded telescopes in which the optical length is maintained by the double prisms with the added advantage that there are an optical system for each eye. Binoculars are customarily specified with two number, say 8×30. The first number gives the magnification and the second number gives the diameter of the objective lens in millimetres, which indicates the amount of optical power captured. Since the magnification is also the ratio of the diameter of the objective lens to the diameter of the eyepiece, the ratio of the second number to the first number gives the diameter of the eyepiece. In the case of an 8×30 pair of binoculars, the eyepiece has a diameter of 3.75 mm.

The larger the objective lens in a telescope, the greater the resolution and the greater the light gathering power. Large diameter lenses are, however, bulky, difficult to support, difficult to make accurately and the glass can contain inhomogeneities. This is where the other type of telescope, the *reflecting telescope* can be used. Large reflectors or mirrors can be made accurately by grinding and coating glass or other material. When the main mirror is paraboloidal in shape, a near perfect images will be formed at the focal point of the paraboloid. If the focal length is long the paraboloidal mirror can be replaced by a spherical mirror, as shown in Section 9.5. Mirrors do not have problems with distortions due aberrations because there is no refraction. They also have the important advantage that they operate at any wavelength above the wavelength which causes the surface to appear rough.

The image in a reflecting telescopes is formed at the focal plane of the paraboloid which is in front of the mirror. This means that the beam must be diverted so that it can easily be viewed. The first person to invent a scheme for doing this was Newton who inserted a plane mirror into the reflected beam, Figure 9.29(a). This does not change the magnification. Both Gregory and Cassegrain realised that the use of a second conic section mirror would increase the magnification. The Cassegrain telescope, Figure 9.29(b), uses a convex hyperboloidal mirror to give a secondary focus behind the main para-

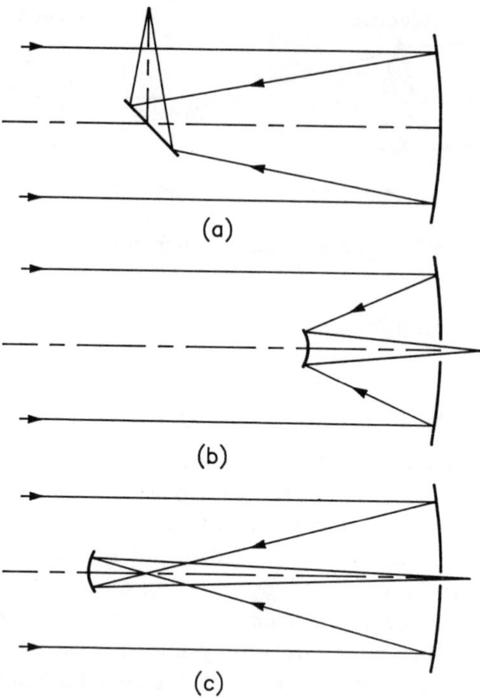

Figure 9.29. Reflecting telescopes. (a) Newtonian with plane mirror. (b) Cassegrain with convex hyperboloidal mirror. (c) Gregorian with concave ellipsoidal mirror

boloidal mirror. The Gregorian telescope, Figure 9.29(c), does the same task by using a concave ellipsoidal mirror. The Cassegrain is more popular because the secondary mirror is easier to support and is widely used for high magnification systems. It is also widely used at microwave frequencies and at millimetrewaves for both radio telescopes and communication antennas.

MICROSCOPES

The microscope is designed to magnify the image of a small object so that it can be viewed by the eye. The basis of the microscope is a simple *magnifier*, Figure 9.30, which is a positive convex lens. The nearest focus distance of the average eye is assumed to be 250 mm. If a small object at this distance is seen by the eye, it will make an angle α_0 at the eye. Inserting a magnifying lens and moving the object to the focal distance of the lens, s_0, will form a virtual image which subtends a larger angle α_M at the eye. The angular magnification, M, is then defined as the ratio of α_M to α_0. In the paraxial region, $\alpha_M = h/250$ and $\alpha_0 = h/s$. s is equal to the focal length of the lens f, when the image is viewed at infinity, so

$$M = \frac{250}{f} \tag{9.48}$$

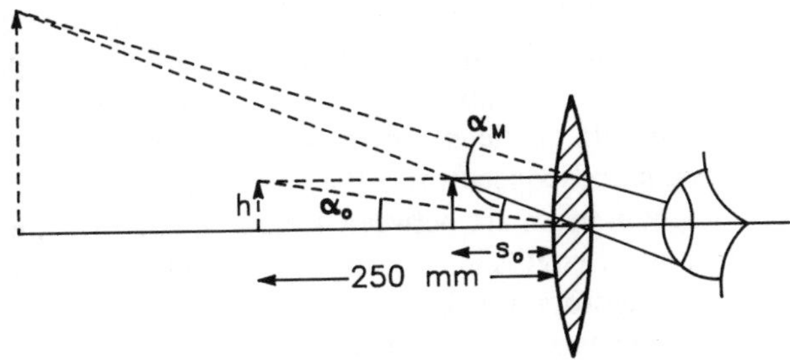

Figure 9.30. Optics of a simple magnifier

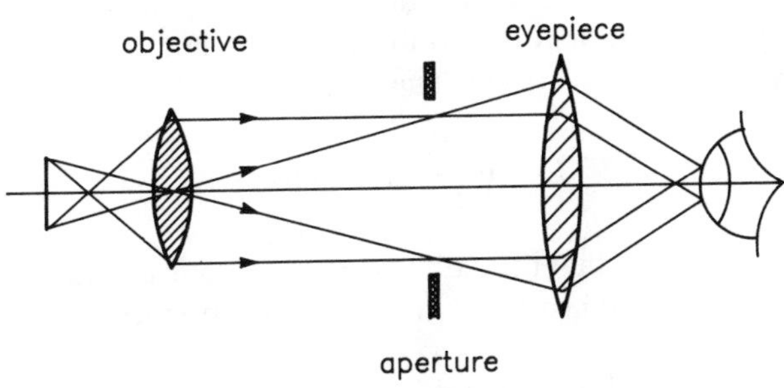

Figure 9.31. Optics of a compound microscope

where f is in millimetres. A magnification of five times thus requires a lens with a focal length of 50 mm.

The *compound microscope* increases the magnification of the simple magnifier by adding a second positive lens which, in the notation of the telescope, is the objective lens, Figure 9.31. This has a short focal length so that the object is close to but just outside the focal length of the lens. The objective lens forms a real image inside the microscope tube, between the two lenses. This is seen by the eyepiece lens which operates as a magnifier. An aperture is placed at the intermediate focal plane so that a sharp boundary is seen by the observer. The magnification is given by equation (9.48) with f replaced by the combined focal length of the two lenses separated by the intermediate distance.

9.11 LED TRANSMISSION SYSTEMS

The light emitting diode is a compact solid state source of incoherent light which operates with low voltages and low currents. The photodiode and phototransistor are similarly compact solid state detectors. These devices have given rise to a large number of short distance communication and signalling applications.

In principle any artificial source of incoherent light would be suitable, but most of the other types described in Section 9.8 are relatively large, inefficient and require large power supplies. It was only with the large scale production of cheap LEDs, photodiodes and phototransistors that applications started to emerge which could benefit from small sources and detectors. Both the LED and the photodiode are usually packaged for this type of application with an integral plastic lens, Figure 9.32, which is typically between 2 and 5 mm in diameter. As with any lens, the shape of the surface changes the radiation pattern of the source beam and concentrates the power in the forward direction. The subtended angles for the main beam are between 40° and 80°. LEDs are available which produce optical power at wavelengths in the range 550 nm (green) to 1000 nm (infrared). Photodiodes and phototransistors are more sensitive in the 800 to 1000 nm range of wavelengths so infrared systems are preferred where power has to be conserved.

An LED free space communication or signalling system uses a transmitter and a receiver in much the same way as any of the communication systems described in Chapter 7. Figure 9.32 shows a schematic system. Spherical waves are radiated by the LED which become plane waves after a short distance. These are picked up by the detector. There will be a path loss due to the spreading out of the source power in the same way as with a microwave system. This is given by,

$$L_p = \left(\frac{\lambda}{4\pi R}\right)^2 \quad \text{Nepers} \tag{9.49}$$

The short wavelengths means that the path loss is very significant and only short distance systems are feasible. For instance $R = 1$ m at $\lambda = 800$ nm gives $L_p = 4 \times 10^{-15}$ Nepers or -144 dB.

The range of applications is considerable and some are listed below:

- Card and tape readers
- End of tape sensors
- Optical tachometers
- Optical shaft position encoders
- Bar code scanners
- Liquid level monitors
- Smoke detectors
- Remote control links

The first four all involve very short links which are designed merely to detect the presence or absence of optical power from the source. This signal is then used as the input to a control system to determine position or input data. Bar code readers are in widespread use in shops to provide stock control and sales data. In this case the source and detector are adjacent and the detector receives on or off signals by reflection off the black and

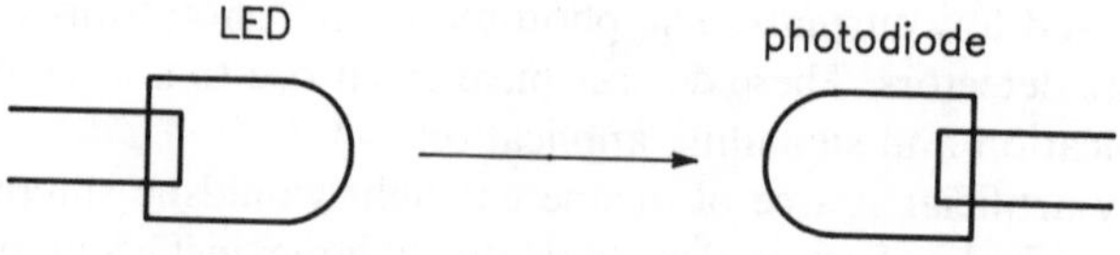

Figure 9.32. Transmission system with an LED and photodiode incorporating integral lens

white bars of the bar code. A code is used which minimises the chances of reading errors. Liquid level detectors work on the same reflective principle as bar code readers to detect the amount of liquid in a container.

Smoke detectors use an infrared source and detector placed a short distance apart. The normal situation is that power is continually received by the detector. If smoke is present it will considerably increase the path loss so that the reduction in received signal level can be detected and an alarm sounded.

Short distance infrared remote control links are widely used for controlling domestic audio and video equipment. They are also used where a clear line of sight path exists to send data or sound without the need for interconnecting wires. Infrared wavelengths are used in order to get the maximum sensitivity from the detectors. As mentioned above, the path loss over even a short distance is significant and needs to be minimised. On the other hand the signal bandwidth and data rate are usually low so that a low signal-to-noise ratio is adequate. The power budget for a remote control link can be evaluated in the same manner as described for communication systems in Chapter 7. The source power and the real gains of the transmitting and receiving lenses are added together (in dBs). The path loss is subtracted from this quantity to obtain the received power. This must be greater than the signal-to-noise ratio if the system is to function as specified.

9.12 LASER TRANSMISSION SYSTEMS

The invention of the laser provided the first, and still the only source of coherent optical plane waves. It is used in many applications from material processing to medicine and from printers to communications. Applications where the power is guided along optical fibres will be described in Chapter 13. Here the interest is in laser beam systems utilising the theory in earlier sections of the book.

The laser as a source was described in Section 9.8 and the laser beam as a source of plane waves was described in Section 5.14. In that section, it was noted that the cross-section of a laser beam has a Gaussian distribution described by equation (5.40)

$$E_y = A(z)e^{(-r^2/w^2)} \sin(\omega t - \beta z) \tag{9.50}$$

where r is the radial variable and w the radius of the *beam waste* of the laser beam which is typically 0.3 mm. w is defined as the radius where the electric field has fallen to $1/e$ of the value on axis.

The beam diameter is physically small but large in wavelengths and it can be thought of as being made up of a large number of parallel optical rays with amplitudes which fit the Gaussian envelope pattern. The Gaussian distribution is as the result of the distribution of power across the laser cavity and lens at the exit to the laser.

The lens at the exit of the laser is designed to focus the beam to infinity, which according to the geometric optics theory of lenses would mean that a parallel, non-divergent beam is radiated. In practice this is impossible and geometric optics does not describe the true situation. This must be done with physical optics or with antenna theory which then shows that the beam must always diverge. Similarly geometric optics predicts that rays from a lens can be focussed to a point. If this were correct then the power density at the focal spot would be infinite. Thus there is a minimum beam diameter to which the laser energy can be focused. Laser applications divide into those requiring a narrow radiated beam and those requiring a focused spot so the beam divergence and the minimum focusing diameter are important for laser applications.

A schematic of a beam emerging from a laser and passing through two lenses is shown in Figure 9.33. The cross-section increases and decreases smoothly, but everywhere maintains the Gaussian radial profile of equation (9.50) with the factor $A(z)$ describing the variation of the amplitude in the z direction. The beam waste is at the minimum cross-section or the focus point of the first lens. The second lens is geometric optically focused at infinity which means that the physical optics beam waste is at the lens. After the second lens, the beam diameter diverges continuously. The radiation pattern of the laser beam is a plot of power against normalised angle and is sketched for a Gaussian beam in Figure 9.34. The horizontal axis has units of $(D/\lambda)\sin\theta$ where D is the diameter of the beam waste at the lens, $(D = 2w)$. For the narrow beams radiated by lasers, $(D/\lambda)\sin\theta$ approximates well to $D\theta/\lambda$. The narrowness of the laser beam is inversely proportional to the normalised diameter, D/λ. A convenient measure is the beamwidth between the half power points, which is called the *beam divergence angle*, ϕ, and for a laser beam is given approximately by

$$\phi = \frac{1.3\lambda}{D} \tag{9.51}$$

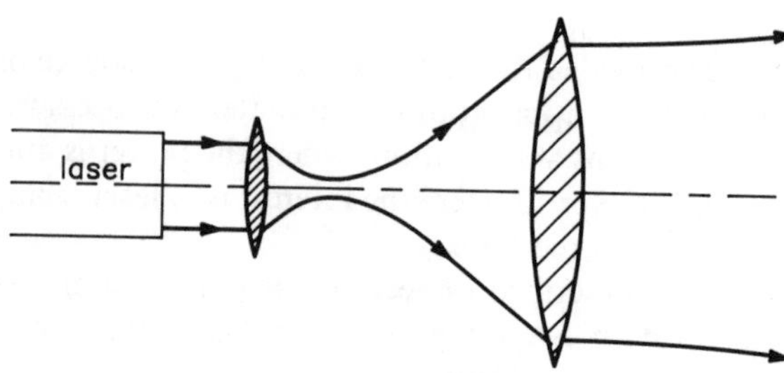

Figure 9.33 Beam from a laser passing through two focusing lenses

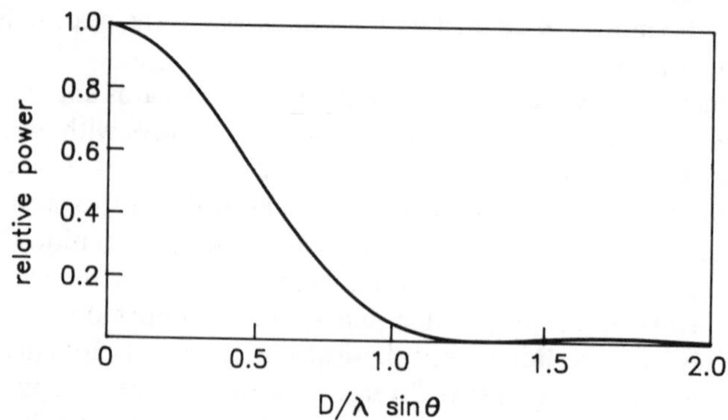

Figure 9.34. Radiation characteristics of a Gaussian laser beam

The beam does not completely fill the lens so the beam divergence angle is a function of the beam diameter. If the beam was larger than the lens, then D would be the diameter of the lens. A typical value for D is 6 mm, so a He–Ne laser at 632.8 nm has $\phi = 1.4 \times 10^{-4}$ radians or 0.5 arcminutes. This means that after 1 m, the beam has a half power diameter of 0.14 mm. The narrowness of the beam waste can be improved to some extent by better laser design, but other factors come into play, such as lens imperfections, which limit the beam angle.

The minimum diameter to which the laser beam can be focussed is called the *diffraction limited spot diameter, d*, and is a function only of the wavelength. It can be determined by a complete physical optics analysis, but conveniently is given by

$$d \approx \lambda \tag{9.52}$$

This is the best that can be achieved and any imperfections in the laser or focusing optics will increase the diameter. Focused laser beams are used for cutting of materials by concentrating power onto a small spot. The power density can be very high. For instance a carbon dioxide laser at 10.6 μm which radiates 80 W of CW power, could be concentrated to a spot diameter of 10.6 μm. If the power was uniformly distributed over that diameter, and there were no losses, the power density would be nearly 10^7 W/mm^2.

The applications for lasers are so numerous that only a few can be briefly described in order to give an indication of the potential for optical plane wave systems. *Material processing* uses high power focused laser beams for cutting, welding, drilling and heat treatment. The advantages are partly the high power densities which give quick results and high efficiencies, but also the sharp definition of the laser beam. This means that high accuracy can be achieved for precision drilling and cutting. A typical system focuses the output from a pulsed laser with peak outputs in the kW range using a lens. Gas is blown into the beam so as cool the operating spot and to keep cut material away from the beam. Cutting speeds of 100 mm/s can typically be achieved when cutting 2 mm thick steel. Holes can be accurately drilled and positioned in the metal. Plastics can be welded using lower power lasers. The flexibility of the high power laser means that thousands of laser are used in industry daily for material processing.

Lasers have proved to by an excellent diagnostic and therapeutic tool in *medicine.* Again it is the fine diameter of the laser beam and the precision with which it is positioned which are the main advantages. There are also other benefits. The power levels can be controlled precisely and because different body cells react to different wavelengths in predictable ways this can be used to benefit. Blue–green argon laser light at 488 nm is selectively absorbed by red blood cells, so that it can be penetrate the upper layers of the skin without harm, but coagulate blood deeper in the body. Water is very lossy in the infrared region so CO_2 laser light at 10.6 μm is almost totally absorbed. Since muscles and tissues contain a high percentage of water, a laser can be used to produce a clean incision. The added benefit occurs that the laser beam cauterises the flesh so that the cutting is almost bloodless. Lasers with wavelengths from ultraviolet to infrared are used for specific applications. The none visible light lasers often have a subsidiary visible light laser attached which produces a low power beam coincident with the high power beam. This is in order that the beam can be seen and more easily handled. Laser beams, which can be sent down optical fibres, give an additional degree of flexibility. These enable investigations or surgery to be carried out deep in the body.

There are a range of applications using mechanically scanned laser beams for information processing. Examples are photocopiers, point of sale bar code readers and laser printers. In all cases a laser beam with a small divergence angle is generated so that precision pointing and scanning are possible. The typical distances involved are only tens of centimetres which means that a beam can be directed to within a few microns of the intended position.

The laser printer is representative of this class of applications. It is a development of the photocopier, Figure 9.35, and uses a modulated laser beam to scan a photo conductive drum. The laser light selectively discharges areas of the drum. Carbon toner with the opposite polarity sticks to these areas. The drum rotates onto the paper and the toner is transferred and heat sealed so that the required image is formed on the paper. Rotationally oscillating mirrors are used to move the beam so that rapid scanning of the drum is achieved. There are usually a number of optical components in the laser beam for focussing and to produce the correct image on the drum. Laser printers produce high quality output from computers at speeds from four pages per minute to two pages per second, depending on the sophistication of the scanning and copying process. This is achieved with minimum acoustic noise so that it is ideal for the office environment.

Laser communications is mainly by means of optical fibre waveguides. Free space communication links have been tried but atmospheric turbulence is a major problem. Even on a clear day the gases in the atmosphere are in continual motion. These are sufficient to mean that, except for distances of a few metres, a narrow laser beam wanders randomly away from the receiver. The air movements in urban areas makes this beam movement even worse. Permanent communication links are very difficult and unsatisfactory. There are however good possibilities for laser communication links in *space*. The absence of the air means that the laser beam is stable and can be used to communicate between space craft or satellites. They are proposed for inter-satellite links particularly between low orbit satellites and geostationary relay satellites. The standard power budget calculations would be used to design the link. The path loss would be very high, but this is counteracted by the narrow beam divergence angle which effectively means a high gain antenna.

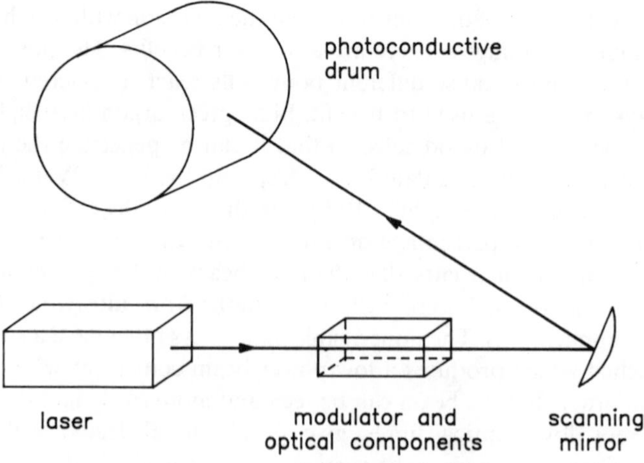

Figure 9.35. Laser optical system in a photocopier

Lasers are used for optical data storage and retrieval because very high packing densities can be achieved. The need for large amounts of permanent or semi-permanent computer data storage have given a big boost to this area of applications. The favourite medium for storage is a rotating disc, because writing and reading heads can be easily moved radially across the surface of the disc whilst the disc rotates. The data is written to the disc by a laser which modifies the surface of the disc. The data are read from the disc using reflected laser light. Optical data discs can hold gigabytes of data.

COMPACT AUDIO DISC SYSTEM

The compact audio disc system is a good example of optical transmission technology. Compact discs (CDs) were introduced in 1982 and have been one of the most successful domestic electronic products in terms of the number of sales in a short period of time. This is a tribute to the ability to mass produce a complex system at a low price, because the CD player contains a very sophisticated optical system as well as using sophisticated data processing. The optical system is worth describing because it illustrates the use of many of the optical components which have been described in this chapter, as well as implementing the basic theory of plane waves.

Compact disc players use semiconductor AlGaAs lasers which produce an output of about 5 mW at a wavelength of 790 nm. This is just outside the visible spectrum. This wavelength is fixed and is crucial to the design of the data storage on the disc. The CD is a digital data storage system using 16 bit words to represent sounds. The digital data is stored as data pits on the disc, which are made with a laser cutter. Each pit is 127 nm deep and oval in shape, about 500 nm wide and the length and the spacing between pits on a track varies between 0.8 μm and 3 μm. Tracks are 1.6 μm apart which means that on a standard 120 mm diameter disc there are about 16 billion bits of data. Only about a third of these are used for audio data, the remainder are used for error correction and synchronisation.

A cross-section of a compact disc is shown in Figure 9.36. It is 1.2 mm thick and made of polycarbonate with a refractive index of 1.55. The laser beam shines on the bottom surface, through the disc and is reflected off a reflective layer which is on the top surface. The plastic disc helps the focussing of the laser beam because the rays are refracted towards the normal meaning that the beam is more parallel as it travels through the disc. The wavelength in the disc is 790/1.55 = 510 nm. When the optical waves arrive at the top surface, they encounter either a pit or a plane surface. The pits are quarter

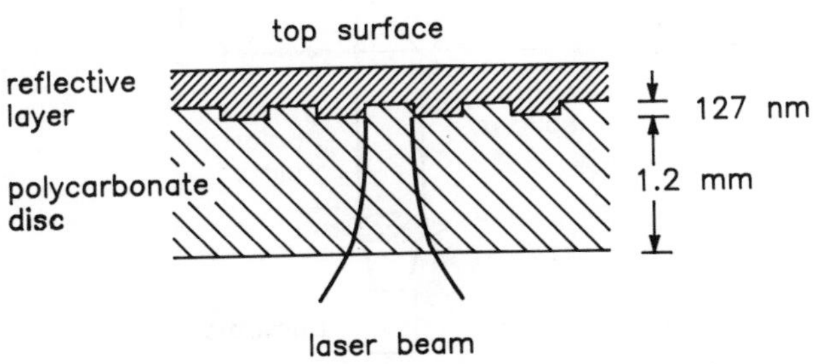

Figure 9.36. Cross-section of a compact disc

of a wavelength deep, i.e. 127 nm, which means that after reflection off the top of a pit, the wave has travelled an extra half wavelength by comparison with the waves reflected off the plane surface. This causes complete cancellation of the waves incident on a pit and represents a '0'. The absence of a pit represents a '1'. This is a much better way of ensuring that the presence of a pit is read than any mechanical method.

The basic optical system of a CD player is sketched in Figure 9.37. The optical beam from the laser is linearly polarised and passes through a collimating lens onto a polarisation beam splitter. This consists of two birefringent prisms back to back with a dielectric spacer in between, as described in Section 9.3. Optical waves with the linear polarisation of the source pass straight through the prism, whilst reflected waves with the orthogonal polarisation are reflected through 90° towards the detector. The transmitted beam next encounters a polarising plate which twists the polarisation vector through 45°. A plane mirror bends the beam through a right angle before it encounters an objective lens which focusses the waves onto the disc. Waves reflected off the disc pass through the polarising plate again so that the result of the double pass through the plate is a polarisation rotation of 90°, as required so that the beam splitter will send the reflected beam to the detector. This will pass through another lens before being incident on a photodiode.

The optical system described above is the part that is needed to read the data off the disc. Two other parts are also needed. Firstly, there must be a dynamic focussing system for the objective lens next to the disc and secondly there must be a control system which keeps the focussed beam onto a track. There are various schemes for doing these control functions. The signal needed to control the objective lens and ensure that the laser beam is focussed onto the disc can be obtained by using a cylindrical lens which produces an elliptical shaped beam. The detector is then a quadrant of photodiodes. The sum of the outputs gives the signal information, the individual diode outputs gives control information. The control data needed to keep the beam onto a track can be obtained by inserting a diffraction grating into the beam. This splits the single beam into a main beam and a series of lower level beams on either side. The beams adjacent to the main

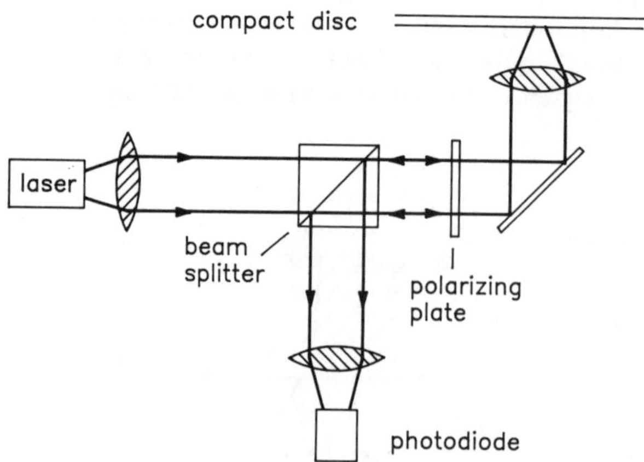

Figure 9.37. Outline of the optical system in a compact disc player

beam are used for tracking. These pass through the optics and should hit the disc so that the three beams are aligned with a track. If the alignment is incorrect the tracking beams in front of, and behind, the main beam will not see data pits and no signal will be recorded from them. This can be used to make an auto-track mechanism. It is clear from this brief description of the CD player, that a complicated optical system is used. The CD player also needs considerable electronic processing capability, only a part of which is concerned with the audio signals which are the final output.

PROBLEMS

1. Use Fermat's principle to derive the geometric optics law of reflection at a boundary.

2. Three layers of glass have refractive indices of 1.4, 1.5 and 1.6. A ray is incident on the layer with the highest index at an angle of 60°. Find the optical path length in each layer and obtain the position at which the ray leaves the layers of glass.

3. The paraxial approximation can be used for rays arriving at a narrow angle with respect to the optical axis. If the paraxial values must be within 5% of the true values, what is the maximum angle at which the paraxial approximation can be used?

4. A wave is incident on the centre of one surface of a glass prism with internal angle 60° and $n = 1.5$. Trace the ray paths taken by rays which enter at angles of incidence of 22.5° and 45°.

5. Plot a curve of angular deviation against angle of incidence for a prism with internal angles of 60° and $n = 1.55$.

6. A right angle prism is made of fused silica glass with $n = 1.456$ at $\lambda = 656$ nm (red), $n = 1.459$ at $\lambda = 588$ (yellow) and $n = 1.463$ at $\lambda = 486$ nm (blue). White light is incident on one of the shorter faces of the prism at the angle for minimum deviation of the red light. If the point of incidence is 20 mm from the apex find the distance apart of the three colours on the exit face of the prism.

7. Derive a relationship between the incident angle, prism angle and angular deviation for a 60° reflection prism. Hence show that reflection prisms do not suffer from dispersion.

8. Find the range of incident angles over which a ray incident on one of the shorter faces of a Porro prism (with $\alpha = 90°$) is reflected back in the same direction as the incident ray.

9. A ray of light is incident on the broad face of a right angle Porro prism with $n = 1.5$. Find the incident angle at which *all* the power is reflected back in the direction of the incident ray.

10. Confirm the polarisation directions shown for the prisms in Figure 9.8.

11. Plane mirrors can be used to amplify rotational motion of a shaft by reflecting light off a mirror attached to the shaft. If a shaft rotates by 10 arcminutes and a laser beam is reflected off the mirror, by how much does the beam move at a distance of 300 mm from the shaft?

12. A spherical mirror has a radius of curvature of 20 mm and a diameter of 40 mm. Find

the difference in the focal length between a ray incident along the axis and a ray incident parallel to the axis but on the edge of the mirror.

13. Show that a spherical mirror will give a magnification of $M = R/(2s_0 + R)$ where R is the radius of curvature and s_0 is the object distance.

14. Design the shape and focal length of a mirror to magnify by two times the height of an object which is 50 mm from the mirror.

15. Derive the refraction equation at a single spherical surface (equation (9.24)) using Fermat's principle.

16. If the object and image distances of a thin lens minus the focal length are z_0 and z_i (i.e. $z_0 = s_0 - f$), show that the lens maker's equation can be expressed as $z_0 z_i = f^2$.

17. Determine the focal length in air of a planar–convex lens which has a radius of curvature of 50 mm and is made of glass with $n = 1.50$.

18. Design a planar–concave lens to have a focal length of 100 mm with glass of $n = 1.50$. Find the image distance and magnifications for objects at infinity, 500 mm, 200 mm, 100 mm, 75 mm and 50 mm. Compare the results with those of Example 9.3.

19. Calculate the focal length of a thin bi-convex lens with $n = 1.55$ which has radii of curvature of 100 mm and 200 mm. Find the image position and magnification for an object which is 20 mm from the lens.

20. A planar–convex lens has a radius of curvature of 100 mm and is made of the fused silica glass specified in Problem 6. Find the amount of axial movement of the image for an object at infinity between red light and blue light.

21. The planar–convex lens designed in Example 9.3 is placed in water with $n = 1.33$. What will happen to the image distances and magnifications for the specified object distances?

22. Confirm that the ray matrix equations for a thin lens is as given in equation (9.36).

23. Derive the ray matrix for a spherical mirror.

24. Find the combined focal length of two identical thin lenses with focal lengths of 50 mm separated by a distance of 20 mm, using ray matrices.

25. Find the focal length of a thick bi-convex lens with $n = 1.55$, radii of curvature of 100 mm and 200 mm and thickness 30 mm.

26. Find the focal length of the cornea of the eye assuming that it can be treated as a single spherical surface of radius of curvature of 8 mm. If the eye is in water ($n = 1.33$) how much refraction takes place?

27. Calculate the focal length of the crystalline lens in the eye assuming that it can be treated as a thin lens of refractive index 1.40 in a medium of refractive index 1.34. The front surface has a radius of curvature of 10 mm and the back surface of 6 mm.

28. Find the dioptric power of the thin lens required to correct myopia in a person whose object distance for focus on the retina is 1 m instead of infinity.

29. The power received by a camera is proportional to the time of exposure and inversely proportional to the f-number. If a photograph of a moving object is correctly exposed, but blurred, when taken for 1/60th second at $f/16$, what must be f-number be if an exposure of 1/250th second is required to freeze the movement?

30. A camera uses a convex lens of focal length 100 mm. What is the size of the image of a person who is 1.8 m tall and standing 20 m away from the camera?

31. An astronomical telescope has an objective lens of focal length 200 mm and an eyepiece of focal length 50 mm. The telescope looks at the Moon which is 380 000 km distant and has a diameter of 3480 km. Find the image size on the eye assuming that the virtual image distance of the eyepiece is 250 mm.

32. An infrared remote control unit operates at 1 μm with an LED of output power 4 mW. The LED radiates uniformly over a hemisphere. Find the received power over a distance of 3 m.

33. An optical communication link is to be used to communicate between two geostationary satellite positioned at 0° W and 120° W. The wavelength of operation is 800 nm. What source EIRP is needed if the input power at the receiving lens must be at least -180 dBW?

34. The diameter of the laser beam from a ruby laser ($\lambda = 694$ nm) at its beam waste is 6 mm. The laser has a peak output power of 100 W. Find the beam diameter and power density at a distance of 10 km.

10 Guided waves

10.1 INTRODUCTION

The study of microwave and optical transmission now turns to methods of guiding the waves along a fixed path between conductors or along dielectrics. The general name for the structure which guides the waves is a *waveguide*. The fields inside the waveguide are confined in directions transverse to the direction of travel. In the direction of travel the waves behave in the same manner as plane waves in free space which have been the subject of earlier chapters. This chapter describes the basic principles of all waveguides by studying the parallel plate waveguide and using this waveguide to understand the basic properties of the waves which are guided along the waveguide.

At microwave frequencies the waveguides are usually systems of open or closed metal conductors. At optical frequencies, conductors are impractical but dielectric waveguides make excellent guiding structures. These are based on the principles of total internal reflection studied in Chapter 6. A metal waveguide confines the waves to the space between the conductors and the guiding system forms a transmission line for carrying signals between two points. The earlier chapters were concerned with waves travelling in free space which are used for free space communications. A free space system of plane waves wastes the electromagnetic spectrum since only a few signals can be transmitted at any one frequency. Waveguides confine the signals so that a large number of waveguides can be laid close to each other without the signals in one waveguide interacting with the signals in another waveguide.

All microwave and optical systems will include some length of waveguide. Optical communications are based on the cylindrical optical fibre waveguide and there will be shorter lengths of rectangular-shaped dielectric waveguide in the transmitter and receiver, Figure 10.1. The history of optical waveguides is quite short and started with the discovery of the possibilities for long distance transmission with optical fibres in the 1960s. At that time the loss in available glasses was very high but intense research soon produced extremely low loss glasses which now enable economic, long distance optical waveguides to be manufactured. One of the attractions of the optical waveguide is the potentially very large bandwidth available in a very small diameter. This together with the development of solid state optical sources and detectors soon led to the growth of optoelectronics as a subject in which optical signals are processed similar to electronic signals. For this purpose a variety of shapes of waveguide, usually rectangular are used. All the optical waveguides are characterised by the use of at least two dielectric media. One of the media acts as the guiding or core structure, whilst the other acts as a surrounding or cladding structure. Optical waveguides will be studied in Chapter 13.

Most microwave waveguides are used over relatively short distances, but even for free space systems there will be sections of waveguide between the source and the transmitting antenna and between the receiving antenna and the detector. Microwave waveguides have a variety of shapes, Figure 10.2. They include the common flexible coaxial cable, but the losses in the coax become quite high as frequency is increased so

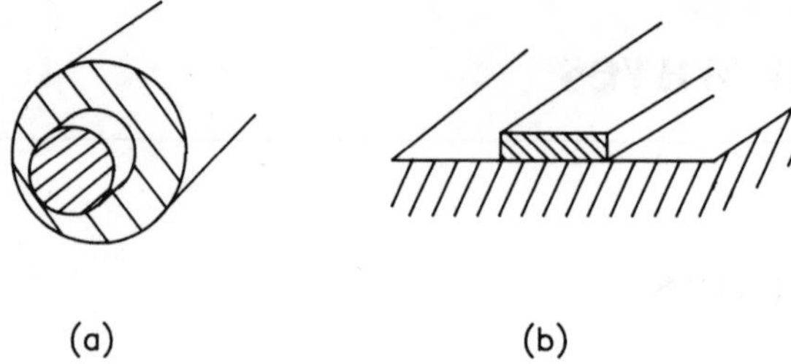

Figure 10.1. Types of optical waveguides. (a) Optical fibre..(b) Dielectric slab waveguide

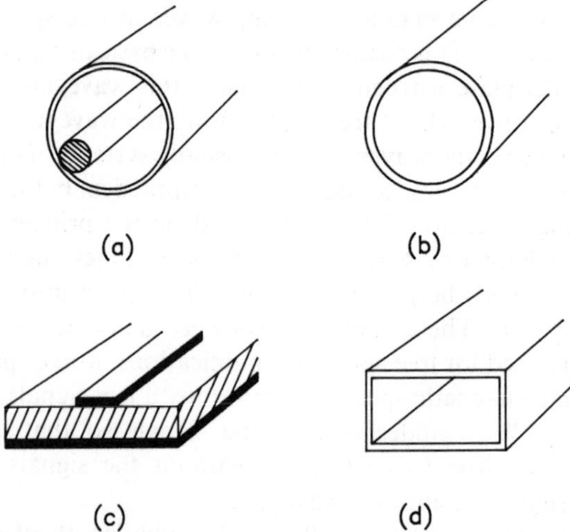

Figure 10.2. Types of microwave waveguide. (a) Coaxial cable. (b) Circular waveguide. (c) Microstrip. (d) Rectangular waveguide

other methods are preferred. The first microwave waveguides were always hollow rectangular or circular metal tubes in which the waves travel along the hollow tube. These were developed in the 1930s for use with microwave radar and communication systems. For many years these, together with the coaxial waveguide, were the only guiding systems used. They have low attenuation and good polarisation properties but they are often bulky and expensive to make due the high precision required. In the 1970s, a new type of waveguide, called *microstrip* emerged. This basically consists of two flat conductors with a dielectric in between the conductors. The advantage of microstrip is its small size and relative simplicity and it has become one of the main types of microwave waveguide. The details of the practical types of microwave waveguide will be described in Chapter 12.

10.2 PRINCIPLES OF GUIDED WAVES

The simplest type of guided wave is a constrained plane wave. This can be demonstrated by considering a plane wave which is travelling in free space, Figure 10.3. The electric

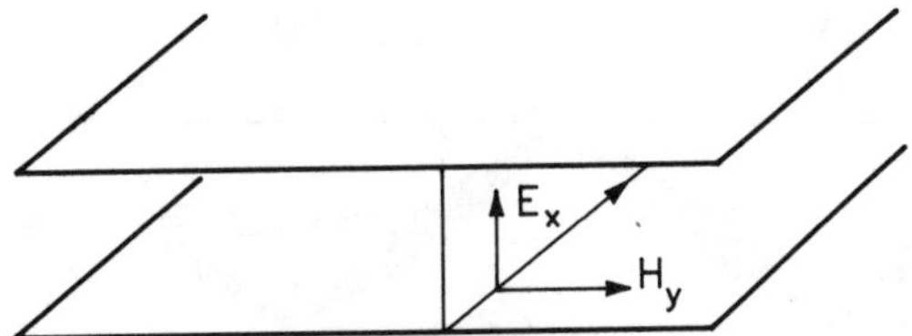

Figure 10.3. Plane wave in free space devolving into a guided wave between parallel plates

field is normal to the plates so that the associated magnetic field is parallel to the plates. The wave is travelling in the z direction, which is into the paper in Figure 10.3. It is possible to bring two parallel conducting plates together with a small gap between the plates, and the plane wave can still exist and will be guided by the plates. The boundary conditions imposed by the conducting plates do not influence the fields because they are normal to electric field. It therefore follows that the plane wave is the E_x and H_y wave. The other linear polarised plane wave with E_y and H_x cannot be supported by the parallel plates because the tangential electric field must be zero on the conducting plates.

The fields of the basic guided wave are the same as for a plane wave:

$$E_x = E_0 \exp[j(\omega t - \beta z)] \tag{10.1}$$

$$H_y = \frac{E_0}{Z_0} \exp[j(\omega t - \beta z)] \tag{10.2}$$

where $\beta = 2\pi/\lambda$, λ is the free space wavelength and Z_0 the wave impedance, $(\mu_0/\varepsilon_0)^{1/2}$ or 377 Ω. This basic wave which is guided by the parallel plates is called a *Transverse ElectroMagnetic Mode or TEM mode.* The word MODE is used to describe a wave which travels along a waveguide. Unlike the free space waves, there can be many modes travelling along a waveguide at a single frequency. Each mode has a different electric and magnetic field configuration and in an ideal waveguide there will be no interaction between different modes. The following sections will show how the modes arise theoretically. The TEM mode is the fundamental type of guided wave and is also the one most used in practice. It is the simplest type of guided waves between plates but enables all the main features of waveguides to be deduced.

The possibility of other modes travelling along the parallel plate waveguide can be demonstrated physically by sketching the picture of rays in the waveguide. Figure 10.4 shows a section through the waveguide along the direction of travel. The TEM mode is represented by a shaped line, or ray, along the axis. The conducting plates suggest the possibility of waves being reflected off the plates so that the rays follow the zig-zag path shown in the figure. In order for this to happen continuously along the waveguide, the electric fields must always be zero at the points of reflection on the plates. This implies that there must be an integer number of half wavelengths between reflection points. In Figure 10.4 one half wavelength is shown between reflection points by a shaped line. There could also be two or three or any integer number of half wavelengths. All these options form modes which are waves that satisfy the boundary conditions. At a fixed frequency, the angle that the rays makes with the z direction will be fixed and determined by the wavelength and the distance apart of the plates.

The physical picture of modes demonstrated in Figure 10.4 does not depend on the

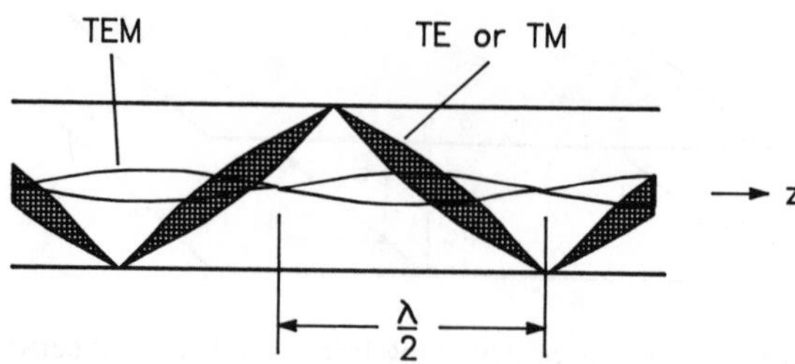

Figure 10.4. Rays travelling along waveguide showing reflections off walls

polarisation of the plane waves which are reflected off the conducting plates. The waves can be perpendicular polarised or parallel polarised. This shows that there will be double the number of modes suggested in the ray picture because for each integer number of half wavelengths between plates there can be a perpendicular polarised wave and a parallel polarised wave. In waveguides, these two polarised waves are given new names. The perpendicular polarised wave has electric fields which are always transverse to the z direction. There is no E_z electric field. These modes are therefore called *Transverse Electric or TE modes.* The modes with parallel polarised waves are conversely called *Transverse Magnetic or TM modes.* These are arbitrary definitions but are universally used in electromagnetic wave studies.

The ray picture discussed above used two parallel conducting plates as the guiding mechanism. The study of reflections at dielectric boundaries in Chapter 6 indicates that exactly the same ray picture will occur if the conducting plates are replaced by dielectric boundaries and total internal reflection takes place. For this to happen a dielectric sheet with a higher refractive index than the surrounding medium is required. Similar guiding action can then be expected. This is the principle of operation of the dielectric, or optical, waveguide. There are however differences from the conducting plate waveguide. The boundary conditions at the dielectric walls state that the total tangential electric field is non-zero in the waveguide, but has some other value. This means that there will not be an exact number of half wavelengths between the reflection points. In addition the need for total internal reflection imposes constraints determined by the relative permittivities or refractive indices of the waveguide and surrounding medium.

As the conducting plate waveguide is simpler to analyze than the dielectric sheet waveguide, the parallel plate waveguide will be used in the rest of this chapter to describe the properties of waveguides. The general principles also apply to dielectric waveguides as will be seen in Chapter 13.

10.3 PROPERTIES OF MODES IN PARALLEL PLATE WAVEGUIDE

In this section the parallel plate waveguide with two conducting plates as shown in Figure 10.5, will be analysed. The starting points for a study of the behaviour of modes are,

(a) the electromagnetic wave equation,

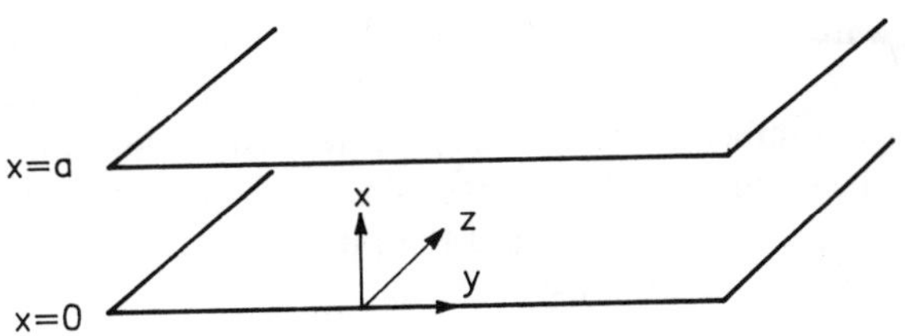

Figure 10.5. Parallel plate waveguide

(b) Maxwell's equations,

(c) the boundary conditions.

The procedure that was used in Chapter 5 to study plane waves will be reproduced for the case of guided waves. The analysis will incorporate all the possible modes which can exist in the waveguide. The electromagnetic wave equation gives the basic form of the solution for waves in the waveguide. Maxwell's equations give the fields which can exist in the waveguide and the boundary conditions give the precise values of the propagation constants and the components of the electric and magnetic fields.

The geometry to be analysed is shown in Figure 10.5 with one of the plates along the yz plane and the other plate at $x = a$. In the plane wave case, the restriction was made that the wave had no variation in the x or y directions which meant that $\partial/\partial x = \partial/\partial y = 0$ and this considerably simplified the electromagnetic equations. For the parallel plate guide, the fields are confined in the x direction, so $\partial/\partial x \neq 0$. The assumption will be made that the plates are infinite in the y direction so that $\partial/\partial y = 0$. This is clearly, an assumption which does not agree with practice, but it turns out to be a reasonable approximation.

The electromagnetic wave equation for the time periodic case with a time dependence $\exp(\mathrm{j}\omega t)$ and air between the plates is

$$\nabla^2 E_x + \omega^2 \mu_0 \varepsilon_0 E_x = 0 \tag{10.3}$$

In component form and applying the assumption that $\partial/\partial y = 0$, equation (10.3) becomes

$$\frac{\partial^2 E_x}{\partial x^2} + \frac{\partial^2 E_x}{\partial z^2} + \omega^2 \mu_0 \varepsilon_0 E_x = 0 \tag{10.4}$$

The waves travel in the z direction with the same form as for plane waves in free space. The z dependence will therefore, be $\exp(-\mathrm{j}\beta z)$ where the negative sign indicates that the wave is travelling in the positive z direction. This assumption will be used in all waveguide studies. It means that $\partial^2 E_x/\partial z^2 = -\beta^2 E_x$ and equation (10.4) becomes

$$\frac{\partial^2 E_x}{\partial x^2} + (\omega^2 \mu_0 \varepsilon_0 - \beta^2) E_x = 0 \tag{10.5}$$

The term in brackets appears in all waveguide analysis and is called the *wavenumber* and given the symbol K_c, i.e.

$$\boxed{K_c^2 = \omega^2 \mu_0 \varepsilon_0 - \beta^2} \tag{10.6}$$

For plane waves in free space and for the TEM mode in waveguides the wavenumber is zero. This is because $\omega^2\mu_0\varepsilon_0 = (\omega/c)^2 = (2\pi/\lambda)^2 = \beta^2$. In general, however, for waveguides, the wavenumber is finite. It has a value which depends on the geometry of the waveguide and the frequency of operation. The value of K_c gives a unique relationship between frequency (represented by ω) and the phase constant (β).

Equation (10.5) can be rewritten as

$$\frac{\partial^2 E_x}{\partial x^2} + K_c^2 E_x = 0 \tag{10.7}$$

A general solution to this equation will have the form

$$E_x = [A\cos(K_c x) + B\sin(K_c x)]\exp[\mathrm{j}(\omega t - \beta z)] \tag{10.8}$$

where A and B are constants to be determined from the boundary conditions. The appropriate boundary condition states that the tangential electric field is zero on the two conducting plates. E_x is not, however, tangential to the plates so there must be at least one more electric field component in order that the boundary conditions can be satisfied. To see which component exists, Maxwell's equations will be expanded using the assumptions that $\partial/\partial y = 0$ and $\partial/\partial z = -j\beta$. The first Maxwell equation gives

$$\nabla \times \boldsymbol{E} = -\mathrm{j}\omega\mu_0 \boldsymbol{H}$$

$$\mathrm{j}\beta E_y = -\mathrm{j}\omega\mu_0 H_x \tag{10.9a}$$

$$-\mathrm{j}\beta E_x - \frac{\partial E_z}{\partial x} = -\mathrm{j}\omega\mu_0 H_y \tag{10.9b}$$

$$\frac{\partial E_y}{\partial x} = -\mathrm{j}\omega\mu_0 H_z \tag{10.9c}$$

The second Maxwell equation gives

$$\nabla \times \boldsymbol{H} = \mathrm{j}\omega\varepsilon_0 \boldsymbol{E}$$

$$\mathrm{j}\beta H_y = \mathrm{j}\omega\varepsilon_0 E_x \tag{10.10a}$$

$$-\mathrm{j}\beta H_x - \frac{\partial H_z}{\partial x} = \mathrm{j}\omega\varepsilon_0 E_y \tag{10.10b}$$

$$\frac{\partial H_y}{\partial x} = \mathrm{j}\omega\varepsilon_0 E_z \tag{10.10c}$$

Examination of these components shows that

(a) These are two independent solutions just as for plane waves.

(b) One of these solutions has components E_x, H_y and E_z. So by comparison with plane waves there is an additional component, E_z, which is in the direction of travel. The wave is no longer completely transverse to the direction of propagation. This solution is the *transverse magnetic mode* or TM mode.

(c) The other solution has components E_y, H_x and H_z and is the *transverse electric mode* or TE mode.

The E_x wave is part of the TM mode and the component which is tangential to the plates will be the E_z field. Using equations (10.10a) and (10.10c)

$$E_z = \frac{1}{j\beta}\frac{\partial E_x}{\partial x} \tag{10.11}$$

so that inserting equation (10.11) into equation (10.8) gives

$$E_z = \frac{K_c}{j\beta}[-A\sin(K_c x) + B\cos(K_c x)]\exp[j(\omega t - \beta z)] \tag{10.12}$$

Applying the boundary condition that $E_z = 0$ at $x = 0$ gives that the coefficient $B = 0$ and the E_x field of equation (10.8) simplifies to

$$E_x = A\cos(K_c x)\exp[j(\omega t - \beta z)] \tag{10.13}$$

The other boundary condition is that $E_z = 0$ at $x = a$, so that

$$0 = -\frac{AK_c}{j\beta}\sin(K_c a) \tag{10.14}$$

This equation is satisfied if

$$K_c a = n\pi \tag{10.15}$$

and determines the relationship between ω and β.

10.4 PROPAGATION CHARACTERISTICS AND CUT-OFF FREQUENCY

Equation (10.15) indicates in principle that the integer n can take on any value from $n = 0$ to $n = \infty$. There are thus an infinite number of TM modes which are designated as TM_n modes where n is the *order* of the mode. If $n = 0$, $E_z = 0$ so that the only field components are E_x and H_y and the result is the basic guided wave of equation (10.1), which is designated the transverse electromagnetic mode, thus the TEM mode is the same as the TM_0 mode.

If the transverse electric or TE modes were analysed using the same procedure as in the last section, the same relationship between ω and β given by equation (10.15) would be found to exist, with one important difference. If $n = 0$ the fields vanish, as explained in Section 10.2. Thus the lowest order TE mode is the TE_1 mode. Combining equations (10.6) and (10.15) gives

$$\left(\frac{\omega a}{c}\right)^2 - (\beta a)^2 = (n\pi)^2 \tag{10.16}$$

This is called the *propagation equation.* A sketch of equation (10.16) is shown in, Figure 10.6. This is called an $\omega - \beta$ diagram and shows the *propagation characteristics*

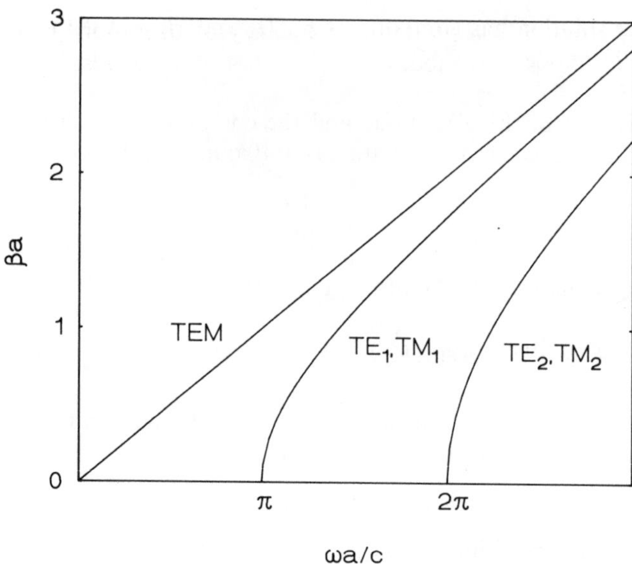

Figure 10.6. Normalised propagation curves for parallel plate waveguide

for TM or TE modes. By plotting βa against $\omega a/c$, the graph applies to all separation distances, a, between the plates. It shows the first five modes. The TM_0 mode propagates from DC and exists at all frequencies. The TM_n and TE_n modes ($n \neq 0$) are on top of each other. Modes other than the TEM mode all start to propagate at a fixed frequency which is called the *cut-off frequency*. Below this frequency, the particular mode does not exist. Above the cut-off frequency the mode will propagate for all frequencies. Thus if the waveguide is used at a frequency below $\omega a/c = \pi$ only one mode exists. For frequencies between $\omega a/c = \pi$ and $\omega a/c = 2\pi$, three modes can exist and so on. This general behaviour of the propagation characteristics appears in all types of waveguides, although the exact form varies from type to type, depending on the boundary conditions.

The cut-off frequency, f_c, is given by equation (10.16) with $\beta = 0$, that is

$$f_c = \frac{cn}{2a} \tag{10.17}$$

This can also be expressed in terms of the cut-off wavelength λ_c

$$\lambda_c = \frac{2a}{n} \tag{10.18}$$

For the TM_1 or TE_1 modes, $a = \lambda_c/2$, and one half wavelength will just fit between the plates. For the TM_2 or TE_2 modes, one wavelength will fit between the plates at cut-off.

In a practical microwave waveguide the modes are excited by a probe which is designed to be matched to the field configuration for the desired mode. Any discontinuities in the waveguide will, however, tend to excite other modes if they are capable of propagating. For this reason it is desirable to restrict all the power to just one mode. This is easiest to achieve by restricting the maximum frequency of operation to below the cut-off frequency of the TM_1 or TE_1 modes. This is not an onerous restriction because of the large amount of bandwidth available at microwave frequencies, as can be demonstrated with an example.

Example 10.1 Cut-off frequencies for parallel plate waveguide

Find the cut-off frequencies for a parallel plate waveguide spaced 5 mm apart containing a dielectric with $\varepsilon_r = 2.5$.

The analysis performed above was done for air between the parallel plates. It needs to be modified for a finite dielectric. This is straightforward and propagation equation, equation (10.16), becomes

$$\left(\frac{\omega a}{c}\right)^2 \varepsilon_r - (\beta a)^2 = (n\pi)^2 \tag{10.19}$$

The cut-off frequencies are thus

$$f_c = \frac{cn}{2a\varepsilon_r^{1/2}} \tag{10.20}$$

The cut-off frequencies are lower in the dielectric by comparison with the air. For $a = 5$ mm and $\varepsilon_r = 2.5$, equation (10.20) gives the cut-off frequency for the TM_1 and TE_1 modes as 18.97 GHz and for the TM_2 and TE_2 modes as 37.95 GHz, thus restricting operation to just the TEM mode, still gives nearly 19 GHz of bandwidth, which is much larger than would normally be needed in practice.

Below the cut-off frequency, the modes become evanescent. A solution can be found for this situation by starting from the assumption that the waves have a z dependence $\exp(-\alpha z)$ where α is the attenuation constant. The result will be waves which do not travel along the waveguide, but are very rapidly attenuated. For most practical purposes, the evanescent modes can be ignored.

10.5 GUIDED WAVES VIEWED AS TWO INTERACTING PLANE WAVES

It is possible to view the fields between parallel plates as made up of two interacting plane waves. This provides useful physical insight into their behaviour. In order to develop this view, the solution for the E_x wave in equation (10.13) will be rewritten in terms of components in the direction of travel, the z direction and components in the transverse, or x, direction. This can be done by writing the $\cos(K_c x)$ term in exponential form:

$$\begin{aligned} E_x &= A[\exp(jK_c x) + j\exp(-jK_c x]\exp[j(\omega t - \beta z)] \\ &= A\{\exp[j(K_c x - \beta z)] + \exp[-j(K_c x + \beta z)]\}\, e^{j\omega t} \end{aligned} \tag{10.21}$$

The first exponential term in the curly brackets describes a plane wave which is travelling in the negative x direction and the positive z direction. The second term describes a plane wave which is travelling in the positive x direction and the positive z direction. Each term thus describes a plane wave which is travelling at an angle to the z axis of $\theta = \pm \tan^{-1}(K_c/\beta)$. The ray picture of the two plane waves is sketched in Figure 10.7 against distance in the z direction. The two waves are continuously reflected off the waveguide walls. This is the description which was used qualitatively in Section 10.2 to describe the physical principles of guided waves.

The same result can be obtained by starting with two plane waves in free space. If a sketch is drawn of the wavefronts of two plane waves travelling in space at oblique angles to each other, the result is shown in Figure 10.8. The solid lines represent peaks of the wave and the dotted lines represent troughs of the waves. At points where the solid lines and the dotted lines cross, the fields will cancel and the resultant field will be zero. These points lie along horizontal lines. If conducting sheets are placed along any of the horizontal lines, the fields will not be disturbed and the boundary conditions along the conductor will be satisfied. Then the fields between two plates will be the same as that derived for the parallel plate waveguide.

The wavefronts shown in Figure 10.8 and the rays shown in Figure 10.7 are orthogonal to each other, thus if the angle, θ, is large so that the rays bounce rapidly between the plates, the wavefronts are widely spaced out along the axis of the waveguide. At cut-off, $\beta = 0$ and $\theta = \tan^{-1}(\infty) = 90°$. In this case the rays do not travel along the waveguide, but are reflected back and forth between the plates. At very high frequencies, $\beta \Rightarrow \infty$ and $\theta = 0°$. The waves are now travelling along the z axis. The contrasting picture of rays near to cut-off and at very high frequencies is shown in Figure 10.9. For increasing intermediate frequencies between cut-off and very high frequency, the waves gradually change from rapid reflection with distance to very few reflections.

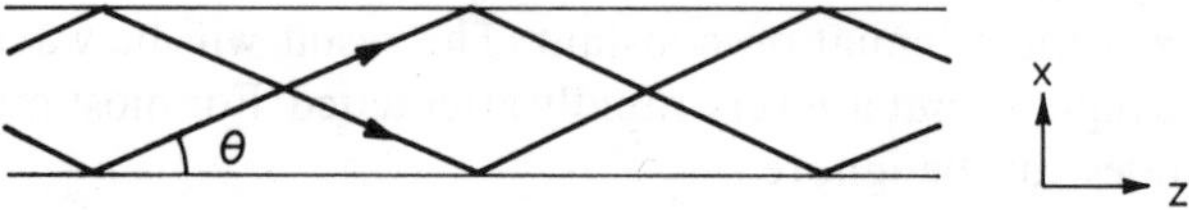

Figure 10.7. Rays for two interacting waves along a parallel plate waveguide

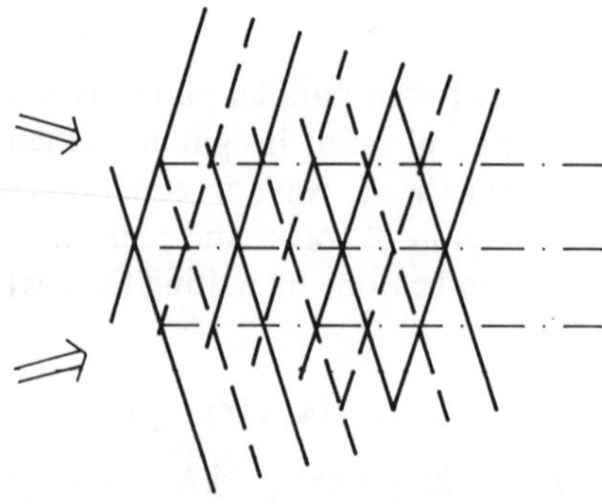

Figure 10.8. Interaction of two plane waves in free space showing the wavefronts. ——— peaks of waves, ___ troughs of waves

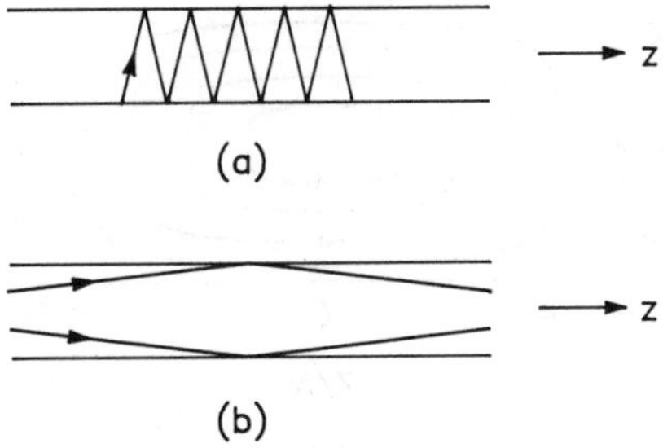

Figure 10.9. (a) Rays near cut off in waveguide, (b) Rays at high frequencies

10.6 FIELDS OF TEM, TM AND TE MODES

The field components for the TM mode are derived from equations (10.9) and (10.10) as

$$E_x = A\cos\left(\frac{n\pi x}{a}\right)\exp[\mathrm{j}(\omega t - \beta z)] \tag{10.22a}$$

$$H_y = \frac{\omega\varepsilon_0}{\beta}A\cos\left(\frac{n\pi x}{a}\right)\exp[\mathrm{j}(\omega t - \beta z)] \tag{10.22b}$$

$$E_z = \frac{-K_c}{\mathrm{j}\beta}A\sin\left(\frac{n\pi x}{a}\right)\exp[\mathrm{j}(\omega t - \beta z)] \tag{10.22c}$$

$E_z = 0$ for the TEM mode and the field components are equations (10.22a) and (10.22b) with $n = 0$. These are the same as equation (10.1) since $\omega\varepsilon_0/\beta = f\lambda\varepsilon_0 = \varepsilon_0/(\varepsilon_0\mu_0)^{1/2} = 1/Z_0$. Figure 10.10 sketches the contour lines for the electric field for the TM_1 mode in the xz plane. The half loops of electric field come from the combination of the $\exp(-\mathrm{j}\beta z)$ dependence in the axial direction and the $\cos(\pi x/a)$ or $\sin(\pi x/a)$ dependence in the transverse direction. The electric field at the plates is only in the x direction whilst the electric field on the axis is only along the axial direction. This looping behaviour is characteristic of waveguide modes. The associated magnetic field has a single component, H_y, and forms lines at right angles to the plane of the paper in Figure 10.10. The magnetic field is tangential to the plates so the magnetic field boundary condition indicates that

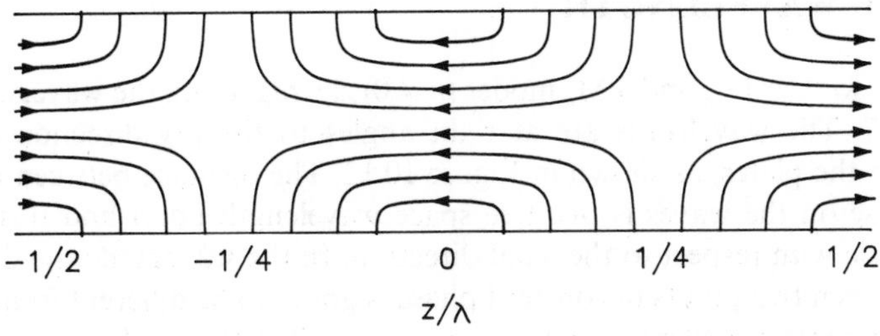

Figure 10.10. Electric field lines of TM_1 mode between plates of waveguide

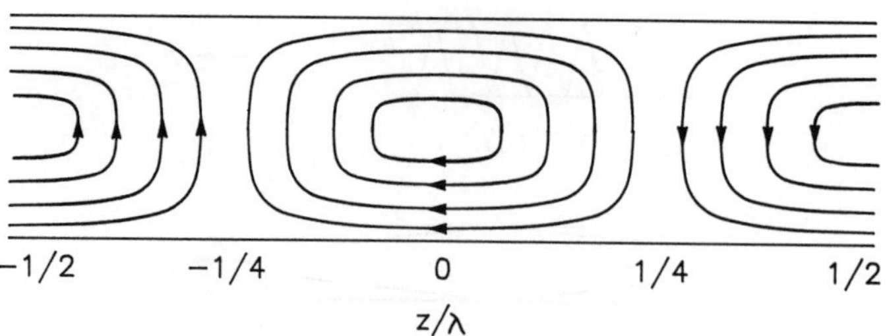

Figure 10.11. Magnetic field loops of TE_1 mode between plates of waveguide.

there will be a current induced in the plates with a value $J_z = H_y$. This is along the direction of travel and along the waveguide.

The fields for the transverse electric modes can be obtained by the same procedure as used in Section 10.3 to derive the fields for TM modes. TE modes have E_y, H_x and H_z field components given by

$$E_y = A \sin\left(\frac{n\pi x}{a}\right) \exp\left[\mathrm{j}(\omega t - \beta z)\right] \tag{10.23a}$$

$$H_x = \frac{-\beta}{\omega\mu_0} A \sin\left(\frac{n\pi x}{a}\right) \exp\left[\mathrm{j}(\omega t - \beta z)\right] \tag{10.23b}$$

$$H_z = \frac{-K_c}{\mathrm{j}\omega\mu_0} A \cos\left(\frac{n\pi x}{a}\right) \exp\left[\mathrm{j}(\omega t - \beta z)\right] \tag{10.23c}$$

The cut-off relationships for TE modes are identical to those for TM modes. Examination of equation (10.23) indicates that if $n = 0$, both E_y and H_x are zero and therefore there is no solution. The lowest order mode in the TE mode set is the TE_1 mode and there is no analogy of the TEM mode.

In contrast to the TM modes, there are two magnetic field components and one electric field component for TE modes. This means that the equivalent contour sketch of the fields of the TE_1 mode has to show the magnetic field lines, Figure 10.11. These form loops with the peak value along the axis of the waveguide. The associated electric field forms lines at right angles to the plane of the paper. The current in the conducting plates is due to H_z as the only tangential magnetic field component. This induces a current $J_x = H_z$ which is at right angles to the direction of travel.

10.7 GUIDE WAVELENGTH

The ray paths for the TE_n and TM_n modes ($n \neq 0$) zig-zag along the waveguide as shown in Figure 10.7. The wavefronts are at right angles to the ray direction and are also reflected from the plates, as shown in Figure 10.12. The distance between two points of constant phase on the waves is one free space wavelength apart and this is measured at a finite angle with respect to the axial direction. In the waveguide the distance along the z axis between two points of constant phase is going to be different from, and greater than, the free space wavelength. This distance is called the *guide wavelength*, λ_g. The guide wavelength is the distance over which the phase changes by 2π. The phase

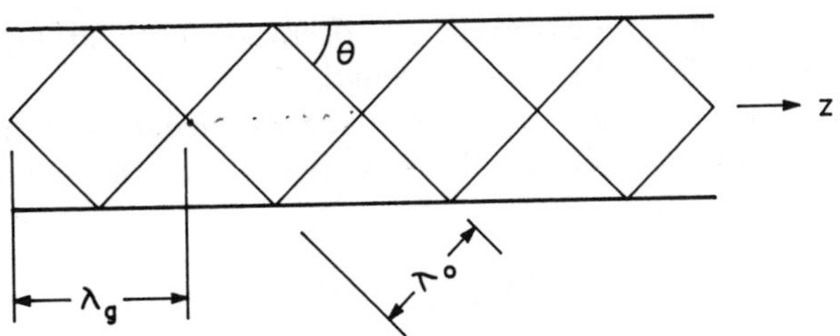

Figure 10.12. Wavefronts along parallel plate waveguide

coefficient is β, therefore

$$\lambda_g = \frac{2\pi}{\beta} \tag{10.24}$$

The guide wavelength is important in practical waveguides because it is a quantity which can be measured in much the same way as the free space wavelength can be measured in air. The easiest way is to put an obstacle in the path of the waves so that some of the signal is reflected. A standing wave pattern will be established in exactly the same way as described for plane waves in Chapter 6. A small probe is inserted between the plates and the distance apart of two nulls in the standing wave pattern is measured. This will be half the guide wavelength, or $\lambda_g/2$. The probe should be small enough that it samples the fields, but does not perturb them.

The variation of guide wavelength for a mode is shown in Figure 10.13. At the cut-off frequency, $\beta = 0$ and the guide wavelength is infinite. As the frequency increases above

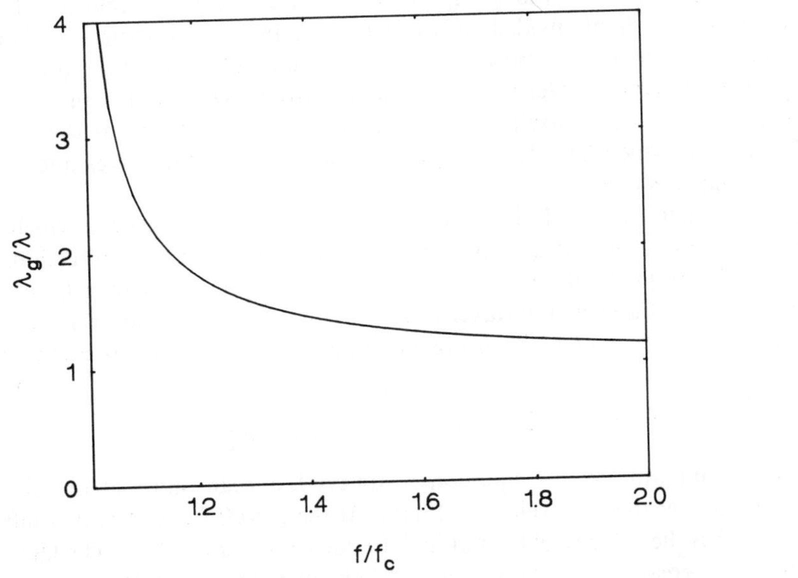

Figure 10.13. Normalised guide wavelength against normalised frequency for TE_n and TM_n modes.

the cut off frequency, the guide wavelength decreases until at high frequencies it approaches the value of the free space wavelength.

Sometimes it is convenient to express the guide wavelength in terms of the cut-off wavelength, λ_c. Equation (10.16) can be rewritten using the definition of cut-off wavelength, equation (10.18) as

$$\left(\frac{\omega}{c}\right)^2 - \beta^2 = \left(\frac{2\pi}{\lambda_c}\right)^2 \tag{10.25}$$

Replacing ω/c by $2\pi/\lambda$ and β by $2\pi/\lambda_g$, respectively, gives

$$\frac{1}{\lambda^2} - \frac{1}{\lambda_g^2} = \frac{1}{\lambda_c^2} \tag{10.26}$$

An alternative expression which gives the phase constant β in terms of frequency is obtained from equation (10.25) by replacing λ_c by $2\pi c/\omega_c$ as

$$\beta = \frac{\omega}{c}[1 - (\omega_c/\omega)^2]^{1/2} = \frac{\omega}{c}[1 - (f_c/f)^2]^{1/2} \tag{10.27}$$

This equation is only valid for frequencies above the cut-off frequency. The square root term appears in many waveguide formula, such as velocities and impedances.

10.8 PHASE AND GROUP VELOCITY

In the study of plane waves in free space, two velocities were defined. These were the phase velocity which is the speed at which a constant phase point on the wavefronts travels along and the group velocity which is the velocity of energy transport. For plane waves both the phase and group velocities are equal to the velocity of light in air or the velocity of light divided by the refractive index in a dielectric. The situation in waveguides is in general more complicated because of the zig-zag path travelled by rays along the waveguide. Neither the phase nor group velocities is equal to the velocity of light in an air-filled parallel plate waveguide. The exception to this statement is the TEM mode which has characteristics similar to the plane waves and the same phase and group velocity.

For a general TE_n or TM_n mode the phase velocity is the speed at which the wavefront of the waves in the waveguides appears to move along in the z direction. Assuming an air-filled waveguide, the velocity of the wave along its ray path will be the velocity of light, c. From the sketch of wavefronts in Figure 10.12, the phase velocity is $v_p = c/\cos\theta$. Now $\cos\theta = \lambda/\lambda_g = c\beta/\omega$. So using equation (10.27) gives an expression for v_p as

$$v_p = \frac{\omega}{\beta} = \frac{c}{[1 - (\omega_c/\omega)^2]^{1/2}} \tag{10.28}$$

The denominator of this equation is always less than unity so the phase velocity is always greater than the velocity of light. At the cut-off frequency, it is infinite and then approaches the velocity of light at high frequencies, Figure 10.14. The fact that the phase velocity is greater than the velocity of light does not imply that energy travels faster than light. The effect is similar to a wave of water striking a wall at an angle. The wave

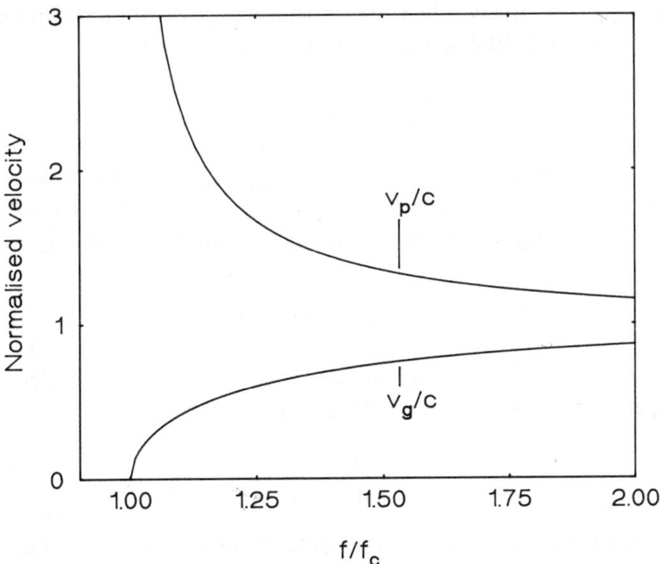

Figure 10.14. Normalised phase velocity (v_p/c) and group velocity for TE_n and TM_n modes

will appear to move along the wall much faster than the speed at which the actual wave is travelling.

The velocity of energy flow is the group velocity and this is the velocity of the rays as they travel the zig-zag path along the waveguide. These follow a longer path than the direct path along the z axis. Consequently the group velocity will always be slower than the velocity of light. It is given by $v_g = c\cos\theta$, or using the definition given in Section 5.14, $v_g = d\omega/d\beta$. From equation (10.28) this gives

$$v_g = \frac{d\omega}{d\beta} = c[1 - (\omega_c/\omega)^2]^{1/2} \tag{10.29}$$

The group velocity is also shown on Figure 10.14. At the cut-off frequency it is zero, because there is no energy travelling along the waveguide. As the frequency increases, the group velocity increases until it tends towards the velocity of light. This shows that at high frequencies, all modes tend towards the velocity of light, irrespective of their cut-off frequency.

10.9 WAVE IMPEDANCE

A wave impedance was defined in Chapter 5 for a plane wave as the ratio of the electric field to the magnetic field. In the case of plane waves there is only one electric field component and one magnetic field component. A wave impedance can still be defined for modes in a waveguide but the definition needs to be modified to the ratio of the *transverse* electric field and the *transverse* magnetic field. The field components of the modes differ so there is one expression for the wave impedance of TEM modes, another expression for TM modes and a third expression for TE modes. These will be designated

Z_{TEM}, Z_{TM} and Z_{TE}, respectively. The wave impedance for TEM modes in a dielectric-filled parallel plate waveguide is the same as that for plane waves

$$Z_{TEM} = \eta = \left(\frac{\mu_0}{\varepsilon_0 \varepsilon_r}\right)^{1/2} \tag{10.30}$$

For convenience, the TEM wave impedance is given te symbol η. The wave impedance for TM modes is the ratio of the E_x electric field to the H_y magnetic field. From equation (10.22), and equation (10.28) modified for a dielectric, this gives

$$Z_{TM} = \frac{E_x}{H_y} = \frac{\beta}{\omega \varepsilon_0 \varepsilon_r} = \eta[1 - (\omega_c/\omega)^2]^{1/2} \tag{10.31}$$

Similarly the wave impedance for TE modes is the ratio of the E_y electric field to the H_x magnetic field. From equations (10.23) and (10.28) this gives

$$Z_{TE} = \frac{E_y}{H_x} = \frac{\omega \mu_0}{\beta} = \frac{\eta}{[1 - (\omega_c/\omega)^2]^{1/2}} \tag{10.32}$$

None of the wave impedances are functions of transverse position (x, y) in the waveguide. The wave impedances for the general TM and TE modes are modified by comparison with the wave impedance of the medium. Z_{TM} is always less than the wave impedance of the medium. It is zero at the cut-off frequency and then increases towards the TEM wave impedance, η. Z_{TE} shows the opposite behaviour and is infinite at the cut-off frequency. It decreases towards η at high frequencies. The behaviour of Z_{TE} and Z_{TM} with frequency is analogous to the behaviour of the phase and group velocities, respectively. This can be seen by comparing equations (10.31) and (10.32) with equations (10.28) and (10.29).

10.10 ATTENUATION IN PARALLEL PLATE WAVEGUIDE

The parallel plate waveguide studied so far in this chapter has been assumed to be loss-less. In practical waveguides, there are two sources of loss:

(a) Dielectric loss if the parallel plate waveguide is filled with a dielectric as will often be the case.

(b) Conductor loss due to the finite conductivity of the metal plates.

Expressions for the attenuation caused by these losses will now be developed. The attenuation due to dielectric loss can be treated by replacing the relative permittivity ε_r of a loss-less dielectric by a complex permittivity $\varepsilon_r = \varepsilon' - j\varepsilon''$. The real part, ε', is the same as before. The imaginary part, ε'', represents the small loss in the dielectric. The amount of loss is often expressed by the loss tangent, $\tan\delta = \varepsilon''/\varepsilon'$.

The procedure for analysing the attenuation is the same as that developed in Chapter 8 for plane waves travelling in a lossy medium. By direct analogy with that procedure, (see Section 8.2), the phase coefficient, $j\beta$, is replaced by the complex propagation coefficient, γ, where $\gamma = \alpha + j\beta$ and the wavenumber becomes

$$K_c^2 = \omega^2 \mu_0 \varepsilon_0 (\varepsilon' - j\varepsilon'') + \gamma^2 \tag{10.33}$$

For TEM modes $K_c = 0$, and then

$$\gamma = \alpha + j\beta = j\omega[\mu_0\varepsilon_0(\varepsilon' - j\varepsilon'')]^{1/2} \tag{10.34}$$

If the ratio $\varepsilon''/\varepsilon'$ is small by comparison with unity, a binomial expansion can be used to show that the attenuation constant is approximately

$$\alpha_{TEM} = \frac{\omega(\mu_0\varepsilon_0\varepsilon')^{1/2}}{2}\frac{\varepsilon''}{\varepsilon'} \quad \text{Nepers/m} \tag{10.35}$$

For TM and TE modes, the same procedure can be used, but using the general expression for the wavenumber. This leads to a result which is valid for both TM and TE modes,

$$\alpha_{TM,\,TE} = \frac{\alpha_{TEM}}{[1 - (\omega_c/\omega)^2]^{1/2}} \quad \text{Nepers/m} \tag{10.36}$$

This equation shows that the attenuation due to dielectric loss in TM and TE modes is always higher than that due to the TEM modes.

Example 10.2 Dielectric loss in a parallel plate waveguide

Plot the attenuation in decibels due to dielectric loss in a parallel plate waveguide with separation distance 5 mm and filled with dielectric of relative permittivity 2.5 and loss tangent 0.001. Plot for the TEM, TM_1 and TE_1 modes for frequencies up to 80 GHz.

The cut-off frequencies for the TM_1 and TE_1 modes were found in Example 10.1 as 18.97 GHz. The attenuation per metre is given by equations (10.35) and (10.36) but these values are in Nepers/m. They need to be multiplied by 8.686 to convert them into dB/m. Inserting the values gives $\alpha_{TEM} = 0.144f$ dB/m where f is in GHz. Note that this is independent of the distance apart of the plates. $\alpha_{TM} = \alpha_{TE}$ and is given directly by equation (10.36). The results are plotted in Figure 10.15. The attenuation is relatively large at higher frequencies and shows the importance of choosing a dielectric with a low loss. It is possible to get the dielectrics with $\tan\delta = 0.0001$ which reduces the attenuation by a factor of ten and makes it very small. The attenuation of the TE_1 and TM_1 modes is infinite at cut-off, then reduces to a minimum and then increases again. The values are slightly above those for the TEM mode.

The attenuation due to conductor loss can be found from the fact that the average power in the waveguide is

$$P = P_0 e^{-2\alpha z} \tag{10.37}$$

where the factor 2 in the exponential comes from the square of the electric field. Differentiating equation (10.37) with respect to distance, z, and rearranging gives

$$\alpha = \frac{-\,dP/dz}{2P} \tag{10.38}$$

where dP/dz is the average power loss per distance. This gives an expression for the attenuation constant. In order to evaluate it, the average power flow, P, and the power loss per distance must be found. For TEM modes between parallel plates of separation distance a, the average power flow through a cross-section of height a and width w can

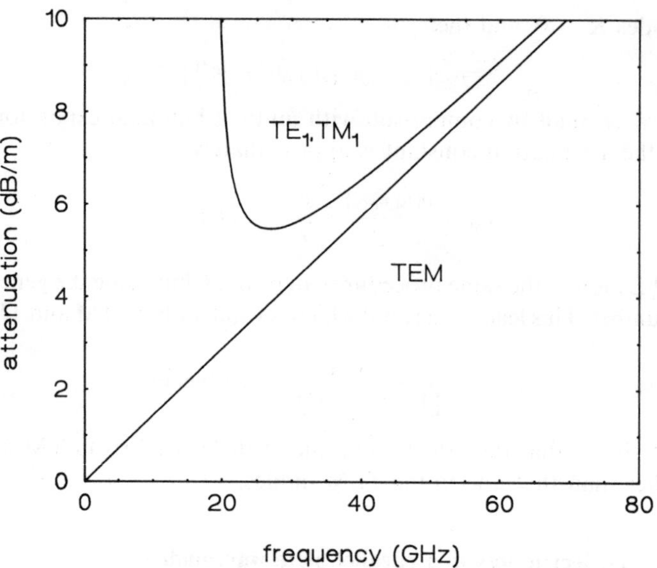

Figure 10.15. Attenuation due to dielectric losses in parallel plate waveguide

be found by analogy with plane waves as

$$P = aw\frac{A^2}{2\eta} \tag{10.39}$$

The power loss per distance comes from the finite conductivity of the plates and is the I^2R loss in the plates. For a plate of width w and unit length, this is

$$\frac{\mathrm{d}P}{\mathrm{d}z} = \tfrac{1}{2}wR_s|J_z|^2 \tag{10.40}$$

where R_s is the surface resistivity. This was found in Section 8.4, equation (8.21) to be $R_s = (\omega\mu_0/2\sigma)^{1/2}$. The boundary condition states that the current density, J_z, is equal to the magnetic field, H_y. For the TEM mode this is also equal to A/η, where A is the amplitude coefficient of the electric field. The power loss then becomes

$$\frac{\mathrm{d}P}{dz} = \tfrac{1}{2}wR_s\left|\frac{A}{\eta}\right|^2 \tag{10.41}$$

There are two plates so the total loss is double that given in equation (10.41). Substituting equation (10.39) and two times (10.41) into (10.38) gives for the attenuation of a TEM mode due to conductor loss as

$$\alpha_{\mathrm{TEM}} = \frac{R_s}{\eta a} \tag{10.42}$$

The surface resistivity is proportional to the square root of the frequency so the attenuation due to conductor loss for the TEM mode is also proportional to the square root of the frequency. The expression is independent of the effective width even though this was needed in the derivation.

The attenuation due to conductor loss for the general TM and TE modes can be found by a similar procedure to above. The power flow, P, must be found from integrating the Poynting vector for the mode. In the case of TM modes, the power flow is

$$P_{\text{TM}} = \frac{w}{2}\int_0^a (E_x H_y)\,\text{d}x \tag{10.43}$$

The power loss takes the same form as equation (10.40). Inserting the fields into equation (10.43) and evaluating the integrals and combining with the power loss in equation (10.38) gives the attenuation for TM modes as

$$\alpha_{\text{TM}} = \frac{2\alpha_{\text{TEM}}}{[1-(\omega_c/\omega)^2]^{1/2}} \tag{10.44}$$

The equivalent result for TE modes is

$$\alpha_{\text{TE}} = \frac{2\alpha_{\text{TEM}}(\omega_c/\omega)^2}{[1-(\omega_c/\omega)^2]^{1/2}} \tag{10.45}$$

The behaviour of these attenuations with frequency is more complicated than that for the TEM mode because frequency dependent terms appear in both numerator and denominators. The denominators are zero at cut off so that the attenuation starts as being infinite. For the TE modes it then falls continuously as frequency increases, but for the TM modes it falls to a minimum and then increases again. This is illustrated with an example.

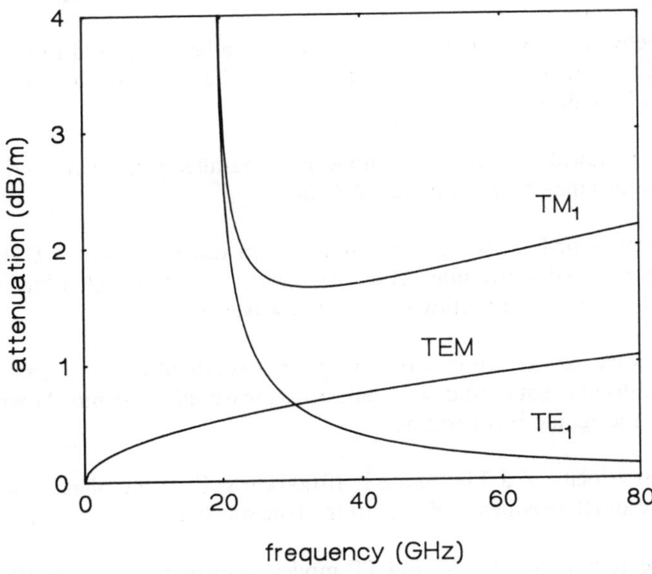

Figure 10.16. Attenuation due to conductor losses in parallel plate waveguide

Example 10.3 Attenuation due to conductor loss

Plot the attenuation in decibels for conductor loss using the parallel plate waveguide of Example 10.2 with an effective width of 3 mm. The conductors are made from brass.

The surface resistivity was derived in Chapter 8 as $R_s = (\omega\mu_0/2\sigma)^{1/2}$. The conductivity of brass from the table in the Appendix is 1.5×10^7 S/m. This gives $R_s = 0.0162 f^{1/2}\,\Omega$, where f is the frequency in GHz. Inserting this and the other values into equation (10.43) gives $\alpha_{TEM} = 0.12 f^{1/2}$ dB/m. The attenuations for the TM_1 and TE_1 modes are given by equations (10.44) and (10.45), respectively with $f_c = 18.97$ GHz. The results are plotted in Figure 10.16. Comparison with the attenuation due to dielectric loss in the previous example shows that the values of conductor loss is less than the dielectric loss. The dielectric loss can, however, be reduced by choosing a dielectric with lower loss whereas the conductor loss cannot be significantly reduced. Using silver conductors instead of brass conductors would reduce the attenuation by $(6.17 \times 10^7/1.5 \times 10^7)^{1/2} = 2.03$ times.

PROBLEMS

1. Derive the propagation equation for a parallel plate waveguide which is filled with dielectric of relative permittivity ε_r.

2. Plot βa against $\omega a/c$ for a dielectric filled parallel plate waveguide with $\varepsilon_r = 4$. Compare with the air-filled case for the first three TM modes.

3. Derive the electric and magnetic field components for TE modes in parallel plate waveguide.

4. A microstrip is formed by a parallel plate waveguide on either side of a sheet $\varepsilon_r = 10$, and thickness 3 mm. Find the cut-off frequency of the TE_1 mode.

5. The air between the Earth and the ionosphere can be considered to be a parallel plate waveguide. If the ionosphere starts at a height of 80 km find the lowest frequency at which a TE mode will propagate.

6. Find the bandwidth of the TE_1 mode in a parallel plate waveguide which has a separation distance between plates of 30 mm.

7. A parallel plate air-filled waveguide has $a = 5$ mm and operates at 40 GHz. If the guided wave is represented as two interacting plane waves, find the angle which the wavefronts make with the z axis and draw to scale the wavefronts.

8. Standing waves are set up in a parallel plate waveguide with $a = 2$ mm. The distance between nulls in the standing wave pattern is measured as 30 mm. At what frequencies could the waveguide be operating?

9. The group velocity of a TE_1 mode is 10 000 km/s. Find the phase velocity, the wave impedance and the frequency of operation of the waveguide with $a = 5$ mm and $\varepsilon_r = 2.5$.

10. Derive the attenuation of TM and TE modes as given in equation (10.36).

11. A TEM mode attenuator is to be made using a parallel plate waveguide filled with lossy foam which has ($\varepsilon_r = 1.05$ and $\sigma = 0.05\,\text{S/m}$). Find the length of the attenuator needed to give a 10 dB attenuation at 10 GHz.

12. Obtain an expression for the power loss for TM modes in parallel plate waveguide and hence confirm equation (10.44) for the attenuation of TM modes due to conductor loss.

13. Copper conductors are deposited onto doped silicon ($\varepsilon_r = 12$, $\sigma = 12\,\text{S/m}$) to form a parallel plate waveguide which is 3 mm thick. Find the attenuation at 20 GHz of the TEM, TE_1 and TM_1 modes.

14. The attenuation due to conductor loss for the TM_1 mode has a minimum at one frequency. Find an equation for this frequency in terms of the cut-off frequency and obtain the corresponding attenuation.

11 Transmission lines

11.1 INTRODUCTION

It is often helpful when designing and analysing waveguides to use a theoretical model which is independent of the actual physical waveguide and is common for all types of waveguide. This can be done by using an equivalent transmission line approach whereby the actual electromagnetic fields are represented by equivalent circuit elements. This approach is widely used in practical microwave system design. The advantages are greatest when active devices and passive components are attached to the waveguide. The devices and components can also be represented as equivalent circuit elements so that the waveguide becomes just another circuit element and the effects of interaction between different parts of the system can be studied using circuit theory.

This method of studying waveguides has been called *electromagnetic circuit theory*. It may be asked why bother to develop a new theory when a perfectly adequate one has already been derived from electromagnetic wave theory. The reason is that a set of simple relationships can be derived which make the design of waveguiding systems easier. Electromagnetic wave theory is perfectly adequate for understanding how the fields and modes in a waveguide behave. It becomes difficult, however, to apply when one waveguide is joined to another waveguide. The junction of two waveguides is a common problem in all practical microwave and optical systems. The transmission line approach allows this problem to be studied and analysed.

This chapter provides an introduction to transmission line theory in order to demonstrate the potential of the technique. The relationship between electromagnetic wave theory which has been developed in previous chapters and electromagnetic circuit theory is described. Then the general transmission line equations are developed. These are directly analogous to those developed for TEM modes in waveguides, but by using impedances they are easier to manipulate when there are multiple reflections. The transmission line approach is particularly useful for understanding what happens when two waveguides are connected together or a waveguide is connected to an arbitrary load. The system designer's aim is usually to match the waveguide to the load so that all the power is transferred along the waveguide into the load. Transmision line theory provides a relatively simple technique for doing this matching using impedance transformations. Consequently impedance measurements are the main practical technique used for the experimental investigation of practical systems. The powerful *Network Analyser* instruments which are available are based on using electromagnetic circuit and transmission line approaches. There is a direct relationship between the TEM mode in a waveguide and plane waves in free space, thus it is also possible to use the transmission line approach to analyse what happens when plane waves impinge on multiple dielectric layers and to find ways of matching the incident plane waves. This will be demonstrated in the last section.

11.2 ELECTROMAGNETIC CIRCUIT THEORY

Electromagnetic circuit theory is based on representing a waveguide by an equivalent circuit. The elements of the circuit can be inductors, capacitors or resistors but since they are representing a continuous wave, they are distributed elements, not discrete elements. The equivalent values can only be found over a length of the waveguide, not at any one point. This is illustrated in the case of a parallel plate waveguide in Figure 11.1. The physical structure is represented by the combination of series resistance and inductance plus parallel conductance and capacitance. As it is cumbersome to draw the individual symbols, the transmission line is drawn as two lines as shown in Figure 11.1(c). It should be remembered that there is continual current flow across the gap between the two lines, unlike low frequency circuits. The following basic elements are used in electromagnetic circuit theory:

A series inductance, L, with units of H/m

A series resistance, R, with units of Ω/m

A parallel, or shunt, capacitance, C, with units of F/m

A parallel conductance, G, with units of S/m

Each of the circuit elements represents a particular effect in electromagnetic wave theory. These have been described in previous chapters. Inductors and capacitors were described in Chapter 2 and the expression derived there give the values of the inductance and capacitance for parallel plate and coaxial waveguides. The relationships will be summarised in the next section. Most practical transmission lines have low loss so the equivalent resistance and conductance will be either negligible or very small. The resistance represents the loss in the conductors, whilst the conductance represents the leakage between conductors through the intervening medium. If a dielectric is present then the conductance represents the loss in the dielectric.

11.3 RELATION BETWEEN FIELD THEORY AND CIRCUIT THEORY

Electric circuit theory is a special case of the more general electromagnetic field theory. This was demonstrated in Chapter 2 for the special case of a conducting wire. In this section the general relationships will be put down which link field and circuit theory.

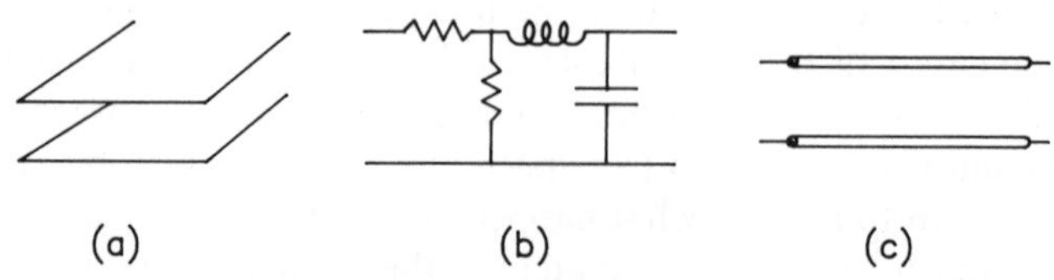

Figure 11.1. Transmission line representation of parallel plate waveguide. (a) Waveguide. (b) Equivalent circuit elements. (c) Transmission line representation

These are equations linking electric and magnetic fields with voltages and currents. They have been derived and used in earlier chapters.

The two equations which link the voltage to the electric field and the current to the magnetic field come from the definition of potential difference and Ampère's law,

$$V = \int \boldsymbol{E} \cdot \boldsymbol{dl} \tag{11.1}$$

$$I = \int \boldsymbol{H} \cdot \boldsymbol{dl} \tag{11.2}$$

The evaluation of the integrals involves unknown constants, thus an additional power conservation relationship is needed which states that the total power in the electromagnetic fields is the same as the total power in the equivalent transmission line:

$$\tfrac{1}{2}\mathrm{Re}(VI^*) = \tfrac{1}{2}\int_s \mathrm{Re}(\boldsymbol{E} \times \boldsymbol{H}^*)\,\mathrm{d}s \tag{11.3}$$

The use of equations (11.1) to (11.3) can be illustrated with an example.

Example 11.1 Equivalent voltage and current for parallel plate waveguide

Find the voltage and current for the parallel plate waveguide shown in Figure 11.2 with plates of width w and height a.

The equivalent voltages and currents are found by starting with the fields of the TEM mode and evaluating the above equations. The electric and magnetic fields are

$$E_x = E_0 \exp[\mathrm{j}(\omega t - \beta z)] \tag{11.4}$$

$$H_y = \frac{E_0}{\eta} \exp[\mathrm{j}(\omega t - \beta z)] \tag{11.5}$$

Applying (11.1) gives the voltage as a function of z as

$$V(z) = C_1 \int_0^a E_x\,\mathrm{d}x = C_1 a E_0 \exp[\mathrm{j}(\omega t - \beta z)] \tag{11.6}$$

where C_1 is a constant and $V = 0$ has been assumed on the bottom plate. Similarly

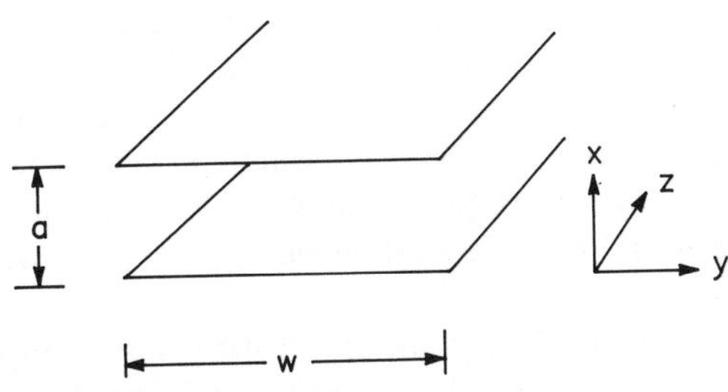

Figure 11.2. Parallel plate waveguide for Example 11.1.

equation (11.2) gives

$$I(z) = C_2 \int_0^w H_y \, dy = C_2 w \frac{E_0}{\eta} \exp[j(\omega t - \beta z)] \tag{11.7}$$

where C_2 is another constant. The total time average power in the TEM wave is

$$\frac{1}{2} \int_s \mathrm{Re}(\boldsymbol{E} \times \boldsymbol{H}^*) \, ds = \tfrac{1}{2} E_x H_y wa = \frac{E_0^2 wa}{2\eta} \tag{11.8}$$

Comparing this result to equation (11.3) shows that equation (11.8) should be equal to $\mathrm{Re}(VI^*)/2$. Therefore $C_1 C_2 = 1$. For convenience, choose $C_1 = C_2 = 1$, so that the voltage and current are given by

$$V(z) = aE_0 \exp[j(\omega t - \beta z)] \tag{11.9}$$

$$I(z) = w \frac{E_0}{\eta} \exp[j(\omega t - \beta z)] \tag{11.10}$$

The equivalent voltages and currents are the electric and magnetic fields multiplied by the height or the width, respectively.

The values of the equivalent circuit elements, the inductance, capacitance, resistance and conductance, are not generally required in practical designs, but it is instructive to understand the relationship between them and the electromagnetic parameters.

In electromagnetic field terms, the inductor describes the amount of stored magnetic energy. This is given by

$$W_{\mathrm{M}} = \frac{\mu_0 \mu_r}{4} \int_s |\boldsymbol{H}|^2 \, ds \tag{11.11}$$

The equivalent stored current in a circuit is $W = L|I|^2/4$, where I is the amplitude of the current flow. Equating the stored energies gives the self inductance as

$$L = \frac{\mu_0 \mu_r}{|I|^2} \int_s |\boldsymbol{H}|^2 \, ds \qquad \mathrm{H/m} \tag{11.12}$$

The capacitor is describing the amount of stored electrical energy and by direct analogy with equation (11.12) for the inductor, the capacitance is given by

$$C = \frac{\varepsilon_0 \varepsilon_r}{|V|^2} \int_s |\boldsymbol{E}|^2 \, ds \qquad \mathrm{F/m} \tag{11.13}$$

All transmission lines will have an equivalent inductance and capacitance, but in many cases the losses will be small and can be neglected. Consequently the resistance and conductance are secondary parameters. The resistor is describing the loss that occurs in the conducting surfaces due to the finite conductivity. The conductance is describing any loss that occurs in between the conductors. For an air-filled waveguide the conductance would be zero. For a dielectric-filled parallel plate waveguide, the loss was found in the last chapter.

For most practical purposes only the equivalent series impedance and parallel admittance are needed. This is partly because the total impedance and admittance can be readily measured. If the transmission line has some loss the series impedance per unit length, Z, will be complex and is given in terms of the series resistance R and inductance L by

$$Z = R + j\omega L \tag{11.14}$$

The parallel or shunt conductance G and capacitance C are similarly given in terms of the shunt admittance Y by

$$Y = G + j\omega C \tag{11.15}$$

The transmission line will have an impedance at any point along its length which is the circuit analogy of the wave impedance derived for guided waves and plane waves. It is called the *characteristic impedance* of a transmission line and is defined as the ratio of the voltage across a line to the current flowing along a line, with no reflected wave:

$$Z_0 = \frac{V(z)}{I(z)} \tag{11.16}$$

The characteristic impedance is the value which is quoted in specifications for coaxial or two wire cables at RF and microwave frequencies.

Example 11.2 Equivalent circuit parameters for a parallel plate waveguide

Find the equivalent circuit parameters for parallel copper plate waveguide with separation distance, $a = 2$ mm and effective width $w = 3$ mm. The waveguide is filled with dielectric of permittivity 2.5 and loss tangent 0.001 and operates at 10 GHz.

The inductance can either be found from equation (11.12) with $\boldsymbol{H} = H_y\hat{\boldsymbol{y}}$ or by the method described in Chapter 2. The result is given in equation (2.45) as $L = \mu_0 a/w$ H/m. For the parameters given this has a value of 0.84 μH/m.

The capacitance can similarly be found either from equation (11.13) or from electric field theory. The result is given by equation (2.28) as $C = \varepsilon_0\varepsilon_r w/a$ F/m. For the parameters given, this has a value of 33.2 pF/m.

The resistance of one plate is the surface resistivity divided by the width of the plates. The surface resistivity is a frequency dependent term given by $R_s = (\omega\mu_0/2\sigma)^{1/2} = 0.026\,\Omega$, where the conductivity of copper, 5.8×10^7 S/m has been used. The total resistance for both plates is $R = 2R_s/w = 17.4\,\Omega$/m.

The conductance is given by $G = \omega\varepsilon_0\varepsilon'' w/a$ where ε'' is the imaginary part of the permittivity of the dielectric and is the relative permittivity times the loss tangent or 0.0025. The conductance therefore has a value of 0.002 S/m.

The characteristic impedance is, from equations (11.9) and (11.10):

$$Z_0 = \eta\frac{a}{w} \tag{11.17}$$

The wave impedance of the parallel plate waveguide is $\eta = 377/\sqrt{2.5} = 238.4\,\Omega$. The characteristic impedance then becomes from (11.17) $Z_0 = 159\,\Omega$.

11.4 INFINITE UNIFORM TRANSMISSION LINE

The development of a theory for transmission lines starts with the case where the line is infinitely long and is uniform along its length. The infinite length implies that the line has no load connected to the end. It is therefore not a practical case but serves to develop a basic theory which is exactly analogous to the waveguide theory developed in the last chapter. The transmission line is illustrated in Figure 11.3(a) and has a voltage between the two lines of V and a current flowing in the line of I. In the analysis in this section a time dependence of $\exp(j\omega t)$ will be implicitly assumed so that it will not appear in the equations.

Consider a short section of the transmission line of length δz. The circuit elements for this short section are shown in Figure 11.3(b). A relationship describing the rate of change in voltage with distance can be derived by summing the voltages around the circuit in Figure 11.3(b) using Kirchhoff's Voltage Law. This gives

$$V - R\delta z I - j\omega L\delta z I - (V + \delta V) = 0 \tag{11.18}$$

or

$$\frac{\delta V}{\delta z} = -(R + j\omega L)I$$

In the limit of δz tending to zero, this becomes

$$\frac{\partial V}{\partial z} = -ZI \tag{11.19}$$

A similar relationship describing the rate of change of current with distance can be derived by summing currents using Kirchhoff's Current Law to give

$$\frac{\partial I}{\partial z} = -YV \tag{11.20}$$

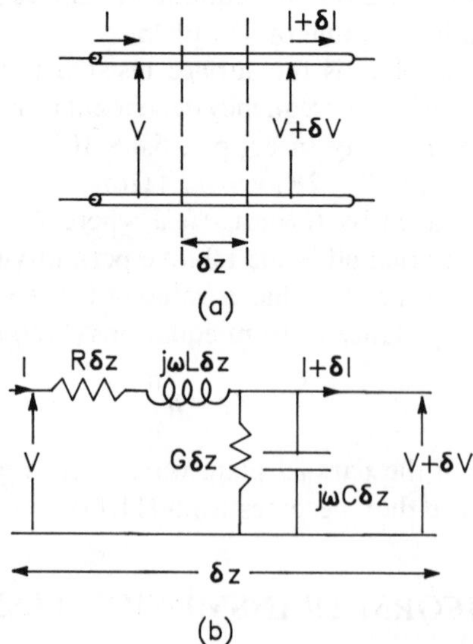

Figure 11.3. Voltage and current on a transmission line. (a) Short section of line. (b) Equivalent circuit of section.

Differentiating equation (11.19) with respect to z and substituting in equation (11.20) gives

$$\frac{\partial^2 V}{\partial z^2} = -I\frac{\partial Z}{\partial z} - Z\frac{\partial I}{\partial z} = -I\frac{\partial Z}{\partial z} + ZYV \tag{11.21}$$

For a uniform line the impedance will be constant along the line, so $\partial Z/\partial z = 0$ and equation (11.21) becomes

$$\frac{\partial^2 V}{\partial z^2} - ZYV = 0 \tag{11.22}$$

By a similarly procedure an analogous equation for the current can be derived. Starting with equation (11.20) and differentiating this with respect to z, substituting in equation (11.19) and using the assumption of constant admittance along the line gives

$$\frac{\partial^2 I}{\partial z^2} - ZYI = 0 \tag{11.23}$$

Equations (11.22) and (11.23) are called the *transmission line equations.* Comparison with the electromagnetic wave equations derived in Chapter 3 show that they have exactly the same form and that the transmission line equations are the basic equation describing the behaviour of waves on transmission lines. As with the wave equations, the transmission line equations for current and voltage have the same constant, ZY, so that only one of the two equations need be solved.

The general solution to the voltage transmission line equation is

$$V = V_{\mathrm{R}}\,\mathrm{e}^{\gamma z} + V_{\mathrm{I}}\,\mathrm{e}^{-\gamma z} \tag{11.24}$$

where V_{R} and V_{I} are amplitude coefficients. The first term represents a wave propagating in the negative z direction and the second term represents a wave propagating in the positive z direction. For later use when a terminated line is considered, the amplitude coefficient associated with the positive wave is designated the incident voltage, V_{I}, and the amplitude coefficient associated with the negative travelling wave is designated the reflected voltage, V_{R}. The full space and time description of the wave would need an exponential $\mathrm{j}\omega t$ term to complete the description.

The constant γ in equation (11.24) is called the *propagation constant* of the transmission line. From equations (11.22) or (11.23) it is given by

$$\gamma = (ZY)^{1/2} = [(R + \mathrm{j}\omega L)(G + \mathrm{j}\omega C)]^{1/2} \tag{11.25}$$

γ is complex and has a real part, α, which is the *attenuation constant*, and an imaginary part, β, which is the *phase constant.* Then $\gamma = \alpha + \mathrm{j}\beta$, just as for plane waves or guided waves. α is equal to the real part of $(ZY)^{1/2}$ and β is equal to the imaginary part of $(ZY)^{1/2}$. Equation (11.24) can then be rewritten as

$$V = V_{\mathrm{R}}\,\mathrm{e}^{\alpha z}\,\mathrm{e}^{\mathrm{j}\beta z} + V_{\mathrm{I}}\,\mathrm{e}^{-\alpha z}\,\mathrm{e}^{-\mathrm{j}\beta z} \tag{11.26}$$

The first term represents a wave which is travelling in the negative z direction and decreasing in magnitude in the negative z direction. Similarly the second term represents a wave which is travelling and decreasing in magnitude in the positive z direction.

If $R=0$ and $G=0$, then $\alpha=0$ because there is no loss and $\gamma=j\beta$ where

$$\beta=\omega(LC)^{1/2} \tag{11.27}$$

If expressions for inductance and capacitance for a TEM waveguide are inserted into equation (11.26), it will be found that $\beta=2\pi/\lambda$. This is the same as found using wave analysis.

The equivalent current on the transmission line can be obtained from equation (11.19) as

$$I=\frac{\gamma}{Z}(V_{\mathrm{I}}\mathrm{e}^{-\gamma z}-V_{\mathrm{R}}\mathrm{e}^{\gamma z}) \tag{11.28}$$

The characteristic impedance of the line, Z_0, is defined as the ratio of the voltage to the current at a point on the line. From equations (11.24), (11.25) and (11.28)

$$Z_0=\frac{V}{I}=\frac{Z}{\gamma}=\left(\frac{Z}{Y}\right)^{1/2}=\left(\frac{R+\mathrm{j}\omega L}{G+\mathrm{j}\omega C}\right)^{1/2} \tag{11.29}$$

For a loss-less line, this becomes

$$Z_0=\left(\frac{L}{C}\right)^{1/2} \tag{11.30}$$

This value will also apply approximately for the case of waves at high frequencies. In this case $\omega L \gg R$ and $\omega C \gg G$ and the real parts of Z and Y can be ignored because of the high value of ω.

Example 11.3 Characteristic impedance of a coaxial cable

Find the characteristic impedance of a coaxial cable filled with polythene ($\varepsilon_r=2.25$) which has an inner diameter of 1.5 mm and an outer diameter of 5 mm.

Equation (11.30) is used to find the characteristic impedance. The capacitance and inductance of coaxial lines were found in Chapter 2 (equations (2.29) and (2.46), respectively). The result is

$$Z_0=\frac{\eta}{2\pi}\ln(b/a) \tag{11.31}$$

η is the impedance of the dielectric and equals $377/\sqrt{2.25}$. Inserting the other parameters gives $Z_0=48.2\,\Omega$. This value is representative of practical coaxial cables.

11.5 TERMINATED LOSS-LESS TRANSMISSION LINE

The last section showed that the form of the waves on a transmission line is the same as in waveguides. Now the problem of a terminated transmission line will be studied, Figure 11.4. The load impedance Z_L is at the origin of the z coordinates. The aim is to relate the impedance at a point on the line, $z=-b$, to the load impedance. The load impedance could be an actual load but it could also be another section of

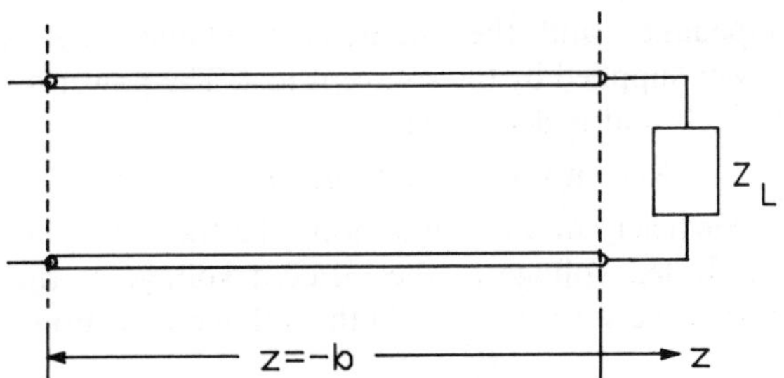

Figure 11.4. Terminated transmission line

transmission line. A loss-less line will be considered so that $\gamma = j\beta$ and the electric field at any point on the line is, from equation (11.24),

$$V = V_I(\rho_L e^{j\beta z} + e^{-j\beta z}) \tag{11.32}$$

where ρ_L is the *voltage reflection coefficient* at the load and equals V_R/V_I. The corresponding current is, from equations (11.28) and (11.29)

$$I = \frac{V_I}{Z_0}(e^{-j\beta z} - \rho_L e^{j\beta z}) \tag{11.33}$$

where Z_0 is the characteristic impedance of the line. The impedance at a point on the line is

$$Z(z) = \frac{V}{I} = -Z_0 \frac{(\rho_L e^{j\beta z} + e^{-j\beta z})}{(\rho_L e^{j\beta z} - e^{-j\beta z})} \tag{11.34}$$

At the load where $z = 0$, the impedance $Z = Z_L$ so equation (11.34) gives

$$\frac{Z_L}{Z_0} = \frac{1 + \rho_L}{1 - \rho_L} \tag{11.35}$$

or

$$\rho_L = \frac{Z_L - Z_0}{Z_L + Z_0} \tag{11.36}$$

It is often useful to work with normalised impedances by dividing the load or impedance at a point in the line by the characteristic impedance. The normalised load impedance is designated $\bar{Z}_L = Z_L/Z_0$.

The reflection coefficient can be calculated given a knowledge of the load and characteristic impedance of the transmission line. Equation (11.35) shows that when the line is short circuit (i.e. $Z_L = 0$), $\rho_L = -1$. When the line is open circuit, $Z_L = \infty$, and $\rho_L = 1$. These results agree with those obtained in Chapter 6 for the reflection of plane waves, and all the results derived for plane waves interacting in space also apply to the waves on transmission lines. Standing waves will be set up and can be detected by sampling the voltages along the transmission line. The *voltage standing wave ratio* (VSWR) is

$$\text{VSWR} = (1 + |\rho_L|)/(1 - |\rho_L|) \tag{11.37}$$

When the load impedance and the characteristic impedance are mismatched, a proportion of the power supplied by the source is lost. The power lost is measured with the *return loss* in decibels and is defined as

$$\text{Return loss} = -20\log_{10}|\rho_L| \quad \text{dB} \tag{11.38}$$

A voltage reflection coefficient at any point along the transmission line can be defined as the ratio of the reflected voltage to the incident voltage at the point. If this is a distance b back towards the source ($z = -b$) the voltage reflection coefficient is

$$\rho_b = \frac{V_R e^{-j\beta b}}{V_I e^{+j\beta b}} = \rho_L e^{-2j\beta b} \tag{11.39}$$

This shows how the reflection coefficient changes as the observation point is moved back from the load.

The most useful expression for transmission line design is one which relates the impedance at the load to the impedance at any point along the line. This is obtained from equation (11.34) by replacing z with $-b$

$$Z_b = Z_0 \frac{(e^{j\beta b} + \rho_L e^{-j\beta b})}{(e^{j\beta b} - \rho_L e^{-j\beta b})} \tag{11.40}$$

Inserting ρ_L from equation (11.36) and replacing the exponentials by sines and cosines leads to

$$Z_b = Z_0 \frac{(Z_L 2\cos\beta b + Z_0 2j\sin\beta b)}{(Z_L 2\sin\beta b + Z_0 2j\cos\beta b)} \tag{11.41}$$

or

$$\boxed{Z_b = Z_0 \frac{Z_L + jZ_0 \tan\beta b}{Z_0 + jZ_L \tan\beta b}} \tag{11.42}$$

This is a fundamental relationship in transmission line theory and relates the impedance at the load to the impedance at any point down the line. It depends on the load impedance, the characteristic impedance of the line and the distance in wavelengths ($\beta b = 2\pi b/\lambda$).

An important feature which is apparent from equation (11.42) is that the impedance is periodic along the transmission line and repeats every half wavelength. This is seen in Figure 11.5 which plots the real and imaginary parts of Z_b for a normalised load impedance $\bar{Z}_L$ of $0.5 + j1.0$. The periodicity means that the impedance of a transmission line at a distance of $p\lambda$ from the load ($p < 0.5$) also occurs at a distance $p\lambda + n\lambda/2$ ($n = 0, 1, 2\ldots$). This is a useful feature when adding components to the line because they can be added at a convenient physical distance from the load.

It is sometimes more convenient to deal with admittances in transmission line design. For instance it is easier to add sections of line in parallel rather than to add sections of line in series. Consideration of metal waveguides shows that adding something in series is not easy, but adding a section of waveguide in shunt is possible. The transmission line equation, equation (11.42) has the same form for admittances if $Y_0 = 1/Z_0$, $Y_L = 1/Z_L$ and $Y_b = 1/Z_b$,

$$\boxed{Y_b = Y_0 \frac{Y_L + jY_0 \tan\beta b}{Y_0 + jY_L \tan\beta b}} \tag{11.43}$$

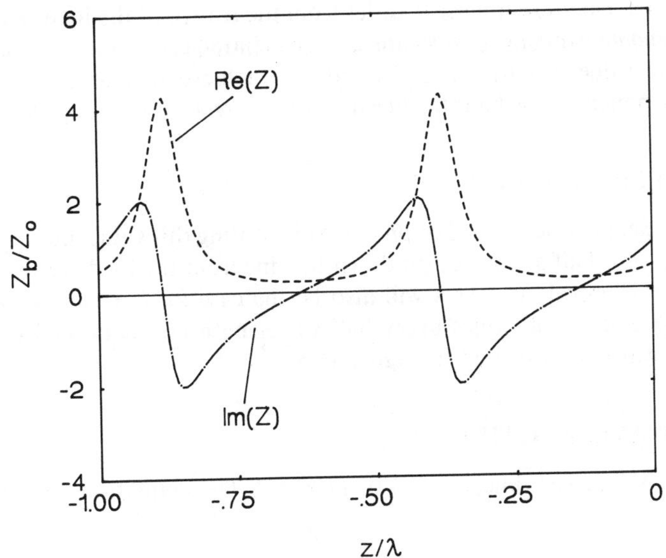

Figure 11.5. Real and imaginary parts of impedance, Z, along a transmission line terminated in a load with $Z_L/Z_0 = 0.5 + j1.0$

These impedance or admittance transformation equations involve complex quantities in both the numerator and denominator, except for certain special cases which will be discussed in the next section. Manual evaluation of the transformations is tedious and not, in general, to be recommended. Fortunately it is relatively straightforward to compute them on computer using one of the software packages designed to handle mathematical or numerical equations. Before computers were widely available the preferred approach was to use a graphical method. The most popular of these is the Smith Chart, which will be described in Section 11.8. Even now that the transformations can be handled easily on a computer, a graphical technique is sometimes useful to provide insight into the behaviour of the impedances.

11.6 SPECIAL CASES FOR TERMINATED LINES

Some special cases of the impedance transformation expression, equation (11.42) are particularly relevant in practical transmission lines. These will now be described.

MATCHED LINE

If the load impedance is matched to the characteristic impedance of the line so that $Z_L = Z_0$, then the impedance is the same at all points along the line. This is fairly obvious but is worth stating explicitly. Normally waveguides are designed to have low loss so that the characteristic impedance of a transmission line is effectively purely real. A perfect match to the load can then only occur if the load itself is a real resistance.

In the general case of an arbitrary, complex, load with both resistive and reactive components, the maximum power transfer from the source to the load takes place when there is a *conjugate match*, which means that the characteristic impedance at the source must have the value $Z = R_L - jX_L$. In order to achieve this situation, an additional matching component must be introduced at the correct point along the line.

HALF WAVELENGTH LINE

For a half wavelength line, $b = \lambda/2$ or $\beta b = \pi$. Substituting this value into Equation (11.42) gives $Z_b = Z_L$, thus half a wavelength down the line from the load the impedance is the same as the load impedance. This will also be the case for $b = n\lambda/2$ so the impedance cycles through a series of values every half wavelength and continually repeats along the line. This was demonstrated in Figure 11.5.

QUARTER WAVELENGTH LINE

For a quarter wavelength line, $b = \lambda/4$ or $\beta b = \pi/2$. Then equation (11.42) gives

$$Z_b = \frac{Z_0^2}{Z_L} \tag{11.44}$$

A quarter wavelength line is called an *impedance transformer* because the impedance is transformed from the load impedance to the impedance Z_b by the characteristic impedance of the line. Since the characteristic impedance is determined by the physical characteristics of the waveguide, this can be chosen to give a range of values. Consequently equation (11.44) gives a powerful and simple method of matching from one impedance to another impedance. It will be demonstrated later in the chapter.

If equation (11.44) is written using normalised impedances, it becomes $\bar{Z}_b = 1/\bar{Z}_L = Y_L$. This shows that the normalised impedance at the end of a quarter wavelength line is equal to the normalised load admittance.

SHORT CIRCUIT LINE

When the transmission line is short circuited, Figure 11.6, then $Z_L = 0$, and equation (11.42) gives

$$Z_b = jZ_0 \tan \beta b \tag{11.45}$$

The impedance along a short circuit section of line is hence imaginary or purely reactive.

The reflection coefficient of a short circuit line is $\rho_L = -1$, as was found for plane waves incident on a conducting sheet. The voltage and current along a short circuit line can be obtained from equations (11.32) and (11.33) as

$$V = -2jV_I \sin \beta z$$
$$I = \frac{2V_I}{Z_0} \cos \beta z \tag{11.46}$$

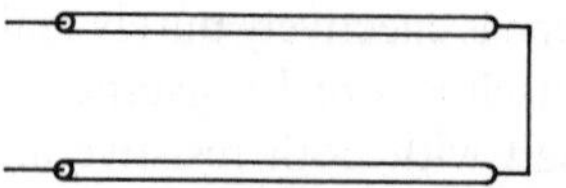

Figure 11.6. Short circuit transmission line

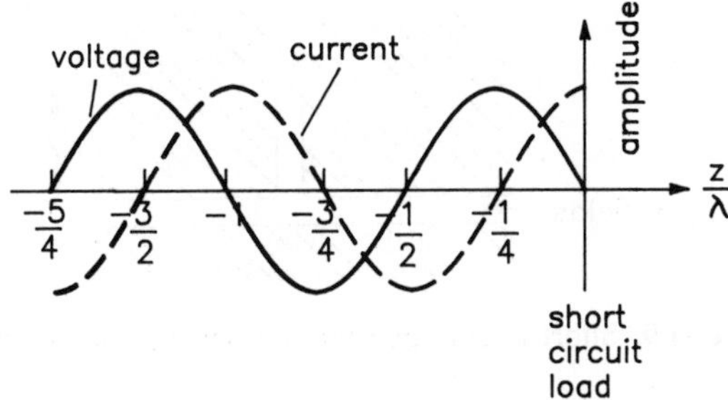

Figure 11.7. Voltage and current on a short circuit transmission line

The variation of voltage and current along a line which is short circuit is shown in Figure 11.7. The voltage varies as a sine wave with zero at the load. The current varies as a cosine wave with the peak at the load. The voltage and the current mirror the behaviour of the electric field and the magnetic field, respectively, of plane waves incident on a conducting sheet. The impedance is the ratio of V to I so it varies as a tangent function. This takes on all values from $Z_b = +\infty$ to $Z_b = -\infty$ which implies that a length of short circuit line can always be chosen to provide any desired reactance. This feature can be used to match an arbitrary impedance as will be demonstrated in the next section.

OPEN CIRCUIT LINE

This is the opposite case to the short circuit line. When the transmission line is open circuited, Figure 11.8, then $Z_L = \infty$, and equation (11.42) gives

$$Z_b = -jZ_0 \cot \beta b \tag{11.47}$$

The impedance along an open circuit section of line is also imaginary or purely reactive, but the inverse of the short circuit values.

The variation of voltage, current and reactance along a line which is open circuit is therefore also the inverse of the short circuit case. The voltage is a maximum at the load and the current is zero at the load. An open circuit line is not easy to achieve with a practical waveguide, because physically ending the waveguide means that the travelling waves radiate into space and *see* the impedance of free space. An open circuit can, however, be simulated with the next special case.

SHORT CIRCUIT, QUARTER WAVELENGTH LINE OR CHOKE

If a short circuit line is made a quarter of a wavelength long then equation (11.47) shows that $Z_b = \infty$. This means that the short circuit has been transformed to perfect open

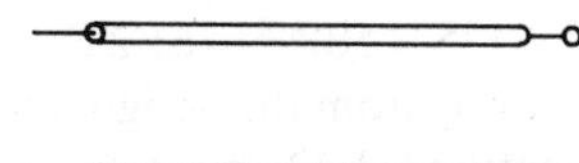

Figure 11.8. Open circuit transmission line

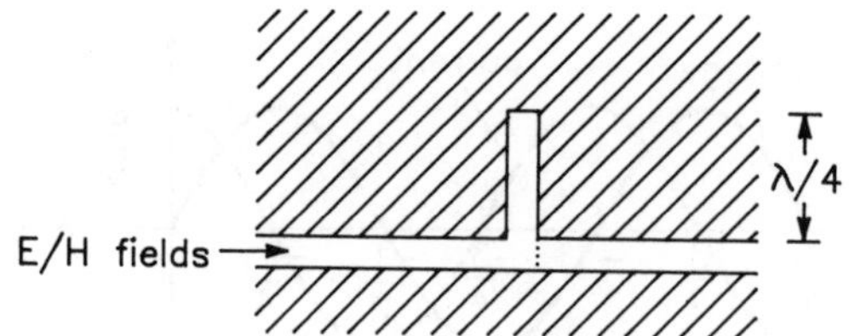

Figure 11.9. Short circuit, quarter wavelength line, or choke

circuit. This is called a *choke* and is widely used at microwaves to simulate an open circuit. In practice it is implemented simply by having a piece of short circuit waveguide which is quarter of a wavelength long. This is sketched in Figure 11.9. The fields in the gap between the two plates *see* an infinite impedance and cannot travel past the choke. The current flow along the conductors is stopped at the gap. This scheme is implemented in many practical systems. For instance it is used on the doors of microwave cookers to ensure that no energy leaks out of the cooker cavity. The mechanical advantage of using a choke is that no physical contact between metal needs to take place.

Note that a choke will only create an open circuit condition at one frequency, where it is exactly quarter of a wavelength long. For the microwave cooker example, this is no problem because all the energy is only at one frequency. For other applications, it implies that away from the design frequency, the choke will stop working perfectly.

11.7 MATCHING OF A TRANSMISSION LINE

Matching implies that for all points between the source and the matching section, the impedance of the line will be the same as the characteristic impedance. A transmission line which is terminated in an arbitrary load can be matched by a number of techniques. The use of quarter wave sections and adding a section of short circuit transmission line at an appropriate point along the line will be described in this section.

QUARTER WAVELENGTH SECTION

The principle of a quarter wavelength section of transmission line was described in the last section. It can be used to transform one real impedance to another impedance, as illustrated with an example.

Example 11.4 Matching a transmission line using a quarter wavelength line

Match a cable of characteristic impedance 75 Ω to a cable of characteristic impedance 50 Ω using a $\lambda/4$ section and find the input impedance and reflection coefficient at 20% away from the design frequency.

The 75 Ω cable forms the load impedance to the 50 Ω cable, Figure 11.10. Using equation (11.44) the impedance of a quarter wavelength long cable must be $(Z_L Z_b)^{1/2} = (75 \times 50)^{1/2} = 61.2\,\Omega$.

At 20% away from the design frequency, the matching line will be $0.25\lambda + 0.05\lambda = 0.3\lambda$ long. Equation (11.42) must be used in this case, with $Z_L = 75\,\Omega$ and $Z_0 = 61.2\,\Omega$.

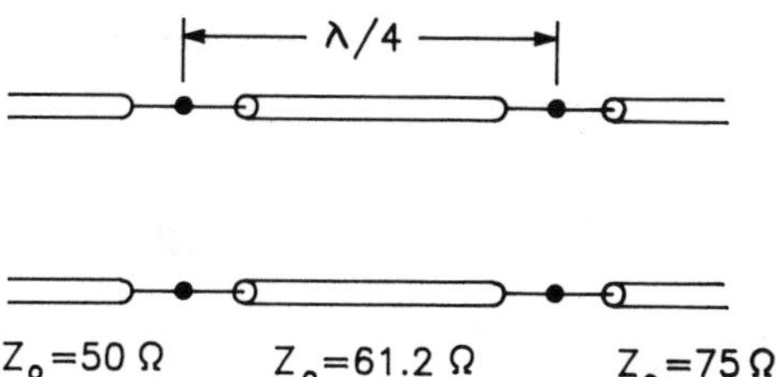

Figure 11.10. Matching using a quarter wavelength line

After evaluation this gives $Z_b = 51.4 + j6.6\,\Omega$. Thus the impedance is no longer matched to the $50\,\Omega$ line, mainly due to the appearance of an reactive component. The reflection coefficient, ρ, is given by equation (11.36) as $|\rho| = 0.063$.

MATCHING USING A STUB TUNER

Matching using a section of short circuit line is usually called matching with a stub tuner. The principle of the method is to find a point along the line from the load where the real part of the impedance is equal to the characteristic impedance. At this point a section of short circuit transmission line is added, Figure 11.11. Since the input impedance of a short circuit line is purely reactive, the length of the short circuit line can be chosen to exactly cancel the reactive part of the impedance of the main transmission line. Equation (11.45) shows that the reactive impedance of a short circuit line can take on any value between positive and negative infinity. The procedure can be demonstrated with an example. Impedances will be used but the same principles would apply with admittances.

Example 11.5 Matching using a stub tuner

Find the position and length of a stub tuner to match a load which has an impedance $Z_L = 0.5Z_0 + jZ_0$ where $Z_0 = R_0$.

The real (R_b) and imaginary (X_b) parts of the impedance down the line from the load has the form shown in Figure 11.5. Figure 11.12 shows an enlarged portion of Figure 11.5. At points marked with an X the impedance of the line is $Z_b = R_0 + jX_b$.

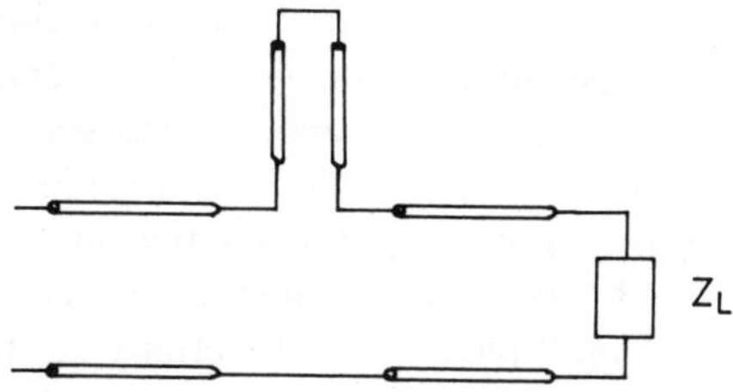

Figure 11.11. Matching using a stub tuner added in series with a transmission line

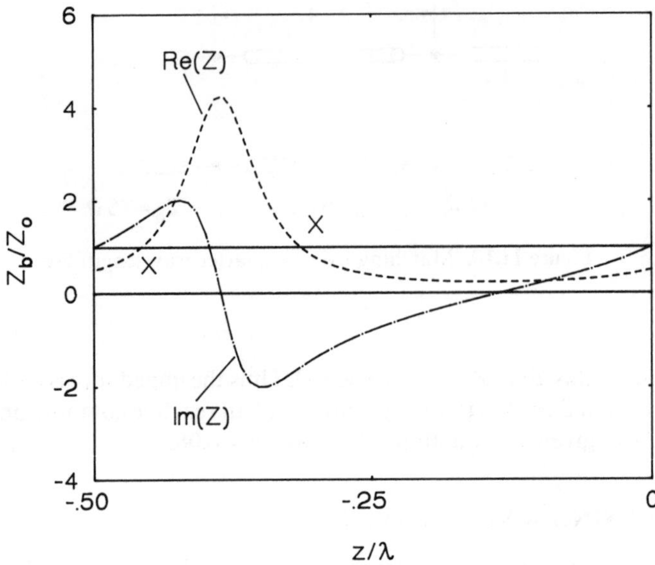

Figure 11.12. Impedance along a transmission line with $Z_L = 0.5\ Z_0 + jZ_0$. Enlarged portion of figure 11.5.

At these points adding, an imaginary impedance of value $-jX_b$ will cancel the imaginary part of Z_b and at all points further down the line from the matching section $Z_b = R_0$. The imaginary impedance $-jX_b$ can be provided by a short circuit section of line chosen of length b_1 where (from equation (11.45), $\beta b_1 = \tan^{-1}(-X_b/Z_0)$.

In this example there are two X points within half a wavelength, at distances of 0.315λ and 0.455λ from the load. The stub tuner could be inserted at either of these points, or at these distances plus n half wavelengths back down the line. At $0.31\,5\lambda$ from the load, $X_b/Z_0 = -1.55$ so $\beta b_1 = 1.0$ and the required length of the stub tuner is $b_1 = 0.16\lambda$. At 0.455λ from the load, $X_b/Z_0 = 1.60$ and the length must be $b_1 = 0.34\lambda$. The length, b_1 could be increased by half a wavelength if this made the physical stub length more acceptable.

MATCHING USING MULTIPLE SECTIONS

The principle of matching is a very powerful tool and is widely used in microwave circuits to match the load to the source to the transmission line. The matching is frequency sensitive since the impedance is a function of $\beta b = 2\pi b/\lambda$, thus the transmission line will only be matched at the design frequency. At frequencies away from the design frequency, equation (11.42) must be used to find the impedance. Much effort has been devoted to finding methods of improving the relative match over a wider band of frequencies. This can be done by using two or more stub tuners instead of the one tuner used in the last example. The multiple matching sections can be designed to cancel the inherent frequency sensitivity so that the bandwidth over which a low reflection coefficient is obtained is increased.

11.8 SMITH CHART

The Smith chart is a graphical aid for dealing with reflection coefficients, voltage standing wave ratios and impedance along a transmission line. It was invented by P. Smith at Bell Telephone Laboratories in 1939. Although the availability of computers and calculators has lessened the need for graphical aids, it is often the case that the output from a computer package will be displayed graphically as a Smith chart. Generations

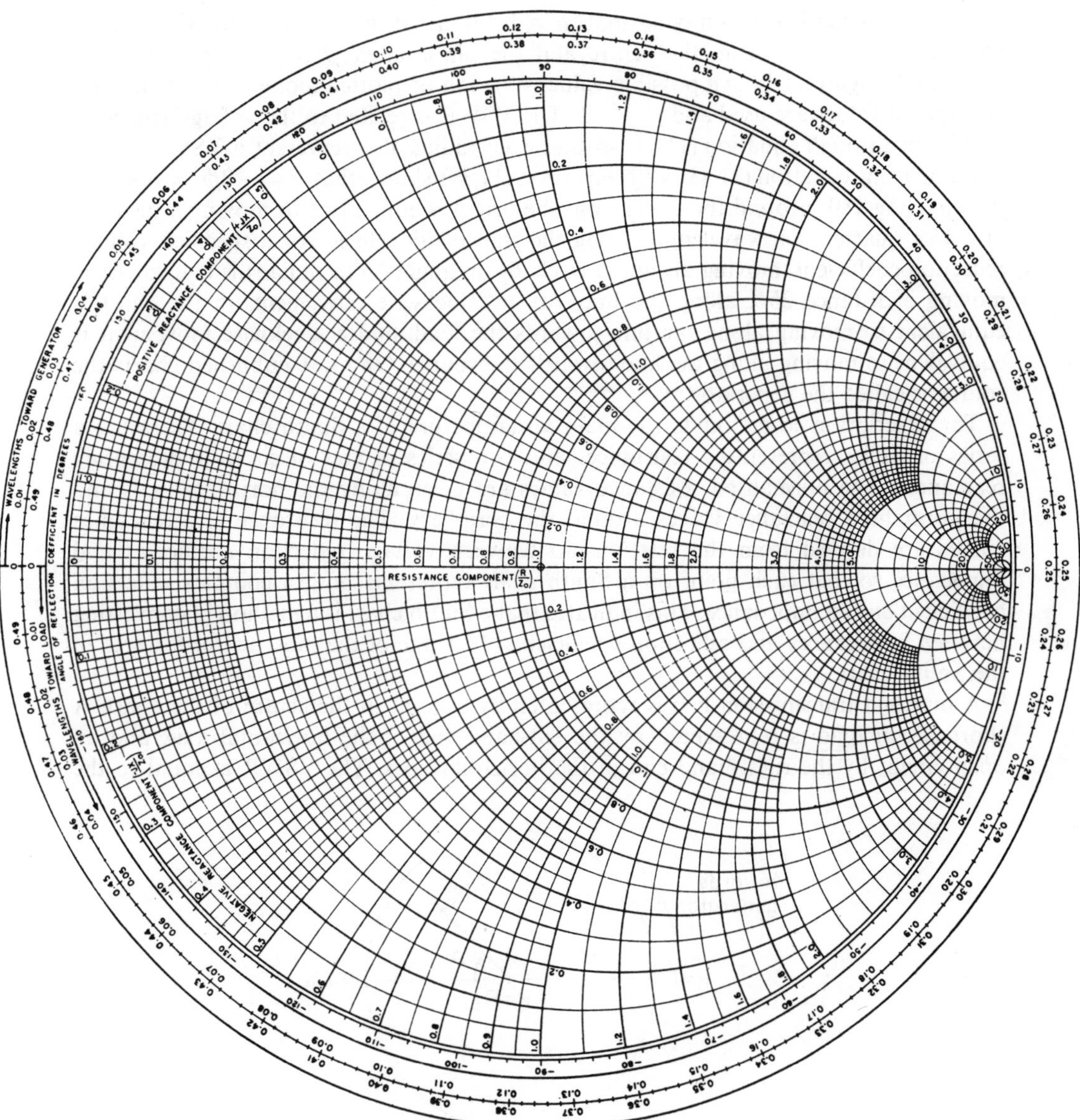

Figure 11.13. The Smith chart

of microwave engineers have grown up with it and it is very widely used in the micro wave industry for the design of matching circuits and observations of the frequency performance of transmission line systems.

A standard Smith chart is shown in Figure 11.13. This appears to be a complicated chart because as much information as possible is condensed onto the chart. It is basically a polar plot of the voltage reflection coefficient. Expressing the voltage reflection coefficient along a line, equation (11.39) in polar form

$$\rho = |\rho| e^{j\theta} \tag{11.48}$$

where ρ is normalised to unity so that $|\rho| \leqslant 1$. On a Smith chart the magnitude $|\rho|$ is plotted as the radius from the centre and the phase angle θ is the angle from the horizontal, Figure 11.14. All possible values of ρ can be plotted along a radius because normalised values are used and $\rho_{max} = 1$. The VSWR is related to the voltage reflection coefficient through equation (11.37) so that each radius also represent a VSWR.

Comparing equation (11.48) with equation (11.39) shows that $\theta = -2\beta b = -4\pi b/\lambda$, thus moving 180° around the circle from the origin is equivalent to moving a distance of $\lambda/4$ or quarter of a wavelength along the transmission line. For a specified voltage reflection coefficient, moving around a circle by an angle $\Delta\theta$ in radians is equivalent to moving a distance $2b/\lambda$ along the line. An anticlockwise movement on the chart is equivalent to moving along the line towards the load, and a clockwise movement is equivalent to moving along the line towards the source. On most published charts the source is called the generator.

The description so far omits the main visual feature of a Smith chart. These are a series of circles representing resistances and two series of arcs of circles representing reactances. The normalised resistance values are given along the horizontal axis and the normalised reactance values around the periphery. The resistance and reactance curves are derived from the relationship between voltage reflection coefficient and impedance so that each point on a Smith chart represents both a complex reflection coefficient *and* an impedance. This indicates that a Smith chart is a method of converting between impedances and reflection coefficients. It is one of the most useful features of the Smith chart.

A matched transmission line is represented by a point in the centre of the chart. A short circuit load is represented by the left-hand end of the horizontal axis, and an open circuit by the right-hand end of the horizontal axis. The VSWR is equal to the

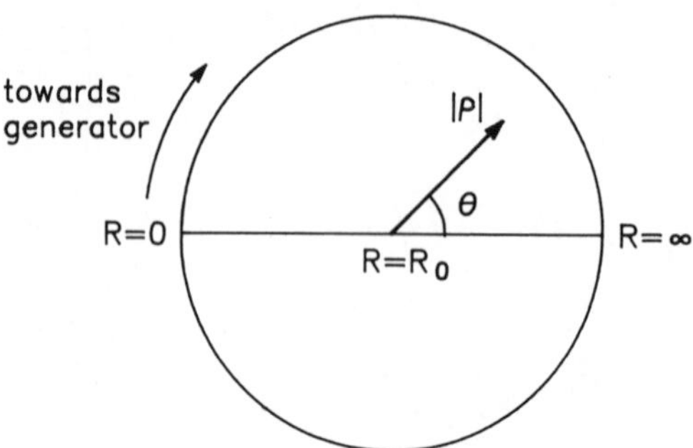

Figure 11.14. Principle of the Smith chart

normalised resistance for a pure resistive load so the normalised resistance values on the right-hand side of the horizontal axis also represent values of VSWR. An example will show the basic principles of use of the Smith chart.

Example 11.6 The Smith Chart

A load of value $20 + j40\,\Omega$ is connected to a transmission line with $Z_0 = 100\,\Omega$. Use a Smith chart to find the load admittance, the VSWR, the voltage reflection coefficient at the load and the impedance at a distance 0.1λ towards the generator.

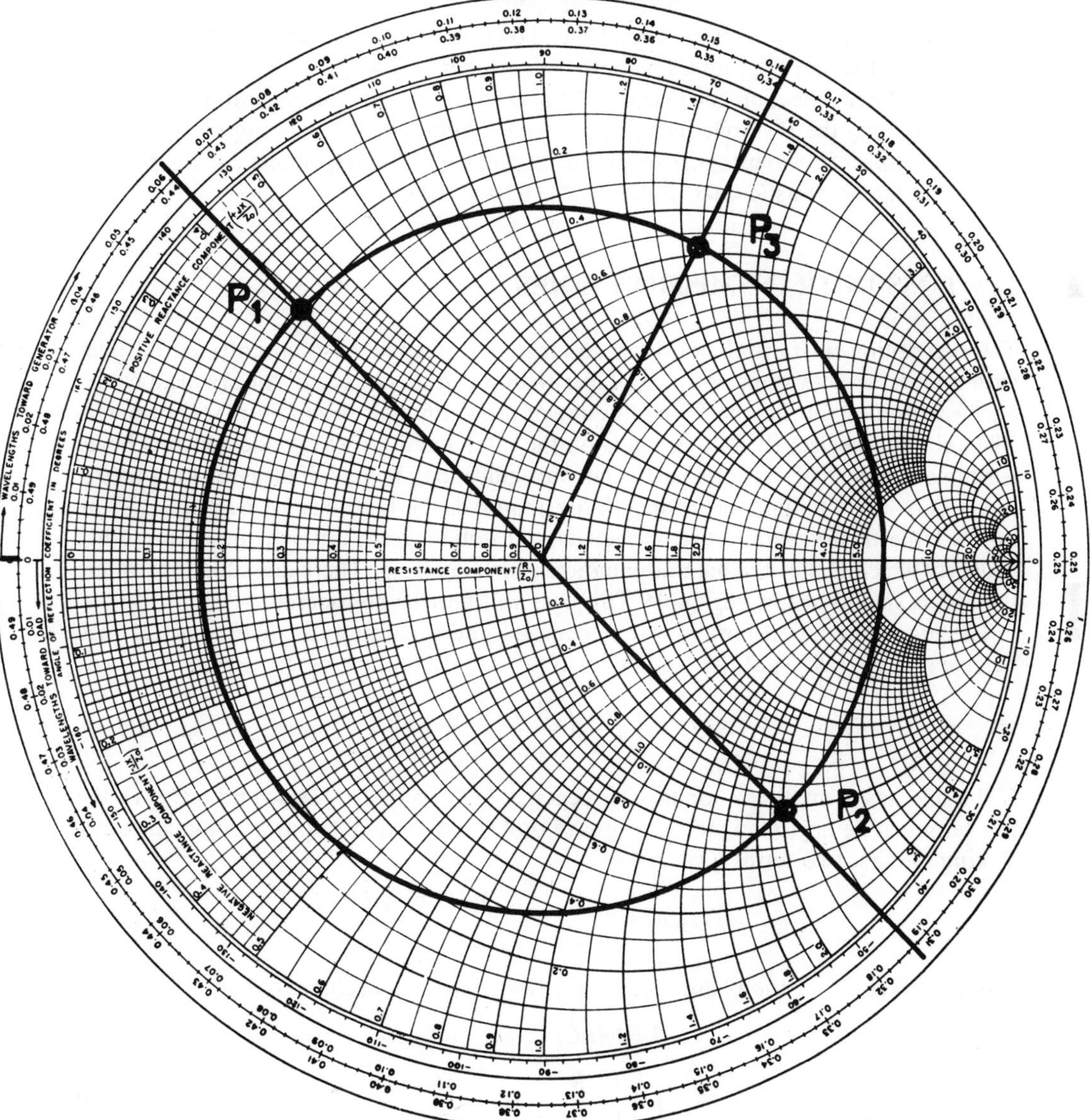

Figure 11.15. Use of the Smith chart in transmission line calculations for Example 11.6

Figure 11.15 shows a Smith chart with the superimposed data for this example. The load impedance is normalized to the characteristic impedance of the line to give $\bar{Z}_L = 0.2 + j0.4$. The impedance circles on the chart are then used to find point P_1 and a circle drawn whose radius is the distance from the centre of the chart to P_1. The admittance can be easily found by noting that an impedance transforms to an admittance through a $\lambda/4$ line. This means moving 180° around the chart on the circle of radius P, to reach P_2. This gives $\bar{Y}_L = 1 - j2$.

The VSWR is given by the point at which the P_1 circle crosses the axis. Reading off the value gives VSWR = 5.7. The magnitude of the normalised load reflection coefficient is the radial length from the centre to P_1 which is $|\rho| = 0.7$. The phase angle of the reflection coefficient can be read off the marked scale around the periphery, at the point where the radial line cuts the periphery. This gives $\theta = 135°$.

The voltage reflection coefficient a distance 0.1 λ towards the generator is given by moving along the P_1 circle by an appropriate angle. The outer periphery is marked with distances in wavelengths so this gives point P_3. The impedance at this point is read off the chart as $\bar{Z} = 0.6 + j1.5$ or $Z = 60 + j150\,\Omega$.

11.9 LOSS IN TRANSMISSION LINES

The description of transmission lines so far in this chapter has concentrated on lines with no loss, or with sufficiently small loss than it can be neglected. All practical cables and waveguides will have a finite amount of loss, although generally it will be small. The ability to neglect the loss depends on the total length of waveguide which is being used in a system. The loss of a particular waveguide may be high if measured per metre, but if only a few millimetres of length are required, the loss may be negligible. If the loss is large then generally the waveguide will not be used, unless it has some other property which makes the high attenuation worth ignoring. There are situations where the loss is small but not negligible. This case will now be considered.

The propagation constant, γ, will be complex and according to equation (11.25) will be given by $\gamma = [(R + j\omega L)(G + j\omega C)]^{1/2}$. If $\omega^2 LC$ is taken out of the expression, then

$$\gamma = j\omega(LC)^{1/2}\left[1 - j\left(\frac{R}{\omega L} + \frac{G}{\omega C}\right) - \frac{RG}{\omega^2 LC}\right]^{1/2} \tag{11.49}$$

If the line has a small loss, or the frequency is high so that $\omega L \gg R$ and $\omega C \gg G$, the last term can be neglected, and the square root can be expanded as a binomial series. Taking the first two terms gives

$$\gamma = j\omega(LC)^{1/2}\left[1 - \frac{j}{2}\left(\frac{R}{\omega L} + \frac{G}{\omega C}\right)\right]$$

Using $Z_0 = (L/C)^{1/2}$, the phase and attenuation constants become

$$\beta = \omega(LC)^{1/2}$$

$$\alpha = \frac{1}{2}\left(\frac{R}{Z_0} + GZ_0\right) \tag{11.50}$$

Only a small loss has been assumed, thus the phase constant, β, has the same value as for the loss-less line. This is dependent on assuming that the characteristic impedance is not modified by any loss. The attenuation is apparently independent of frequency, however, both the resistance and conductance contain frequency dependent terms.

Example 11.7 Phase and attenuation constants for a lossy transmission line

Find the phase constant and attenuation constant for the parallel plate waveguide of Example 11.2.

Example 11.2 evaluated the circuit parameters for a parallel plate waveguide. This gave $L = 0.84\,\mu$H/m, $C = 33.2$ pF/m, $R = 17.4\,\Omega$/m, $G = 0.002$ S/m and $Z_0 = 159\,\Omega$.

Inserting these values into equation (11.50) gives $\beta = 332$ rad/m and $\alpha = 0.21$ Nepers/m = 1.85 dB/m. It is interesting to compare these values with that obtained in Example 10.2 and 10.3 based on electromagnetic wave theory. The dielectric loss is given by equation (10.35) and the conductor loss by equation (10.42). From these equations, or Figures 10.15 and 10.16, the total attenuation is 1.81 dB, which is close to the value found in this example using transmission line theory.

The impedance transformation relationship, equation (11.42), can be easily modified for the small loss case by replacing jβ by γ in equation (11.42). This then gives

$$Z_b = Z_0 \frac{Z_L + jZ_0 \tanh \beta b}{Z_0 + jZ_L \tanh \beta b} \tag{11.51}$$

The only difference compared to equation (11.42) is that $\tan(\beta b)$ is replaced by the complex $\tanh(\beta b)$ function.

11.10 MULTIPLE DIELECTRIC LAYERS IN FREE SPACE

The transmission line approach to solving waveguide problems can also be applied to plane waves in free space. Recall that the characteristics of the TEM waves in parallel plate waveguide are identical to the characteristics of plane waves in free space. Hence an analogy which works for TEM waves in waveguides should also work for plane waves in free space. This is the case and, as far as the plane wave is concerned, it can be represented by a transmission line. This analogy means that the transmission line equations and matching principles can be applied to layers of dielectrics where plane waves are incident on the dielectric. This problem was discussed in Section 6.8 using electromagnetic waves. The manipulation of the multiple electric and magnetic fields was, however, complicated and only the special case of no reflections was studied. The transmission line analogy using the impedance and admittance equations, equations (11.42) and (11.43) can be applied to solve the problem of multiple dielectric layers. The full information of the performance over a band of frequencies is then available.

Consider a sheet of dielectric with a plane wave incident on it, Figure 11.16. The impedance at the first interface can be deduced from equation (11.42) and the special cases. For instance if the dielectric sheet is half a wavelength thick then the impedance at the first interface will be the same as the impedance at the second interface which is the impedance of free space. Hence the plane wave will be matched and there will be no reflections. The physical thickness of the dielectric sheet will depend on the relative

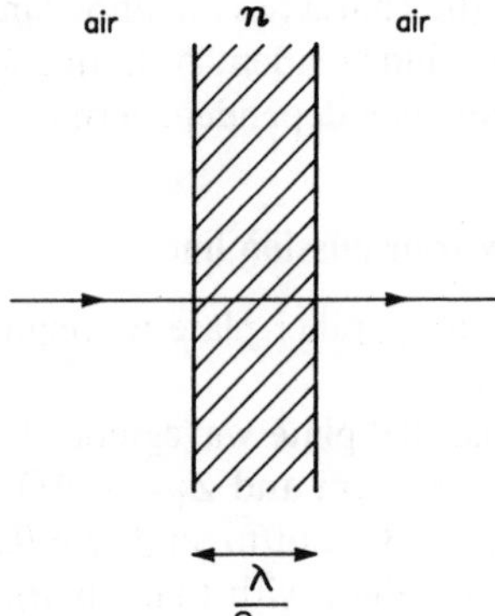

Fig. 11.16. Half wavelength thick sheet of dielectric matched to surrounding air

permittivity of the dielectric and must be $\lambda/2n$ thick, where λ is the free-space wavelength and n is the refractive index of the dielectric.

The quarter wavelength matching principle can be similarly applied to plane wave matching. A common requirement is to match a dielectric slab to free space. This can be done by placing a quarter wavelength thick sheet of dielectric in front of the slab, Figure 11.17. If the refractive index of the slab is n_1 and of the matching sheet is n_2, then at the second interface the impedance of a plane wave will be $Z_L = Z_0/n_1 = 377/n_1\,\Omega$. The impedance at the first interface is given by equation (11.44) so that the characteristic impedance, Z'_0, of the matching sheet needs to be

$$Z'_0 = (Z_0 Z_L)^{1/2} = \frac{377}{n_1^{1/2}} \tag{11.52}$$

but $Z'_0 = 377/n_2$ therefore

$$n_2 = n_1^{1/2} \tag{11.53}$$

If the refractive index of the matching sheet is chosen to be the square root of the refractive index of the dielectric slab, and the sheet is $\lambda/(4n_2)$ thick, the plane waves will

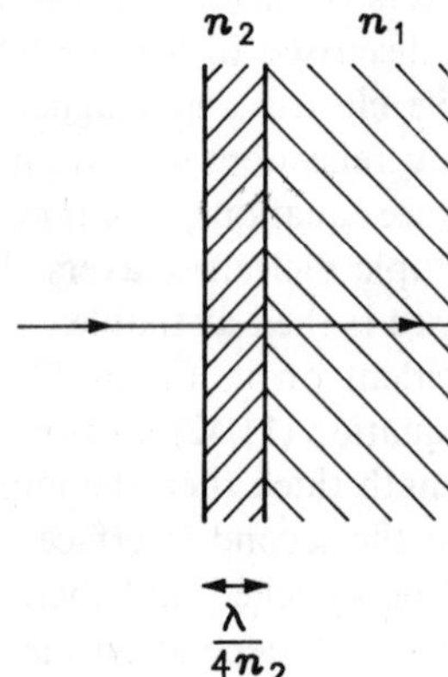

Figure 11.17. Quarter wavelength matching plate

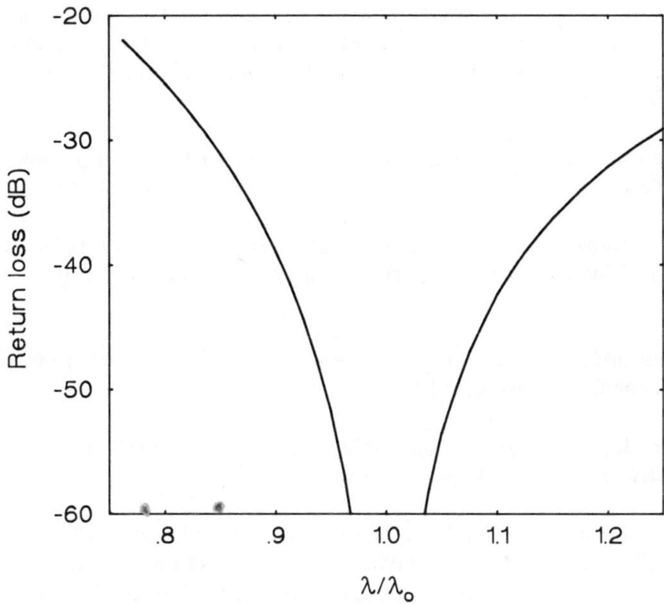

Figure 11.18. Return loss against normalised wavelength for quarter wave matching plate

be matched. This principle is widely used in optics to match lenses to free space. It is then called *blooming* of the lens.

The process is frequency sensitive and an exact match is only obtained at the design frequency. The matched bandwidth can be improved by using multiple layers. This is particular necessary for optical systems using visible light because the bandwidth from red to violet light spans too wide a frequency band to be matched with a single layer.

Example 11.8 Matching plane waves using a quarter wave plate

Design a quarter wavelength matching plate to match a glass lens of refractive index $n_1 = 1.8$ to air for light at a design wavelength of 500 nm. Find the variation in return loss over a 50% bandwidth.

From equation (11.53), the refractive index of the matching plate is $n_2 = 1.342$ and the thickness must be $500/(4 \times 1.342)$ or 93 nm. It can also be $93 + n\lambda_2/2$ thick.

The impedance of the glass lens is $Z_L = Z_0/n_1 = 198.4\,\Omega$. The impedance of the matching plate is $Z'_0 = Z_0/n_2 = 280.9\,\Omega$. The mismatch as the wavelength is changed is obtained from equation (11.42), and the return loss from equation (11.38). This is plotted against λ/λ_0 in Figure 11.18. λ_0 is the design wavelength of 500 nm. The plot shows how the match gradually deteriorates away from the design wavelength. The return loss is below -30 dB from $\lambda/\lambda_0 = 0.83$ to $\lambda/\lambda_0 = 1.23$.

PROBLEMS

1. Find expressions for the equivalent voltage and current along a coaxial cable, using the electric and magnetic field equations in Chapter 2.

2. An air-filled parallel plate waveguide has a separation distance of 5 mm and effective width of 20 mm. The plates are made of aluminium. Find the equivalent circuit inductance, capacitance, resistance, conductance and characteristic impedance at frequencies of 100 MHz, 500 MHz and 1 GHz.

3. Calculate the series inductance and shunt capacitance for a transsmision line with a characteristic impedance of 50 Ω and a velocity of propagation of 1.5×10^8 m/s.

4. A uniform transmission line has elements $R = 10\,\text{m}\Omega/\text{m}$, $G = 0.7\,\mu\text{S/m}$, $L = 1\,\mu\text{H/m}$ and $C = 1\,\text{nF/m}$. At 1 MHz find (a) the characteristic impedance; (b) the velocity ratio v/c.

5. A loss-less 50 Ω transmission line is terminated in 25 + j50 Ω. Find (a) the voltage reflection coefficient; (b) the VSWR.

6. The return loss of a 50 Ω coaxial cable is measured as 20 dB. The cable is connected to a resistive load. Find the load resistance.

7. A loss-less transmission line has a characteristic impedance of 75 Ω. It is connected to a load of (75 + j25) Ω. Plot the amplitude and phase of the reflection coefficient on the transmission line over a half wavelength distance back from the load.

8. Calculate the impedance 0.3λ from a (25 + j50) Ω load of a 50 Ω transmission line. Find the first point along the line where the impedance is purely resistive.

9. Prove that maximum power transfer from a generator of impedance Z_g to a load of impedance Z_L occurs when $Z_g = Z_L^*$.

10. A loss-less transmission line is terminated in a load which gives a normalised reflection coefficient of amplitude 0.75 and phase 45°. Find the distance from the load to the first current minimum at a frequency of 3 GHz.

11. The normalised impedance (Z/Z_0) at a point 0.3 λ from a load is (0.6 − j0.5). Find the value of the normalised load impedance.

12. A 150 mm long section of coaxial cable with an inner diameter of 5 mm and an outer diameter of 16.3 mm is connected to a load of impedance 100 Ω. A second coaxial cable is connected to the end of the coaxial cable. This has the same inner diameter but the outer diameter is 11.5 mm. Find the input impedance seen by the second cable at (a) 500 MHz, (b) 1 GHz, (c) GHz.

13. A line with characteristic impedance R_0 is connected to a load with impedance $Z_L = jX$. Show that the magnitude of the reflection coefficient $|\rho|$ is unity for all points on the line.

14. A short circuit 50 Ω transmission line has an input voltage of j2.35 V and an input current of − 65 mA. Find the possible lengths of the line.

15. Sketch the voltage, current and impedance along a transmission line which is open circuit.

16. A band stop filter is formed by placing a quarter wave ($\lambda_0/4$) choke in series with a transmission line. Find the response, relative to that at λ_0 for wavelengths of λ_0, $\lambda_0 \pm 10\%$ and $\lambda_0 \pm 20\%$.

17. A 150 Ω transmission line is connected through a section of length d and characteristic impedance Z, to a load of $(250 + j100)$ Ω. Find d and Z which match the load to the 150 Ω line.

18. A line length d and characteristic impedance R acts as a transformer to match a load impedance of $(150 + j0)$ Ω to a 300 Ω line at 300 MHz. Find (a) d, (b) R and (c) VSWR on the transformer section at 400 MHz.

19. A quarter wave transformer with a characteristic impedance R is inserted into a 50 Ω line at a distance d from the load of impedance $(100 + j100)$ Ω. Calculate the values of d and R for a match.

20. An antenna with an input impedance of 10 Ω is connected to a coaxial cable with a characteristic impedance of 50 Ω. Design a quarter wave air-filled coaxial section to match the antenna to the cable.

21. A uniform loss-less 400 Ω transmission line is connected to a load of $(200 + j300)$ Ω. A short circuited line of length d_2 of the same transmission line is connected in parallel with the line at distance d_1 from the load. Find the shortest value of d_1 and d_2 for a match.

22. A stub tuner is placed in parallel with a 50 Ω transmission line in order to match a load with an admittance of $(0.1 - j0.05)$ S. Find the length of the stub tuner and its position along the line.

23. A double stub tuner has two short-circuited parallel lines spaced $0.125\,\lambda$ and $0.25\,\lambda$ from a load of impedance $(240 - j320)$ Ω. The characteristic impedance of the lines is 200 Ω. If the length of the first stub is $0.09\,\lambda$, find the length of the second stub for a match.

24. A load of impedance $(80 + j40)$ Ω is connected to a 50 Ω transmission line of length $0.4\,\lambda$. Use the Smith chart to find (a) the VSWR, (b) the voltage reflection coefficient, (c) the input impedance and (d) the distance from the load to the first voltage minimum.

25. Repeat Problem 24 for a load impedance of $(40 - j80)$ Ω.

26. Use the Smith chart to repeat Problem 8.

27. The input impedance of a 300 Ω transmission line is $(90 + j150)$ Ω at 1 GHz. Use the Smith chart to find the resistive load impedance and the length of the line.

28. Find the attenuation of 100 m of the transmission line described in Problem 4.

29. Find the attenuation for the parallel plate waveguide described in Problem 2 at 1 GHz.

30. A quarter wave sheet of foam dielectric is to be used to match a 9 GHz microwave

signal to a dielectric block which has a relative permittivity of 2.5. Calculate the relative permittivity and the thickness of the foam dielectric sheet.

31. A 50 Ω transmission line is matched to a 200 Ω load through two quarter wave sections with characteristics impedances of 70.7 Ω and 141.4 Ω. Show that the bandwidth of the double section is improved over that of a single section by finding the reflection coefficient at 10% above the design frequency for both a single $\lambda/4$ section and the double design described above.

12 Microwave waveguides

12.1 INTRODUCTION

This chapter takes the theory studied in the two previous chapters and applies it to practical waveguides which are used at microwave frequencies. The waves in microwave waveguides are almost entirely guided by metal conductors so that the waves are reflected off the conductors. Dielectric is used in between conducting plates as either a support or as a method of improving the performance. Waveguides made entirely from dielectric which use total internal reflection as the wall reflection mechanism are the subject of the next chapter. Chapter 10 explained the principles of electromagnetic guided waves using parallel plate waveguides. Chapter 11 described the alternative transmission line approach which demonstrated what happens when the waveguide is terminated in a finite load. This showed how it is possible to match out the reflections which take place at junctions between components in waveguide systems. Some of the typical waveguide components are briefly described in this chapter.

Chapter 10 showed that waves propagate along a waveguide in a series of modes which are designated TEM, TM and TE modes. TEM modes exist at all frequencies up from DC. TM and TE modes have a cut off frequency below which they cannot travel along the waveguide. This behaviour was found in a parallel plate metal waveguide, but it applies to all types of waveguide, with one exception. The exception is that in some types of microwave waveguide, the TEM mode cannot exist so that there is a lowest frequency below which no power will propagate. Hollow rectangular and circular waveguides are examples of these types of waveguides. Practical waveguides can be divided into two classes by this distinction. There are waveguides which use the TEM mode of propagation and avoid operating above the cut-off frequency of the TM_1 and TE_1 modes. The two main types within this class are the coaxial cable and microstrip waveguide. The two-wire pair also supports a TEM mode, but this is very rarely used at microwave frequencies because it is an open waveguide where the fields occupy the space around the conductors and are not contained by conductors. It will therefore not be considered in this chapter. The TEM class of waveguides are easy to design and use. They are mainly used at low microwave frequencies partly because of the need to avoid higher order modes and partly because the attenuation due to losses in the conductors and dielectrics increase with frequency. Their inherent simplicity has, however, prompted the development of designs which work at higher microwave frequencies and the upper frequency of use keeps rising.

Rectangular and circular waveguides, which do not support the TEM mode, are more difficult to manufacture because of the high mechanical accuracy required. The size is proportional to the wavelength of the microwaves which means that at low microwave frequencies they can be large, bulky and expensive. They do, however, have good electrical properties and are widely used for microwave systems at frequencies above about 10 GHz. Rectangular waveguide in particular has field patterns which enable a variety of passive components such as attenuators, phase shifters and junctions to be constructed. For this reason rectangular waveguide is the preferred type of microwave waveguide for many microwave system development and measurements.

The electrical parameters which are of interest in microwave waveguides are

Electric and magnetic field patterns

Cut-off frequencies of modes

Impedance of modes

Guide wavelength of modes

Velocities of modes

Attenuation of modes

These parameters were studied in Chapter 10 for parallel plate waveguides and will be discussed for the different types of waveguides below. Not all parameters are relevant for each structure. TEM waveguides have a guide wavelength and phase velocity which are equal to the free space values.

12.2 COAXIAL CABLE

Coaxial cable, Figure 12.1, is used for interconnecting microwave components at low microwave frequencies. It is very widely used at frequencies below about 5 GHz, but is also used where possible at frequencies up to about 20 GHz. The basis of the coaxial cable is the coaxial waveguide and the basis of the coaxial waveguide is two coaxial cylinders. It is possible to analyse coaxial waveguide exactly by solving the full electromagnetic boundary problem using the theory developed in Chapter 10. This is complicated because an annular cylindrical geometry is involved. It is necessary if a detailed study of the higher order modes is needed, but it is not necessary if only the TEM mode is to be used—which is the normal situation. In this case the quasi-static fields on coaxial cylinders which were studied in Chapters 2 and 4 can be used. The results are directly applicable to coaxial cables because they describe the TEM mode of propagation. The transmission line parameters for loss-less coaxial waveguide were

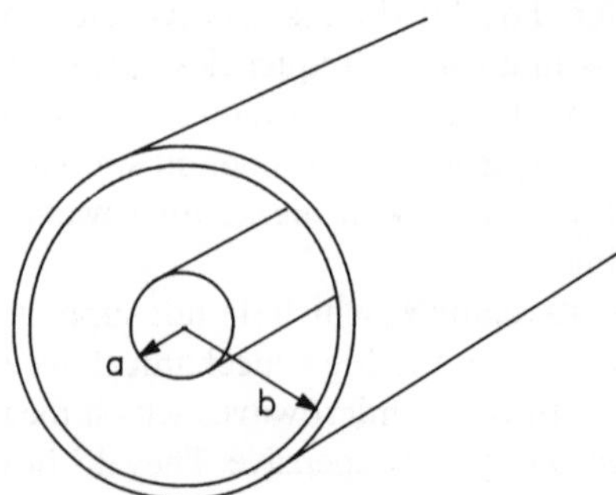

Figure 12.1. Coaxial cable

derived in Chapter 11 using the quasi-static results. These previous results will now be collected together.

The fact that the basic fields in coaxial cable form a TEM mode can be demonstrated physically by taking a simple parallel plate waveguide with E_x and H_y fields, Figure 12.2(a) and bending it around to form a coaxial waveguide, Figures 12.2(b) and 12.2(c). The fields remain qualitatively the same. The electric field becomes E_r and the magnetic field becomes H_ϕ. These two fields form a cylindrical orthogonal pair of fields which correspond to the plane wave fields in rectangular coordinates. For plane waves the electric and magnetic fields were related by $E_x = (\eta_0/\varepsilon_r^{1/2})H_y$. In the coaxial waveguide they are similarly related by $E_r = (\eta_0/\varepsilon_r^{1/2})H_\phi$, where η_0 is the impedance of free space ($= 120\pi\,\Omega$ or $377\,\Omega$) and ε_r is the relative permittivity of the dielectric between the coaxial cylinders. The electric and magnetic fields for a coaxial line with a potential of V_0 between the inner conductor and outer cylinder are

$$E_r = \frac{V_0 \exp[j(\omega t - \beta z)]}{r \ln(b/a)} \tag{12.1a}$$

$$H_\phi = \frac{V_0 \exp[j(\omega t - \beta z)]}{\eta_0 r \ln(b/a)} \tag{12.1b}$$

where a is the radius of the inner conductor and b is the radius of the outer conductor. The fields are sketched in Figure 12.3.

The characteristic impedance was derived in Chapter 11, equation (11.31) as

$$Z_0 = \frac{\eta_0}{2\pi\varepsilon_r^{1/2}} \ln\left(\frac{b}{a}\right) \tag{12.2}$$

Coaxial cables for domestic use at VHF and UHF usually have a characteristic impedance of 75 Ω, but most professional coaxial cable for use of both RF and microwaves has a characteristic impedance of 50 Ω.

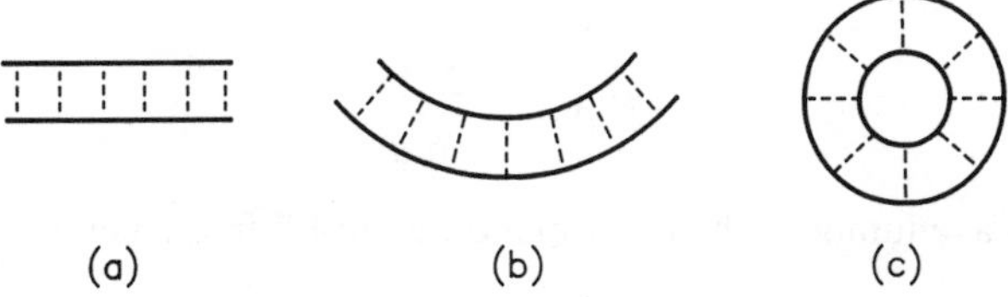

Figure 12.2. Development from (a) parallel plate waveguide to (b) coaxial waveguide

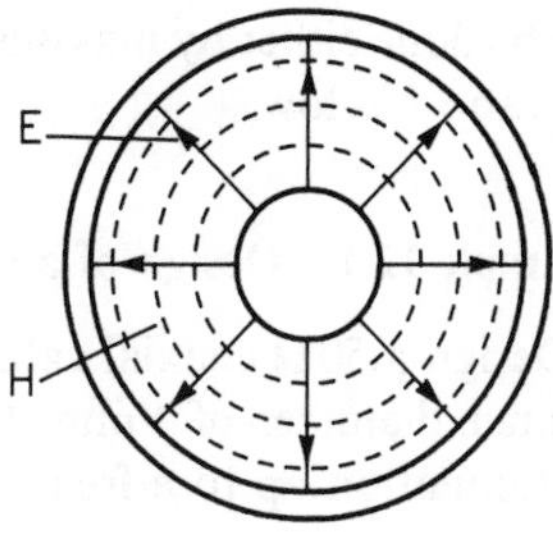

Figure 12.3. Electric and magnetic fields of TEM mode in a coaxial cable

The attenuation of coaxial waveguide can be found using the expression for low loss transmission line, equation (11.50)

$$\alpha = \frac{1}{2}\left(\frac{R}{Z_0} + GZ_0\right) \tag{12.3}$$

The resistance, R, is caused by the loss in the conductors so that it is approximately the surface resistivity, R_s, divided by the area. For unit length the area is the circumference of both cylinders, so

$$R = \frac{R_s}{2\pi a} + \frac{R_s}{2\pi b} \tag{12.4}$$

R_s is given by equation (8.21) as $R_s = (\omega\mu_0/2\sigma)^{1/2}\ \Omega$. The conductance, G, is caused by the loss in the dielectric and is the reactance times the loss tangent, or

$$G = \omega C \tan\delta \tag{12.5}$$

The capacitance, C, was found in Chapter 2, equation (2.29). Combining all the parts into equation (12.3) gives the attenuation of coaxial waveguide as

$$\alpha = \frac{\varepsilon_r^{1/2}}{2}\left[\frac{R_s}{\eta_0 \ln(b/a)}\left(\frac{1}{a} + \frac{1}{b}\right) + \eta_0\omega\varepsilon_0 \tan\delta\right] \tag{12.6}$$

The relative significance of the conductor and dielectric loss parts of the attenuation will be shown in Example 12.1.

The maximum frequency of operation is determined by the cut-off frequency of the next highest order mode. This can be determined with an exact analysis of the coaxial waveguide by solving the wave equation in cylindrical coordinates and applying the boundary conditions on the two cylinders. The expression that results is a combination of Bessel functions which must be solved iteratively on a computer. This will not be stated here because all that is needed is an approximate relationship that enables the cut-off frequency to be calculated. A reasonable relationship that is within a few percent of the correct value is

$$K_c a = \frac{2}{1 + b/a}$$

where K_c is the wavenumber. The corresponding cut-off frequency is

$$f_c = \frac{c}{\varepsilon_r^{1/2}} \frac{1}{\pi a(1 + b/a)} \tag{12.7}$$

It is normally desirable to make f_c as high as possible. Equation (12.7) shows that this can be done either by making the inner diameter small, or to a limited extent, by keeping the ratio b/a low.

Example 12.1 Design of a coaxial cable

Design a $50\,\Omega$ coaxial cable with solid copper conductors, a PTFE dielectric and an inner diameter of 1 mm. Find the cut-off frequency of the next mode and plot the attenuation up to a frequency of 40 GHz.

PTFE has a relative permittivity of 2.1 and a loss tangent of 0.0001, copper has a conductivity of 5.8×10^7 S/m. Equation (12.2) is used to find the ratio b/a needed to give an impedance of $50\,\Omega$, $b/a = \exp[50 \times (2.1)^{1/2} \times 2\pi/377)] = 3.35$. With $2a = 1$ mm this gives the diameter of the outer as $2b = 3.35$ mm. Inserting the values for a, b and ε_r into equation (12.7) gives the bandwidth as 30.3 GHz.

Equation (12.6) is evaluated and plotted in Figure 12.4 for these parameters. This shows that the conductor loss is dominant, partly due to the specified dielectric which has a low loss.

Coaxial cable is made in a variety of packages for microwave use. Flexible coaxial cable has a braided outer conductor made of many helix wound strands of thin copper wire. For maximum flexibility at low microwaves and RF, the inner conductor is also composed of a number of strands of wire. The dielectric is either PTFE or polyethelene. The braided outer conductor increases the losses considerably and imposes a limit on the upper frequency of use. A typical cable, designated RG58, is composed of an inner conductor made up of 19 0.18 mm tinned copper wires surrounded by polyethylene dielectric and a tinned copper braid inside a black PVC sheath. This has a quoted attenuation at 1 GHz of 0.76 dB/m. Its attenuation increases at least proportional to frequency above 1 GHz and limits its use to short distances.

To overcome the losses, the outer conductor must be made from a solid tube of copper. This considerably reduces the flexibility, although by using a helix shape to the outer, some flexibility is retained. At frequencies in the range 5 to 20 GHz, reasonable performance together with bendability can be obtained by making the overall diameter small. The dimensions worked out in Example 12.1 are typical of this type of coaxial cable, which is sometimes called solid jacketted coax. In a type designated RG402, the inner is made of silver coated, copper clad steel. This is surrounded by PTFE dielectric

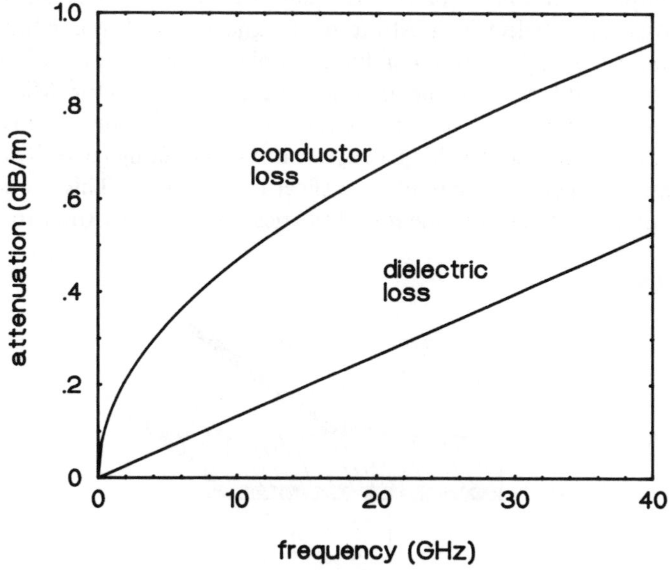

Figure 12.4. Attenuation in coaxial cable with PTFE dielectric for Example 12.1

within a seamless copper tube with an overall diameter of 3.58 mm. The quoted attenuation at 18 GHz is 1.95 dB/m. A smaller diameter version, designated RG403 has a similar construction but with an overall diameter of 2.17 mm. The penalty for the smaller size is that the attenuation goes up to 3.75 dB/m.

The dielectric filling holds the inner and outer in the correct coaxial spacing but limits the power handling capability. For applications which require high powers to be transmitted, such as from a transmitter to an antenna in a microwave radio relay system, air filled coaxial cable is used. The outer is supported by periodic rings of dielectric.

At both ends of a length of coaxial cable there will be a coaxial connector to attach the cable to instruments or equipment. The connector is an important part of the system which is inevitably a compromise between the need to make a component which is mechanically rugged and can withstand repeated handling yet at the same time does not introduce much mismatch to the signals. Many designs have been evolved to try to accomplish these ideals. Some discontinuity is bound to exist so that small mismatch always occurs with typical reflection coefficients of a few percent. Popular types are the N series which has an outer diameter of 19 mm, BNC which is an RF clip on connector and SMA connectors which are small diameter connectors that can be used up to 18 GHz.

12.3 MICROSTRIP

The parallel plate waveguide described in Chapter 10 and shown in Figure 10.5 is obviously impractical because (a) an infinite width was assumed in the y direction and (b) there is nothing to support the plates. Practical implementations of the parallel plate waveguide must have a finite width and this will modify the electrical properties of all the modes. The obvious method of supporting the plates is to use a dielectric to fill the gap. This will have to be of low loss so that it does not introduce significant attenuation.

A popular implementation of the parallel plate waveguide is *microstrip*, Figure 12.5. This is composed of a dielectric substrate of thickness h, with a conductor deposited onto the bottom surface. A strip of conductor is placed on the top surface of width w. The waveguide is then formed in the area below the top conductor. Microstrip can be made by the same technique as low frequency printed circuits by starting with copper covering all the top surface and etching away the non-conducting parts. It is thus possible to make a complete microwave circuit on a sheet of microstrip. This is one reason why microstrip is popular. In addition the total thickness of the microstrip can be made very

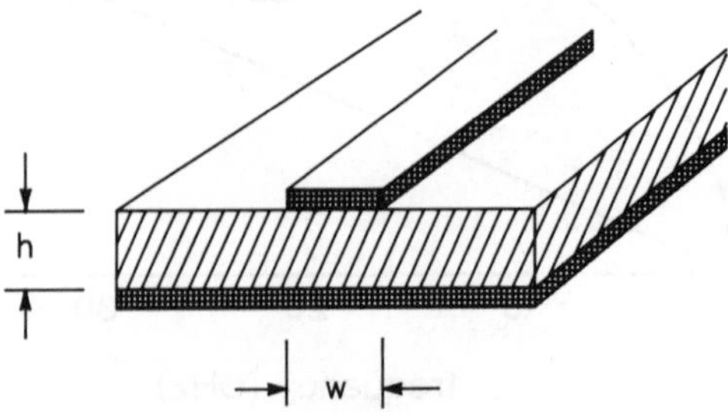

Figure 12.5. Microstrip with copper base, dielectric substrate and copper strip

small so that a low profile system can be designed. A stable, homogeneous, low loss dielectric is required and this explains why microstrip is a relatively recent development in the history of microwaves. It was only in the 1970s that pure dielectrics became available. Prior to this the only TEM waveguide which was easily available was the coaxial cable.

The finite width of microstrip means that there will be fringing fields at the edges of the strip, similar to those discussed in Chapter 4. These are sketched in Figure 12.6. The fringing fields mean that the exact analysis of microstrip is extremely difficult. Consequently considerable effort has been devoted to deriving empirical formulae which will predict the performance. These are based on considering the microstrip as a parallel plate waveguide filled with dielectric, but replacing the actual relative permittivity of the dielectric by an effective permittivity, ε_e, which accounts for the fringing fields. The velocity of propagation in the dielectric becomes $c/\varepsilon_e^{1/2}$ and the propagation constant becomes $2\pi\varepsilon_e^{1/2}/\lambda$. There are many empirical forms for ε_e. A popular one is

$$\varepsilon_e = \frac{1}{2}\left((\varepsilon_r + 1) + \frac{(\varepsilon_r - 1)}{(1 + 12h/w)^{1/2}}\right) \tag{12.8}$$

The characteristic impedance also needs to be modified and this becomes

$$Z_0 = \begin{cases} \dfrac{60}{\varepsilon_e^{1/2}} \ln\left(\dfrac{8h}{w} + \dfrac{w}{4h}\right) & \text{for } w/h \leqslant 1 \\ \dfrac{\eta_0}{\varepsilon_e^{1/2}}\left[\dfrac{w}{h} + 1.393 + 0.667\ln\left(\dfrac{w}{h} + 1.444\right)\right]^{-1} & \text{for } w/h \geqslant 1 \end{cases} \tag{12.9}$$

Two equations are needed to cover the full range of w/h values. The design process is complicated because iteration is needed to find the correct w/h ratio, as will be illustrated in the next example.

There are two parts to the attenuation: that due to the dielectric loss and that due to the conductor loss. The dielectric loss can be found from a modified parallel plate formulae, equation (10.35). The fields around the microstrip are partly in dielectric and partly in air so the modified attenuation due to dielectric loss is

$$\alpha_d = \frac{\omega}{2c}\,\frac{\varepsilon_r(\varepsilon_e - 1)\tan\delta}{\varepsilon_e^{1/2}(\varepsilon_r - 1)} \tag{12.10}$$

The attenuation due to conductor loss will not be altered by the finite width of the top plate, so is given by equation (10.42).

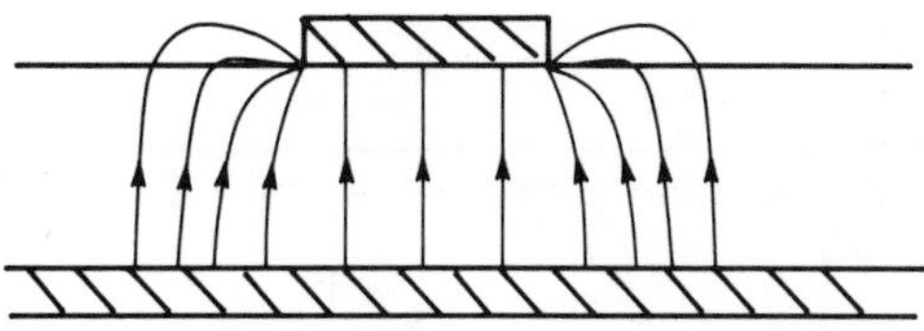

Figure 12.6. Field distribution across a microstrip waveguide showing fringing fields

Example 12.2 Design of a microstrip waveguide

Design a microstrip waveguide having a characteristic impedance of 75 Ω using substrate thickness of 2 mm and a dielectric with relative permittivity $\varepsilon_r = 2.5$ and loss tangent 0.001. It should use copper conductors. Find the attenuation at 10 GHz.

In order to find the width of the microstrip, a graph of equations (12.8) and (12.9) is plotted giving Z_0 against w/h, Figure 12.7. This shows that the approximate ratio required is 1.4. A better estimate can be obtained by iterating the $w/h \geqslant 1$ part of equation (12.9) to give $w/h = 1.455$ or the width $w = 2.91$ mm. The effective permittivity is, from equation (12.8), $\varepsilon_e = 2.0$ compared to the real value of 2.5. The attenuation due to dielectric loss is obtained from equation (12.10) as 0.12 Nepers/m or 1.07 dB/m. In Example 10.2 the attenuation of a parallel plate waveguide of the same parameters as this example was found as 1.44 dB/m. Thus the fringing fields in the microstrip have reduced the loss. The attenuation due to conductor loss is obtained from equation (10.42) as $\alpha_c = R_s \varepsilon_e^{1/2} / \eta_0 a$. This gives a loss of 0.42 dB/m^{-1}.

The higher the relative permittivity of the microstrip substrate, the narrower the spacing between the plates to produce a given characteristic impedance. A narrow spacing between the conductors is usually desirable partly to keep the profile as low as possible and partly to keep the radiation losses low. The radiation losses are an inherent disadvantage of a partly open waveguide. The substrate should also remain stable with temperature variations and ε_r and $\tan\delta$ should not change over the frequency range of interest. Common substrates for microstrip are alumina, sapphire, quartz and RT Duroid. The latter is the registered trade name of a good quality, low loss, substrate which can be easily handled and is widely used. The upper frequency at which microstrip can be

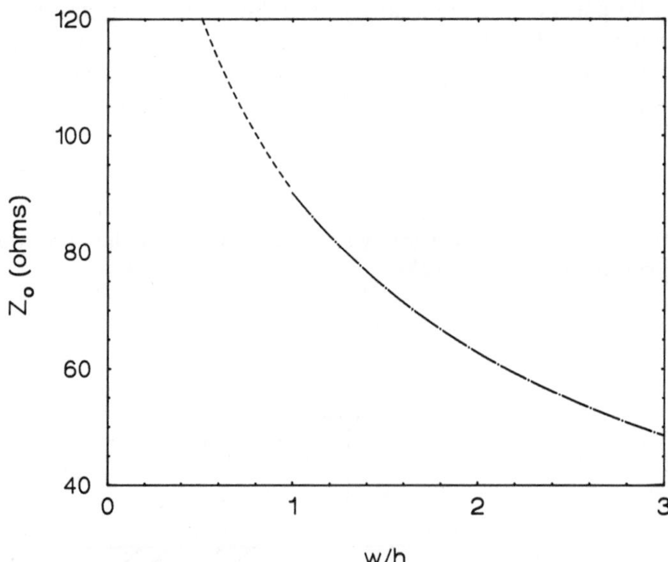

Figure 12.7. Characteristic impedance of microstrip against w/h for $\varepsilon_r = 2.5$

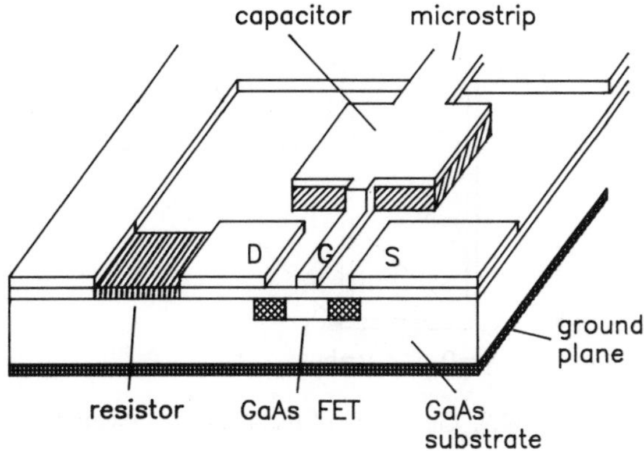

Figure 12.8. Typical layout of components in a monolithic microwave integrated circuit (MMIC)

used is usually determined by the conductor losses. This is currently at about 30 GHz, but advances in design keep pushing the limit to higher frequencies.

Microstrip is not used for long distance transmission. Its major use is within microwave transmitters and receivers for interconnecting components. It is widely used as a transmission line for *microwave integrated circuits* or *MICs*. These are actually hybrid integrated circuits in which discrete components such as semiconductor devices are mounted onto the substrate and microstrip interconnects the components. A complete microwave circuit can be developed and produced with relative ease. For instance, the width of the microstrip lines can be altered to form matching sections. MICs are now widely used in microwave systems operating up to about 18 GHz.

High packing density are not often necessary at microwave frequencies so there has been less need to develop complete integrated circuits. These do, however, exist for high volume applications. The active and passive devices are grown onto a semiconductor substrate such as silicon or gallium arsenide in the same way as digital ICs. The microstrip is also part of the integrated circuit. These are called *monolithic microwave integrated circuits* or *MMICs*; Figure 12.8. The design of MMICs is difficult because the exact shape and position of each component and of each part of the microstrip is important. The whole circuit is either comparable to a wavelength in size or only a fraction of a wavelength in size. This can lead to interactions between nominally separate parts of the circuit.

12.4 RECTANGULAR WAVEGUIDE

Rectangular waveguide, Figure 12.9 has been widely used for the transmission of signals at microwave frequencies. It consists of a hollow rectangular conducting cylinder which is made to precise dimensions. The properties of rectangular waveguides were first studied in the 1930s and then used in microwave radar and communication systems. There are a number of factors which make rectangular waveguide a good transmission systems. It is completely enclosed so there is no interference problem. The hollow waveguide means that high powers can be carried and breakdown is probably set by

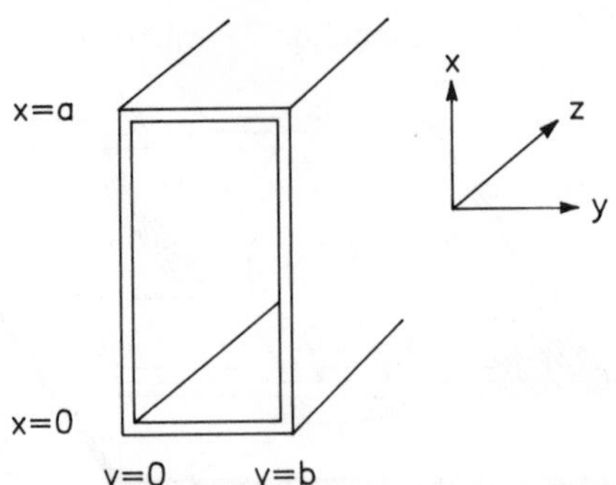

Figure 12.9. Geometry of rectangular waveguide

the components at either end of the waveguide. This is particularly important for high power radars. The fields of the main mode are linearly polarised and easy to excite and detect. This also means that passive components such as attenuators and couplers can be made so that there are available a wide range of rectangular waveguide components. Set against these advantages is the disadvantage that no TEM mode can propagate along a rectangular waveguide. Consequently the cross-section of a rectangular waveguide must be related to the wavelength of operation, in contrast to coaxial cable and microstrip which can be made very small by comparison with the wavelength of operation.

The physical behaviour of the fields of waves travelling in the dominant mode in rectangular waveguide is similar to that of TE modes in a parallel plate waveguide. This can be shown by adding two conducting sheets to a parallel plate waveguide. Consider a transverse electric (TE), mode between parallel plates as shown in Figure 12.10. The field components are E_y, H_x and H_z and there is no variation of field in the y direction. There is only one electric field component so if conducting sheets are introduced normal to the E_y field, as in Figure 12.10(b) the fields will not be disturbed. A closed rectangular box has been created, or in other words, a rectangular waveguide. It can also be seen from this simple explanation that the spacing apart of the two new conducting sheets is not critical to the operation. Any spacing will work.

The above description is a simple physical explanation of how the rectangular waveguide works. As with the parallel plate waveguide, reflections can obviously occur off the new side walls so that potentially there can be more than one mode in the y direction. This implies that the modes will in general depend on both the x and y dimensions. Waves can travel along a rectangular waveguide in an infinite number of modes in the same way as in a parallel plate waveguide. There are again two types of mode.

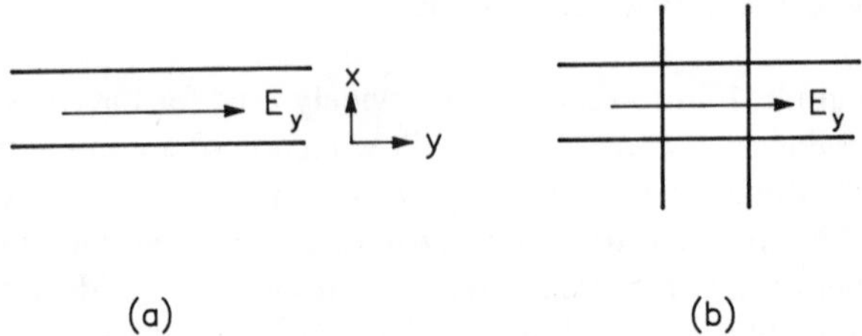

Figure 12.10. Development from (a) parallel plate waveguide to (b) a closed rectangular box

Transverse magnetic (TM) modes with $H_z = 0$ that is, all the magnetic field components are transverse to the direction of travel. Modes are designated TM_{nm} where n and m are integers.

Transverse electric (TE) modes with $E_z = 0$, that is, all the electric field components are transverse to the direction of travel. Modes are designated TE_{nm} where n and m are integers.

In general all field components apart from either H_z or E_z are present in a rectangular waveguide. The integer n refers to field variations in the x direction, the integer m refers to field variations in the y direction. For each variation in one direction, there are an infinite set in the other direction. Thus there are m TM_{1m} modes.

The designation into TM and TE modes refers to a hollow waveguide or a waveguide fully filled with dielectric. If this is not the case, so that the waveguide is partially filled with dielectric or ferrite, then the TM and TE modes are coupled together and produce a *hybrid* mode. It will be assumed in this chapter that the waveguide supports only pure TE or TM modes.

The general procedure for analysing rectangular waveguides follows the same principles as for the parallel plate waveguide but is more complicated because neither $\partial/\partial x = 0$ nor $\partial/\partial y = 0$. The analysis will be done as a series of four steps, which can be summarised as follows.

1. The field components that exist, and the relationship between then can be found from the first two Maxwell equations. When this was done for plane waves it was possible to assume no variations in the x and y directions and when this was done for the parallel plate waveguide no variation in the y direction was assumed. Now there are variations in both x and y directions.

2. The form of the propagating wave is found from the wave equation. This will have harmonics in both directions so a product solution is used. This implies that the x and y directions are separable, which is true for rectangular coordinates.

3. The boundary conditions are used to find expressions for the coefficients.

4. Finally the field components are obtained.

The analysis for TE modes will be done using the H_z wave equation. The analysis for TM modes is identical except that the E_z wave equation is solved.

Step 1. Maxwell's first and second equations, with the assumption that the time and z dependence are $e^{j(\omega t - \beta z)}$, become

$$\begin{array}{lll} & \nabla \times \boldsymbol{E} = -j\omega\mu \boldsymbol{H} & \nabla \times \boldsymbol{H} = j\omega\varepsilon \boldsymbol{E} \\ x: & \dfrac{\partial E_z}{\partial y} + j\beta E_y = -j\omega\mu H_x & \dfrac{\partial H_z}{\partial y} + j\beta H_y = j\omega\varepsilon E_x \\ y: & -j\beta E_x - \dfrac{\partial E_z}{\partial x} = -j\omega\mu H_y & -j\beta H_x - \dfrac{\partial H_z}{\partial x} = j\omega\varepsilon E_y \\ z: & \dfrac{\partial E_y}{\partial x} - \dfrac{\partial E_x}{\partial y} = -j\omega\mu H_z & \dfrac{\partial H_y}{\partial x} - \dfrac{\partial H_x}{\partial y} = j\omega\mu E_z \end{array} \qquad (12.11)$$

the wavenumber is defined as for parallel plate waveguide as $K_c^2 = \omega^2\mu\varepsilon - \beta^2$. The four transverse field components from equation (12.11) are,

$$H_x = -\frac{j\beta}{K_c^2}\frac{\partial H_z}{\partial x} + \frac{j\omega\varepsilon}{K_c^2}\frac{\partial E_z}{\partial y} \tag{12.12}$$

$$H_y = -\frac{j\beta}{K_c^2}\frac{\partial H_z}{\partial y} - \frac{j\omega\varepsilon}{K_c^2}\frac{\partial E_z}{\partial x} \tag{12.13}$$

$$E_x = -\frac{j\omega\mu}{K_c^2}\frac{\partial H_z}{\partial y} - \frac{j\beta}{K_c^2}\frac{\partial E_z}{\partial x} \tag{12.14}$$

$$E_y = \frac{j\omega\mu}{K_c^2}\frac{\partial H_z}{\partial x} - \frac{j\beta}{K_c^2}\frac{\partial E_z}{\partial y} \tag{12.15}$$

In equations (12.11) to (12.15), ε has been written as a short hand for $\varepsilon_0\varepsilon_r$ and μ has been written as a short hand for $\mu_0\mu_r$. These equations apply to both TE and TM modes with either $E_z = 0$ or $H_z = 0$, respectively.

Step 2. The wave equation for TE modes in terms of H_z is

$$\frac{\partial^2 H_z}{\partial x^2} + \frac{\partial^2 H_z}{\partial y^2} + K_c^2 H_z = 0 \tag{12.16}$$

where K_c contains the time and z dependence. To solve the wave equation, a product solution is needed of the form

$$H_z = f_x(x) f_y(y) \exp[j(\omega t - \beta z)] \tag{12.17}$$

where $f_x(x)$ is a function only of x and $f_y(x)$ is a function only of y. Substituting into equation (12.16) and writing $K_c^2 = K_x^2 + K_y^2$ gives

$$\left(\frac{1}{f_x}\frac{\partial^2 f_x}{\partial x^2} + K_x^2\right) + \left(\frac{1}{f_y}\frac{\partial^2 f_y}{\partial y^2} + K_y^2\right) = 0 \tag{12.18}$$

Each large bracket must yield a separate solution, so

$$\begin{aligned} f_x &= A_1 \cos(K_x x) + A_2 \sin(K_x x) \\ f_y &= B_1 \cos(K_y y) + B_2 \sin(K_y y) \end{aligned} \tag{12.19}$$

where A_1, B_1, A_2 and B_2 are constants to be determined from the boundary conditions.

Step 3. The boundary conditions are that the tangential electric fields must be zero at the metal walls, that is (see Figure 12.9)

$$E_x = 0 \text{ at } y = 0 \text{ and } y = b$$

$$E_y = 0 \text{ at } x = 0 \text{ and } x = a$$

The E_x and E_y fields can be obtained from equations (12.14) and (12.15) with equations (12.19) and (12.17). Applying the boundary conditions gives for E_x that $B_2 = 0$ and $K_y b = m\pi$. For E_y it gives that $A_2 = 0$ and $K_x a = n\pi$ where m and n are integers $0, 1, 2, 3, \ldots$.

Then equation (12.17) becomes

$$H_z = A_{nm} \cos(K_x x) \cos(K_y y) e^{j(\omega t - \beta z)} \tag{12.20}$$

where A_{nm} is A_1B_1. Also

$$K_c^2 = \left(\frac{n\pi}{a}\right)^2 + \left(\frac{m\pi}{b}\right)^2 \tag{12.21}$$

The parallel plate waveguide had $K_c = n\pi/a$ so that the closed waveguide has led to an additional, and similar factor, for the b dimension.

The cut-off frequency is determined by the condition $\beta = 0$, which gives

$$\frac{\omega_c}{c} = \left[\left(\frac{n\pi}{a}\right)^2 + \left(\frac{m\pi}{b}\right)^2\right]^{1/2} \tag{12.22}$$

This can also be expressed in terms of the cut-off wavelength

$$\lambda_c = \frac{2ab}{[(nb)^2 + (ma)^2]^{1/2}} \tag{12.23}$$

If $n = 0$ and $m = 0$ then H_z is constant, which means that no mode exists. Thus it is not possible to have TE_{00} modes. By convention $a > b$ so the lowest order mode is the $n = 1$, $m = 0$ mode or TE_{10} mode.

Step 4. The transverse fields of the TE_{nm} modes are determined by substituting equation (12.20) into equations (12.12) to (12.15) with $E_z = 0$,

$$H_x = \frac{j\beta K_x}{K_c^2} A \sin(K_x x)\cos(K_y y) \tag{12.24a}$$

$$H_y = \frac{j\beta K_y}{K_c^2} A \cos(K_x x)\sin(K_y y) \tag{12.24b}$$

$$E_x = \frac{j\omega\mu K_y}{K_c^2} A \cos(K_x x)\sin(K_y y) \tag{12.24c}$$

$$E_y = \frac{-j\omega\mu K_x}{K_c^2} A \sin(K_x x)\cos(K_y y) \tag{12.24d}$$

where $A = A_{nm}$. This completes the analysis of the TE modes. The TM modes follows a similar procedure and leads to an E_z field of the form

$$E_z = B_{nm} \sin(K_x x)\sin(K_y y)\exp[j(\omega t - \beta z)] \tag{12.25}$$

The wavenumber and cut-off frequencies obey the same equations as for TE modes. However, because of the double sine function in equation (12.25) there can be no mode with either $n = 0$ or $m = 0$. Consequently the lowest order mode is the TM_{11} mode which has a cut-off frequency the same as the TE_{11} mode. The transverse electric and magnetic field components for the TM_{nm} modes are

$$H_x = \frac{j\omega\varepsilon K_y}{K_c^2} B \sin(K_x x)\cos(K_y y) \tag{12.26a}$$

$$H_y = \frac{-j\omega\varepsilon K_x}{K_c^2} B \cos(K_x x)\sin(K_y y) \tag{12.26b}$$

$$E_x = \frac{-j\beta K_x}{K_c^2} B \cos(K_x x)\sin(K_y y) \tag{12.26c}$$

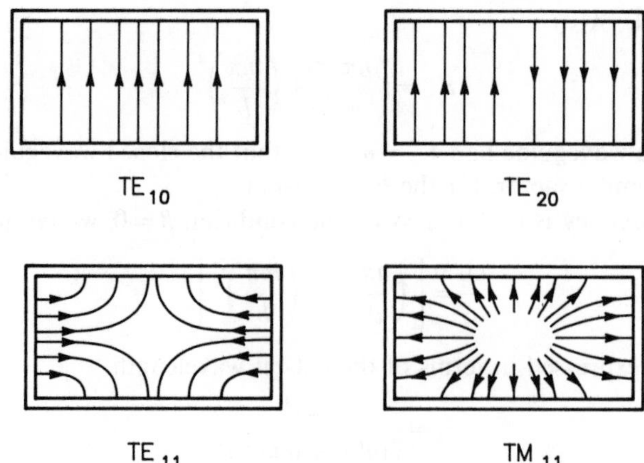

Figure 12.11. Transverse electric field patterns for modes in rectangular waveguide

$$E_y = \frac{-j\beta K_y}{K_c^2} B \sin(K_x x) \cos(K_y y) \tag{12.26d}$$

The electric field patterns for the TE_{10}, TE_{20}, TE_{11} and TM_{11} modes are sketched in Figure 12.11. Notice how the number of cycles of fields in the x and y directions reflects the order of n and m respectively. Of particular note is the TE_{10} mode. The electric field lines go straight from one broad wall of the waveguide to the other broad wall and there is no variation of field in this direction. The TE_{10} mode is called the dominant mode and is the one which is used most in practice. The main characteristics of this mode will be described below.

The TE_{10} mode has $\omega_c/c = \pi/a$ or $\lambda_c = 2a$. The next highest mode depends on the ratio a/b. For the TE_{01} mode, $\lambda_c = 2b$; for the TE_{20} mode, $\lambda_c = a$; for the TE_{11} and TM_{11} modes, $\lambda_c = 2ab/(a^2 + b^2)^{1/2}$.

Example 12.3 Cut-off frequencies and modes in rectangular waveguide

Find the cut-off frequencies and cut-off wavelengths for the first five modes in WR90 waveguide and plot the propagation characteristics.

WR90 is one of a series of standard rectangular waveguide sizes with standard codes. It has $a = 22.9$ mm (0.9 inches) and $b = 10.2$ mm (0.4 inches). Using the above equations gives the order of cut-off as TE_{10}, TE_{20}, TE_{01}, TE_{11}/TM_{11} with the following frequencies and wavelengths

Mode	Cut-off frequency(C_1H_2)	Cut-off wavelength (mm)
TE_{10}	6.56	45.7
TE_{20}	13.10	22.9
TE_{01}	14.76	20.3
TE_{11}	16.16	18.6
TM_{11}	16.16	18.6

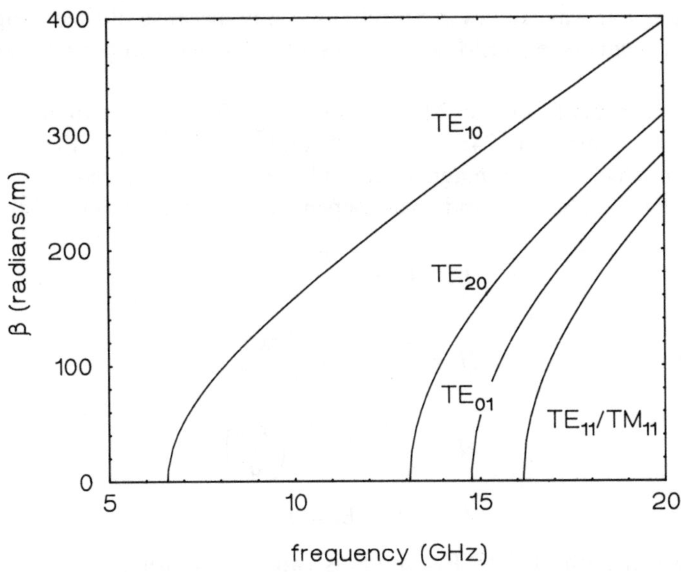

Figure 12.12. Propagation characteristics of WR90 rectangular waveguide

The propagation characteristics of β against frequency are plotted in Figure 12.12. It is usual to operate rectangular waveguides with only one mode so that problems do not arise due to power converting inadvertently into a higher order mode. The restriction to one mode still leads a very large frequency bandwidth available for use. In the case above, the potential single mode bandwidth is from 6.56 GHz to 13.10 GHz. In practice, the attenuation is high near to cut-off so the usable bandwidth is less than the dominant mode bandwidth.

The standard sizes for rectangular waveguide were developed in the era of imperial units so the internal dimensions are usually exact multiples of hundredths of an inch. In the WR designation the number refers to the wide internal dimension in 1/100ths of an inch. All the waveguides in the series, with the exception of the WR90 waveguide, have internal dimensions with $a = 2b$. Some representative waveguides with their designated operating bands are given below.

Designation	Inside dimension (mm)	Useful TE_{10} frequency range (GHz)
WR340	86.36 × 43.18	2.2–3.3
WR90	22.86 × 10.16	8.2–12.4
WR28	7.11 × 3.56	26.5–40.0
WR10	2.54 × 1.27	75 –110
WR4	1.02 × 0.508	170 –260

Large rectangular waveguides are rarely used now because of their weight and high cost. At lower frequencies TEM waveguides such as microstrip or coaxial cable are preferred.

The field components of the TE_{10} mode in air-filled rectangular waveguide are obtained from equation (12.24) with $n = 1$ and $m = 0$. It has the following field components in which the two magnetic fields have been normalized to the only electric field component E_y. The time and space dependent factor $\exp[j(\omega t - \beta z)]$ is assumed.

$$E_y = E_0 \sin\left(\frac{\pi x}{a}\right) \tag{12.27a}$$

$$H_x = \frac{-\beta}{\omega\mu_0} E_0 \sin\left(\frac{\pi x}{a}\right) \tag{12.27b}$$

$$H_z = \frac{j\pi}{\omega\mu_0 a} E_0 \cos\left(\frac{\pi x}{a}\right) \tag{12.27c}$$

$$E_z = H_y = E_x = 0$$

The cut-off wavelength of the TE_{10} mode is twice the width of the broad dimension. The E_y field peaks at the centre of the waveguide and then falls to zero at the narrow walls according to a sine function. In the y direction the electric field is constant. The longitudinal magnetic field is zero along the axis and rises to peaks at the narrow walls. The current on the walls is at right angles to the magnetic field and is sketched in Figure 12.13. Notice that along the centre line of the broad wall, the current is only along the z direction. There is no x directed current along the centre line. This distribution enables a slot to be cut along the axis of the broad wall and the fields inside the waveguide will not be distorted. This is implemented in *standing wave detectors* where the field inside the waveguide is sampled with a small probe which is on a movable carriage.

The transverse wave impedance is defined as the ratio of the amplitude of the E_y electric field to the H_x magnetic field. From equation (12.27) the wave impedance is

$$Z_{TE} = \frac{\omega\mu_0}{\beta} = \frac{\eta_0}{[1 - (\omega_c/\omega)^2]^{1/2}} \tag{12.28}$$

The right-hand side is obtained by expressing β in terms of the cut-off frequency using equation (10.27). Equation (12.28) is exactly the same as the wave impedance for TE modes in parallel plate waveguide, equation (10.32). Similarly the velocities and the guide wavelength are the same as developed in Section 10.8, equations (10.28) and (10.29).

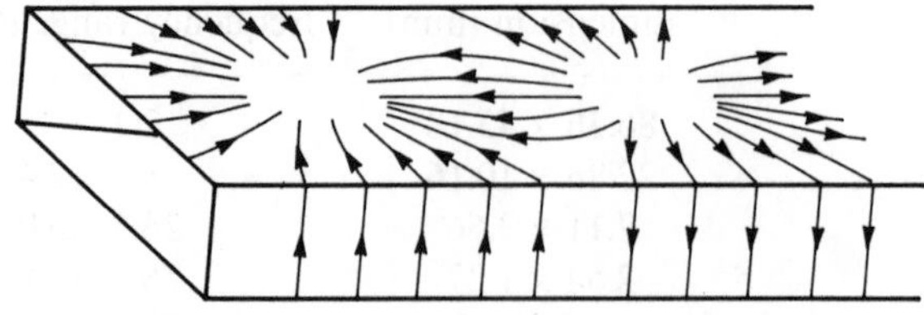

Figure 12.13. Current flow on conducting walls of rectangular waveguide

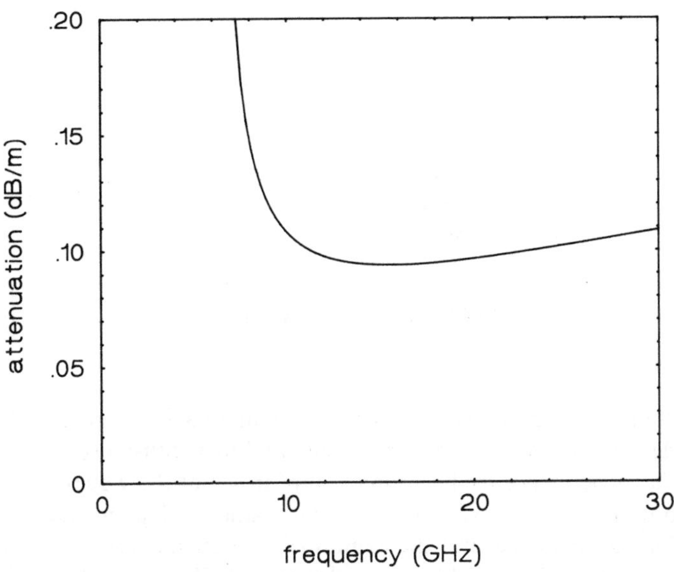

Figure 12.14. Attenuation of TE_{10} mode in WR90 rectangular waveguide

This also applies to the attenuation due to dielectric loss because K_c for the TE_{10} mode is the same form as for the TE_1 modes in parallel plate waveguide. Hence equation (10.36) gives the attenuation due to dielectric loss.

The attenuation due to conductor loss is different because there are four walls instead of two walls. The procedure used in Section 10.10 can be followed. The currents on the side walls are due to H_z and the currents on the broad walls are due to both H_z and H_x. The derivation is left as a problem. The result is

$$\alpha_c = \frac{R_s}{\eta b[1-(\omega_c/\omega)^2]^{1/2}}\left[1+\frac{2b}{a}\left(\frac{\omega_c}{\omega}\right)^2\right] \quad \text{Nepers/m} \qquad (12.29)$$

This equation shows a multiple dependence on frequency because the surface resistivity is proportional to the square root of frequency. The form is sketched in Figure 12.14 for WR90 copper waveguide. The form is the same as for parallel plate waveguide, but the value is higher due to the extra walls in the rectangular waveguide. Equation (12.29) is inversely proportional to the width of the narrow wall, b. So that as the operating frequency increases and the waveguide gets smaller, the attenuation increases. This is a limiting factor to the use of rectangular waveguide at millimetrewaves. The attenuation theory used to derive equation (12.29) assumes that the walls are perfectly smooth. In practice this is not possible and the measured attenuation can by two or three times the theoretical value at millimetrewaves.

12.5 CIRCULAR WAVEGUIDES

Circular waveguide, Figure 12.15, operates in a very similar way to rectangular waveguides. Power can be carried in both TE and TM modes with cut-off and

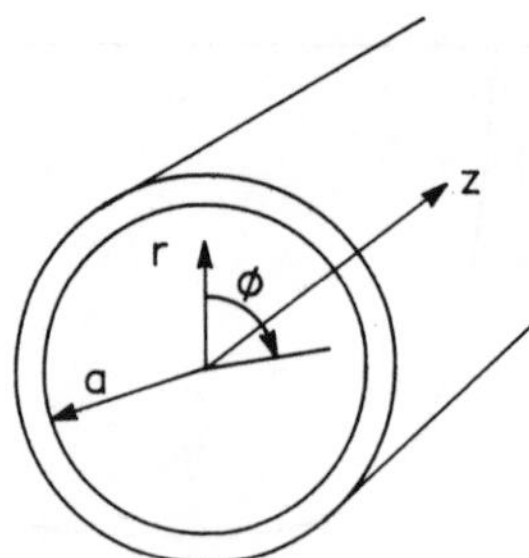

Figure 12.15. Geometry of circular waveguide

propagation characteristics similar in form to rectangular waveguides. Again there is no TEM mode. The analysis procedure is similar but it must use cylindrical (r, ϕ, z), coordinates. The dependence in the z direction of travel of the waves is the same as for rectangular coordinates, i.e. $\exp[\mathrm{j}(\omega t - \beta z)]$. The radial, r, dependence, is equivalent to the x rectangular dependence, and the azimuthal, ϕ, dependence is equivalent to the y rectangular dependence. The mode sets are designated TE_{nm} and TM_{nm} as before where n refers to the radial mode set and m refers to the azimuthal mode set.

Circular waveguide is less used than rectangular waveguide because it does not have a preferred polarisation direction. In TE_{10} rectangular waveguide the electric fields are linear polarised and parallel to the narrow wall. The cylindrical geometry of circular waveguide means that there is no preferred azimuthal direction for the electric fields. This has, however, the benefit that two orthogonal polarised modes can be sent through the same circular waveguide, or a circular polarised wave can be propagated. In neither case is this possible with rectangular waveguide, thus circular waveguide is used in applications associated with antennas which must transmit or receive two polarisations or circular polarisation.

The cylindrical geometry leads to more complicated analysis than for rectangular waveguides. Since the objective here is to highlight the main features, only the general form of the results will be described. The wave equation in cylindrical coordinates for H_z is

$$\frac{1}{r}\frac{\partial}{\partial r}\left(r\frac{\partial \phi}{\partial r}\right) + \frac{1}{r^2}\frac{\partial^2 \phi}{\partial \phi^2} + K_c^2 \phi = 0 \tag{12.30}$$

The wavenumber K_c has the same form as for the other waveguides, i.e.

$$K_c^2 = \omega^2 \varepsilon_0 \varepsilon_r \mu_0 \mu_r - \beta^2 \tag{12.31}$$

The wave equation is separable into a radial term and an azimuthal term, so that a product solution similar to equation (12.18) is used. The radial solutions are Bessel functions and the azimuthal solutions are trigonometric functions.

The axial magnetic field of TE_{nm} mode is described by

$$H_z = A J_n(K_c r) \cos n\phi \exp[\mathrm{j}(\omega t - \beta z)] \tag{12.32}$$

where $J_n(K_c r)$ is a Bessel function of the first kind and order n. Bessel functions are the cylindrical equivalent of rectangular trigonometric functions. The form of $J_n(x)$ for $n = 0$ and 1 are plotted in Figure 12.16, from which it is observed that Bessel functions are

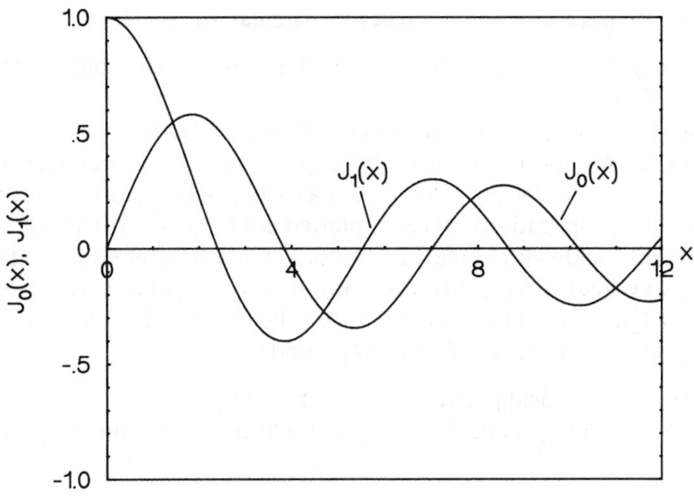

Figure 12.16. Bessel functions of first kind, $J_0(x)$ and $J_1(x)$

not periodic with x so that the roots do not occur at regular intervals of x. The amplitude of the function also decays as the argument increases.

The axial electric field of TM_{nm} modes is similarly described by

$$E_z = BJ_n(K_c r)\cos n\phi \exp[j(\omega t - \beta z)] \tag{12.33}$$

The boundary condition states that the tangential electric fields are zero at $r = a$, where a is the radius of the waveguide. This leads to the values of the wavenumber for a particular mode as follows

$$\begin{aligned} K_c a &= j_{nm} \quad \text{where} \quad J_n(j_{nm}) = 0 \quad \text{for TM modes} \\ K_c a &= j'_{nm} \quad \text{where} \quad \frac{dJ_n}{dx}(j'_{nm}) = 0 \quad \text{for TE modes} \end{aligned} \tag{12.34}$$

The order of cut-off frequencies for the modes depends on the values of j_{nm} and j'_{nm}. The first five modes have wavenumbers as follows

Mode	$K_c a$
TE_{11}	1.841
TM_{01}	2.405
TE_{21}	3.054
TM_{11}	3.831
TE_{01}	3.831

The last two modes are said to be degenerate because they lie on top of each other. This does not mean than both modes will automatically propagate along the waveguide. That depends on the method used to excite the modes.

Example 12.4 Propagation characteristics for circular waveguide

Plot the propagation characteristics for a hollow circular waveguide with an internal diameter of 26.0 mm.

Equation (12.31) is used to find the cut-off frequency, $K_c a = \omega_c a/c$, or $f_c = (K_c a) c/2\pi a$. Using the values in the table above gives the cut off frequency for the TE_{11} mode as 6.76 GHz, and for the other modes as 8.833 GHz, 11.21 GHz and 14.07 GHz, respectively. The propagation curves are plotted in Figure 12.17. This example appears to show that the bandwidth of the TE_{11} mode is much smaller than for the equivalent rectangular waveguide. In practice this is not the case because it is a lot more difficult to excite the TM_{01} and TE_{21} modes so that the real bandwidth of the TE_{11} mode is set by the cut-off frequency of the TM_{11} mode.

The transverse electric field patterns for the TE_{11}, TE_{01} and TE_{21} modes are sketched in Figure 12.18. The TE_{01} mode has a pattern which is independent of the azimuthal

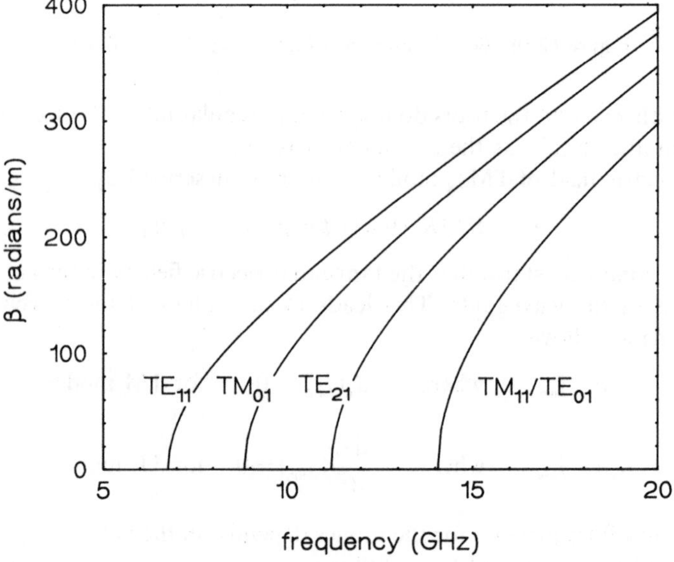

Figure 12.17. Propagation characteristics for modes in circular waveguide with internal diameter of 26 mm

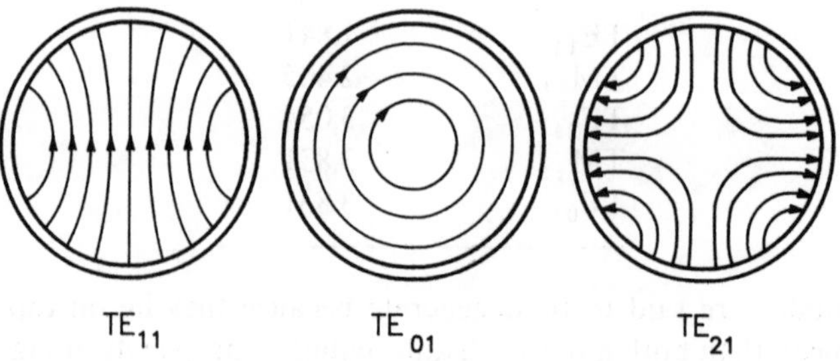

Figure 12.18. Transverse electric field patterns of modes in circular waveguide

orientation. The other modes all have a preferred azimuthal orientation for the electric fields. The dominant mode is the TE_{11} mode. The fields are similar to those of the TE_{10} mode in rectangular waveguide and it is usual to excite the circular TE_{11} mode by first generating a TE_{10} mode in rectangular waveguide and then attaching a short semi-conical transforming section to convert the cross-section to circular waveguide. Circular waveguide is only practical over relatively short lengths as it is difficult to maintain the circular cross-section to be precisely circular over long distances. The waveguide has to be very accurately made with no deformations in the waveguide walls.

The attenuation of the dominant TE_{11} mode has a frequency dependence which is very similar to that of the TE_{10} mode in rectangular waveguide, Figure 12.14. The TE_{01} mode has a different behaviour with frequency. It has the property that the attenuation is proportional to (frequency)$^{-3/2}$. It hence decreases continuously as frequency increases.

The low attenuation property of the TE_{01} mode led in the 1950s and 1960s to the proposal to develop a long haul low-loss telecommunication system using circular waveguide at millimetrewaves. The telecommunication authorities invested considerable effort in developing a system and a number of successful trials were conducted. For instance the British Post Office had a 14.2 km test system which was measured to have an attenuation of between 1.5 and 2.5 dB/km using the frequency range 30 to 110 GHz. This wide spectrum would have carried an enormous amount of telecommunication traffic. The projects were, however, over ambitious. The waveguide system was very expensive and that amount of bandwidth was not needed. More importantly, optical fibre came along at about that time and it became clear that optical fibre had a much greater potential at a much lower cost.

12.6 OTHER WAVEGUIDE TYPES

The preceding sections have described the main types of microwave waveguide which are used in practical systems. There are in addition a large number of specialised types of waveguide, each of which has one or more desirable attribute for a particular applications. New types are continually being developed to improve performance or to meet some new requirement. A few types are now briefly described and are shown in Figures 12.19 and 12.20.

Circular waveguide suffers from the fact that there is no preferred orientation. This can be overcome with *elliptical waveguide* which can be made with thin walls to give a

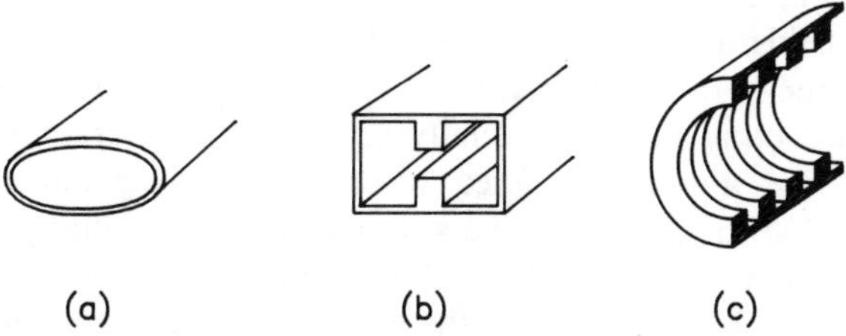

Figure 12.19. Types of microwave waveguide. (a) Elliptical waveguide. (b) Ridged waveguide. (c) Cross-section of corrugated waveguide

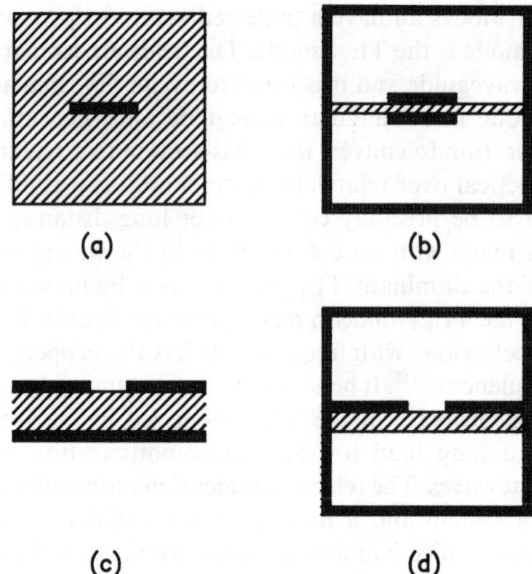

Figure 12.20. Types of stripline waveguide. (a) Stripline. (b) Triplate. (c) Slotline. (d) Finline

certain degree of flexibility. Elliptical waveguide is, however, expensive to manufacture and very difficult to analyse.

There are some applications where a very broad frequency band of operation is needed and the bandwidth of standard rectangular waveguide is inadequate. *Ridged waveguide* has been developed to meet this problem. In this waveguide a central ridge is introduced so that the height of the waveguide is narrowed. This has the effect of lowering the cut-off frequency of the TE_{10} mode by a considerable amount. Bandwidth ratios of between 2:1 to 4:1 are available. The losses go up as the bandwidth increases and may limit performance.

Circular waveguide is amenable to being modified in order to improve the field patterns. This can be done by inserting a dielectric into the cylinder, or by modifying the walls. Making the walls corrugated in which the corrugations are quarter of a wavelength deep gives an electric field pattern which is nearly perfectly linearly polarised. This is used in a type of antenna, called a *corrugated horn*, to give a very high performance system.

Section 12.3 described microstrip. This is only the most popular of a large number of related designs, Figure 12.20. *Stripline* has a strip of conductor enclosed within a box of dielectric. The fields are similar to microstrip but the characteristic impedance is halved because of the two parallel plate waveguides. *Triplate* is a version of the same thing inside a conducting box and usually air filled in order to reduce losses. This waveguide looks like a partially filled rectangular waveguide, but the mode of propagation is a TEM mode so that the size limitations of rectangular waveguide are avoided. The dimensions can be made small by comparison with a wavelength. *Slotline* is similar to microstrip but the strip on the dielectric substrate is replaced by a slot in a conducting sheet. The waves travel in the slot between the two conducting sheets. The fields are closely confined to the slot. It is a useful alternative to microstrip, particularly for microwave integrated circuits because it is possible to connect devices across the slot to put them in parallel with the transmission line. In microstrip it is easier to put devices in series with the transmission line. The fields are closely confined to the slotline, but are more complicated to study. The *finline* is a sort of slotline in an enclosed box. The enclosure reduces the radiation from discontinuities and junctions which is a problem with microstrip. Finline works well at high frequencies and a number of variations have been developed for millimetrewave applications.

12.7 WAVEGUIDE COMPONENTS

Components are needed in microwave waveguide to control the power and signals which are travelling along the waveguide. There are many different designs of components for the different types of waveguide—coaxial, microstrip or hollow waveguide. The principle of component design is to use a particular electromagnetic field configuration to interact with another element so as to control the flow of power, for instance the phase of signals can be changed by introducing a dielectric which slows down the waves. Some of the main waveguide components will be described here to illustrate the principles. Components can be divided into passive devices which are self contained and depend for their operation only on the waves in the waveguide, and active devices which use semiconductor devices with some external input or output of signal and power. Only the main passive components will be described here. Active microwave devices form a large subject on their own right.

Matched loads, Figure 12.21, are required to absorb all the power incident upon them so that there is no reflected signal. This is done in microstrip or coax by inserting a resistance strip in place of one of the conducting strip so that a load of $R = Z_0$ terminates the line. In waveguide a tapered load made of lossy material gradually absorbs the energy. The more gradual the taper, the better the performance, but the length increases so that a compromise is necessary. The design is largely empirical, or *cut-and-try*. In rectangular waveguide, carbon loaded epoxy is used, but even wood will work quite well. The matched load cannot be made perfect and there will always be a certain amount of reflection, particular over a wide band of frequencies.

Transitions perform the function of transforming from one type of waveguide system to another type. Usually this means to and from coaxial cable because this is the best waveguide for interconnecting equipment and instruments. Coax to microstrip

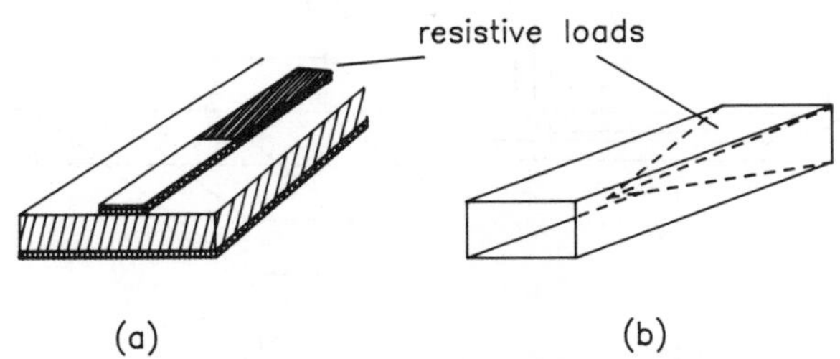

Figure 12.21. Matched loads with (a) microstrip, (b) rectangular waveguide

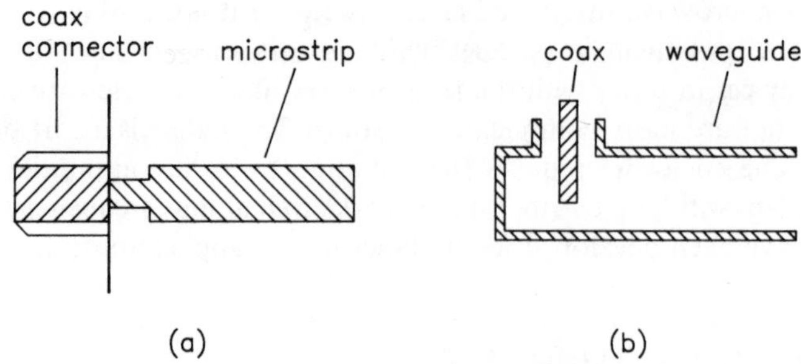

Figure 12.22. Transitions. (a) Coax to microstrip (plan view). (b) Coax to rectangular waveguide

transitions, Figure 12.22(a) are easy to make because both waveguides support TEM modes. The central conductor of the coax is attached to the strip conductor and the outer conductor is connected to the ground plane. Any impedance difference can be compensated by altering the width of the microstrip next to the coax connector. A coax to rectangular waveguide transition is shown in Figure 12.22(b). The objective is to insert a probe into the waveguide which will excite the TE_{10} field pattern. At the same time it should not introduce an impedance discontinuity. A vertical conducting post which is the central conductor of the coax connector is partially inserted into the waveguide. The post excites an electric field which is parallel to the conductor and this is the required E_y field of the TE_{10} mode. The size and depth of the probe can be adjusted to achieve a good impedance transformation. Making the impedances match over a reasonable band is an art and many empirical designs have been used to achieve broad band performance.

Directional couplers are used to couple one transmission line to another line, Figure 12.23. This may be as a means of sampling a proportion of the power in the main line, or as a means of splitting the power into two or more waveguides. The amount of coupling is expressed by the *coupling factor* which is the ratio, in decibels, of the incident power in the main line to the power in the coupled line. A 3 dB coupler splits the power equally between the main and coupled lines. Another term used is the *coupler directivity* which is a measure of how well the coupler distinguishes between waves travelling the forward direction and waves travelling in the backward direction. This is again expressed as a ratio in dBs, of the incident power to the backward travelling power. Ideally the directivity should be infinite (in dBs). In practice values of 30 dB, or

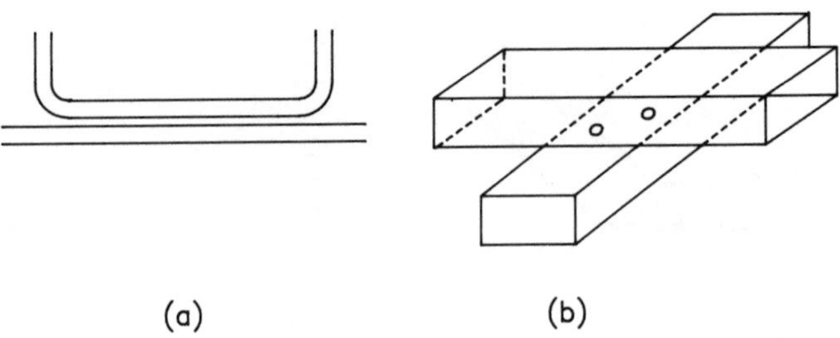

Figure 12.23. Directional couplers. (a) Two microstrip lines (plan view). (b) Two waveguides

one thousandths of the incident power are both required and achieved. Microstrip couplers can be simply made by two striplines close together, Figure 12.23(a). The fringing fields of the two lines will interact and transfer some power from one line to the other line. The amount of coupling is controlled by the spacing apart of the two striplines. Rectangular waveguide couplers, Figure 12.23(b), use holes cut in one of the walls of the waveguide, usually the broad wall. The holes interrupt the current flow in the walls which creates a field across the hole and this radiates into the adjacent waveguide. The amount of coupling is determined by the number of holes, the position of the holes and the size and shape of the holes. The large number of parameters means that many designs have been evolved over time. Each design has some merits which must be matched to a specific application.

Attenuators are designed to control the amplitude of the microwave signal in a waveguide. In microstrip this can be done in a similar manner to the matched load, except that the conducting strip is continued after the resistor. In rectangular waveguide a simple attenuator can be made by inserting a tapered resistive card into the centre of the waveguide so that it is parallel to the electric field and placed where the field is strongest, Figure 12.24. The attenuator can made variable by adding a mechanism to adjust the amount of resistive card in the waveguide. An accurate attenuator can be made by inserting a resistive card into circular waveguide and rotating the card. When the card is parallel to the electric field maximum power will be absorbed. When it is orthogonal to the electric field, very little power is absorbed. At intermediate angles, the amount of power absorbed is exactly proportional to the component of the electric field. This means that the attenuator can be calibrated accurately. It is called a *rotary vane attenuator* and is used for calibration measurements.

Phase shifters are designed to control the phase of the microwave signals. There are two methods by which this can be done, either by altering the length of the waveguide or by inserting a dielectric into the waveguide. The simplest method is to insert a length of waveguide into the transmission line which adds a desired proportion of a wavelength. 90° of phase is added with a quarter wavelength long section of waveguide. The relevant wavelength is the guide wavelength, not the free space wavelength, although in TEM systems, the two wavelengths are the same. This method of generating a phase shift is

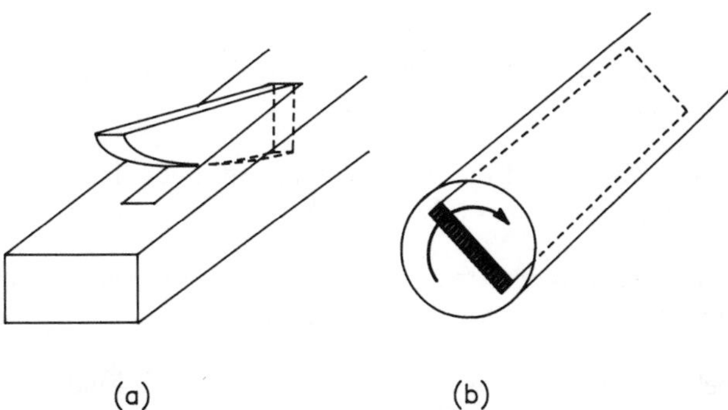

Figure 12.24. Waveguide attenuators. (a) Resistive slab in rectangular waveguide. (b) Rotary vane attenuator

widely used with coaxial cable. It does not give a variable phase shifter, other than by inserting different discrete lengths of cable into the line. Variable coaxial phase shifters have been made using a trombone type system, but this does not work well at RF due to the interrupted current flow. Waveguide phase shifters can be made by the same method as waveguide attenuators. The resistive card is replaced by a tapered slab of low loss dielectric. Instead of absorbing the power, the dielectric slows down the waves and hence effectively introduces a phase shift. An accurate phase shifter can also be made by using the principle of the rotary vane attenuator to construct a rotary vane phase shifter.

PROBLEMS

1. Find the single mode bandwidth of a coaxial cable with an inner radius of $a = 3\,\text{mm}$ and containing polyethylene with (a) $b = 6\,\text{mm}$, (b) $= 9\,\text{mm}$, (c) $b = 12\,\text{mm}$.

2. Design an air filled coaxial cable to have a characteristic impedance of $50\,\Omega$ and to operate up to 50 GHz.

3. Calculate the attenuation in a $75\,\Omega$ polyethylene filled coaxial cable with $a = 2.5\,\text{mm}$, and copper conductors at 500 and 1000 MHz. The loss tangent of the polyethylene is 0.0005.

4. Plot the effective permittivity of a microstrip with $\varepsilon_r = 4$ against w/h for $0.5 < w/h < 10$. Hence find the ratio of w/h which reduces the actual permittivity by half.

5. Simplify the expressions for the effective permittivity and impedance of microstrip for the case of $w/h = 3/2$. Find the impedance for a microstrip line which uses silicon as the substrate ($\varepsilon_r = 12$).

6. Find the attenuation at 8 GHz in a microstrip line with $w = 6\,\text{mm}$, $h = 3\,\text{mm}$, $\varepsilon_r = 2.5$, $\tan\delta = 0.0001$ and copper conductors.

7. A microstrip transmission line of impedance $80\,\Omega$ is to be matched to another microstrip line of impedance $50\,\Omega$ by a quarter wave microstrip transformer section. The substrate is made of alumina ($\varepsilon_r = 10$) with a thickness of 2 mm. Find the width of the three microstrip lines.

8. Derive the electric and magnetic field components for TM_{nm} modes in rectangular waveguide, equation (12.26).

9. Find the order of cut on and the relative dominant mode bandwidth for rectangular waveguide with (a) $a = 4b$, (b) $a = b$.

10. For a rectangular waveguide with sides 34.85 mm by 15.80 mm sketch a curve of guide wavelength against frequency for the TE_{10} mode.

11. A road tunnel has a rectangular cross-section of $7 \times 5\,\text{m}^2$. Find the lowest frequency which will propagate in the tunnel.

12. Design an air-filled rectangular waveguide to be used for the transmission of microwave

power at 2.45 GHz. The frequency should be in the middle of the operating band of the dominant mode.

13. Derive an expression for the surface current density in the broad wall of a rectangular waveguide propagating the TE_{10} mode. Hence confirm that there is no transverse current flow along the axis of the broad wall.

14. Show that the time average power transmitted in the TE_{10} mode along a rectangular waveguide is given by $P = (\beta\omega\mu ab A^2)/4K_c^2$ where A is the amplitude coefficient and the other symbols have their usual meaning.

15. If air breaks down at 3×10^6 V/m, use the results of the last problem to find the maximum power which can be carried in a rectangular waveguide of dimensions 15.8 mm by 7.9 mm at 15 GHz.

16. It is required to send both the Ku up link satellite band (14 to 14.5 GHz) and the down link hand (10.75 to 11.5 GHz) through the same rectangular waveguide. If WR90 (22.86 mm by 10.16 mm) and WR62 (15.80 mm by 7.90 mm) are available, which waveguide would be preferable?

17. Derive the equation for the attenuation in rectangular waveguide, equation (12.29).

18. Find the attenuation, in dBs, of a 2 m length of WR90 brass rectangular waveguide at 8 and 12 GHz.

19. Find the ratio of the dominant mode bandwidths for a rectangular waveguide and a circular waveguide which have the same dominant mode cut-off frequency. Assume that the bandwidth of the circular waveguide is set by the cut-off frequency of the TM_{11} mode.

20. A polyethylene filled circular waveguide operates from 20 GHz to 30 GHz with a 10% guard band above the cut off frequency. Find the waveguide diameter and the number of modes at the top of the band.

21. Waveguide mode filters which filter out certain modes can be constructed by inserting thin metal wires into a waveguide. By examining the field patterns in Figure 12.18, deduce the affects on the three modes of a pair of crossed wires across a circular waveguide, which are at right angles to each other.

13 Optical waveguides

13.1 INTRODUCTION AND POTENTIAL OF OPTICAL WAVEGUIDES

An optical waveguide is a dielectric waveguide and consists of a slab or rod of dielectric surrounded by another dielectric. There are no conducting surfaces. It is this feature which distinguishes optical waveguides from metal waveguides. Conducting surfaces are impractical at optical frequencies so the guiding action of reflections off a metal surface has to be replaced by the guiding action of reflections at a dielectric boundary. In fact there is no reasons why dielectric waveguides should not be used at microwave frequencies, but generally metal waveguides are more convenient. The guiding action of dielectric waveguides is more complicated than that for metal waveguides and the detailed analysis of dielectric waveguides is a lot more complicated than for metal waveguides. Fortunately simplified theories have been evolved which make the task of analysis more manageable. These will be developed in this chapter so that the principles of optical waveguides can be understood.

The history of optical waveguides is more recent than that of microwave waveguides and stems from developments in the 1960s which made guided optical waves feasible. One of these was the development of semiconductor light sources in the form of the LED and the laser. Another one was the discovery that a viable waveguide could be made from two dielectrics which differed in their refractive indices by only a small amount. This was at the time a theoretical prediction and the realisation depended on major developments in material technology. The most suitable dielectric was found to be glass and the need was to produce very pure glasses so that the attenuation in optical waveguides could be made sufficiently low to make an economic system. In fact the success in this area has been so great that the attenuations are now significantly below that available with metal waveguides.

Some of the advantages of optical waveguides can be summarised as

- Low transmission losses.
- Large frequency bandwidth available.
- Small size and weight.
- Flexible cables possible.
- Immunity from electrical interference.
- Security of signal transmission.
- Waveguides can be integrated with other components.
- Low potential cost.

There are also some disadvantages which can be summarised as

- Small size makes coupling to fibre difficult.

- Waveguides are fragile
- Electrical power cannot be carried along waveguide.
- Sources and detectors have low efficiencies and high noise.

The short wavelength at optical frequencies brings both advantages and disadvantages. It means that the waveguide and associated components are very compact, but it also mean that the physical problems of dealing with the waveguide are considerable. This includes the problem of coupling energy into and out of the waveguide. The short wavelength brings the advantage of very large frequency bandwidths when viewed in terms of carrying information, for instance 1 GHz of bandwidth is available between optical wavelengths of 1 μm and 1.000 003 33 μm. This is a percentage bandwidth of only 0.000 33%. In practice this wide bandwidth means that simple but inefficient transmission methods can be used.

The immunity from electrical interference comes from the fact there are no metal parts, so there can be no conduction currents, and none of the static electric, magnetic and current factors which were described in the early chapters of this book. The associated disadvantage with having no conductor is that electrical power cannot be carried along the waveguide. This means that a separate electrical supply for the sources and receivers must be provided.

The fact that the waves in optical waveguides are largely confined to the centeral dielectric means that cross talk between two waveguides is very small and therefore the information in the waveguide is secure and the waveguide cannot easily be *tapped.* This confinement of the waves is lost when the optical waveguide is bent around a tight corner. In this circumstance there can be a loss in signal as it radiates out of the central guiding region.

The potential applications for optical waveguides are considerable and the rate of growth of the subject is high. Whereas microwave waveguides are a mature technology, optical waveguides are a fast evolving subject. This partly stems from the fact that it is only with the availability of good quality optical waveguides and optical semiconductor sources that new uses are being found. Each new use has different requirements and leads to new developments. There are three main areas of applications of optical waveguides:

(a) Long distance telecommunications over land or under the sea. The low attenuation and the wide bandwidth are the most attractive features. Repeaters can be spaced a long distance apart and many channels of information can be multiplexed together.

(b) Short length waveguide applications in industry and medicine for measurement, control and monitoring. The immunity from electrical interference is particularly useful in these applications. The compact nature of the waveguide is exploited in medical applications.

(c) Integrated optics where very short lengths of waveguide are used to interconnect and process optical signals. The ability to manufacture semiconductor dielectric waveguides in integrated circuits is a major advantage.

Each of these areas of applications relies on the principles of optical waveguides which will be described in this chapter. Two geometries are used for optical waveguides. Circular dielectric waveguides are used for short to long distance transmission. These are called *optical fibres*. Rectangular dielectric slab waveguides are used in integrated optics and optoelectronics. These are simpler to analyse and will be used in the next two sections to describe the principles and operation of optical waveguides.

It is clear from the above short description of advantages and applications that the subject of optical waveguides is a large topic in its own right. This chapter provides an introduction which builds on the subject matter already studied in earlier parts of the book.

13.2 PRINCIPLES OF DIELECTRIC WAVEGUIDES

All dielectric waveguides guide waves using the principle of total internal reflection. This was described in Chapter 6 and the ray picture is shown in Figure 13.1. The incident ray on the dielectric boundary is totally reflected if two conditions apply.

(a) The refractive index of the dielectric containing the incident ray must be *greater* than the refractive index of the outer dielectric.

(b) The angle of incidence θ must be greater than a critical angle θ_c which is given by

$$\sin\theta_c = \frac{n_2}{n_1} \tag{13.1}$$

where n_1 and n_2 are the refractive indices of the incident and transmitted medium, respectively.

In Chapter 6, equation (13.1) was derived using relative permittivities as the description of the electrical effects of the dielectric. It is normal in optical studies to work exclusively with refractive indices, so these will be used in this chapter. The relationship between the two is $n = \varepsilon_r^{1/2}$. It is also usual to use wavelength, in microns (μm) or nanometres (nm), rather than frequencies to describe the periodicity of the waves.

A dielectric waveguide consists of two dielectrics, although in the simplest case, one of the dielectrics could be air. The dielectric which guides the waves is called the *core*. The other dielectric is called the *cladding*. These terms assume a cylindrical geometry but are used for convenience with rectangular geometries. The ray picture in a dielectric waveguide is sketched in Figure 13.2. This shows the total internal reflection taking place at two boundary so that the result is a ray picture very similar to that of the

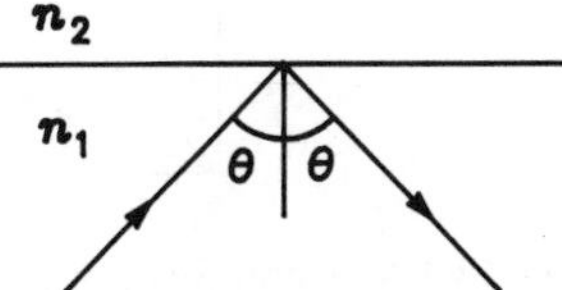

Figure 13.1. Principle of total internal reflection at a dielectric boundary

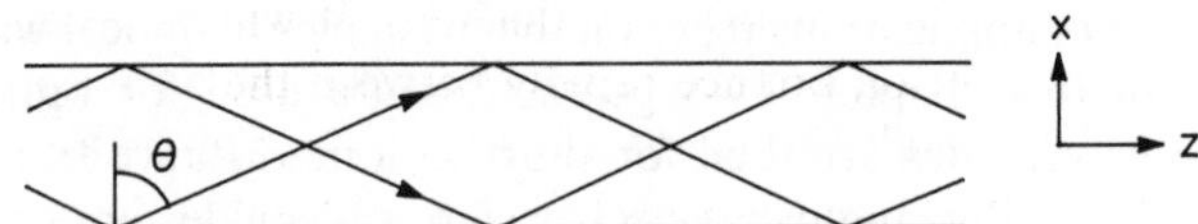

Figure 13.2. Rays in a dielectric waveguide

guided waves between conducting plates described in Chapter 10. There are however two important differences.

(a) The need to satisfy the total internal reflection condition, equation (13.1), at the boundary means that only rays at angles greater than the critical angle can exist.

(b) A finite non-propagating fields must exist outside the dielectric waveguide, as was shown to be necessary in order to satisfy the boundary conditions in Section 6.6.

These two factors have the theoretical consequence that the analysis is much more complicated than for waves guided by conductors. For instance the wavenumber and propagation characteristics cannot be obtained in closed form and must be solved numerically on a computer. This process will be described for a dielectric slab waveguide in the next section.

Apart from the two factors just described, the principles of guided waves which were deduced in Chapter 10 apply to optical waveguides. The ray picture is useful for physical insight, but a complete understanding must use the electromagnetic wave analysis to find the fields in the waveguide. The wave analysis shows that similar characteristics apply as metal waveguides. Power is carried in modes and a number of modes can propagate in the waveguide. The number is determined by the physical size of the waveguide. Each mode has a cut-off frequency and wavenumber which relates the propagation constant β to the radial frequency, ω $(= 2\pi f = 2\pi c/\lambda)$. The propagation

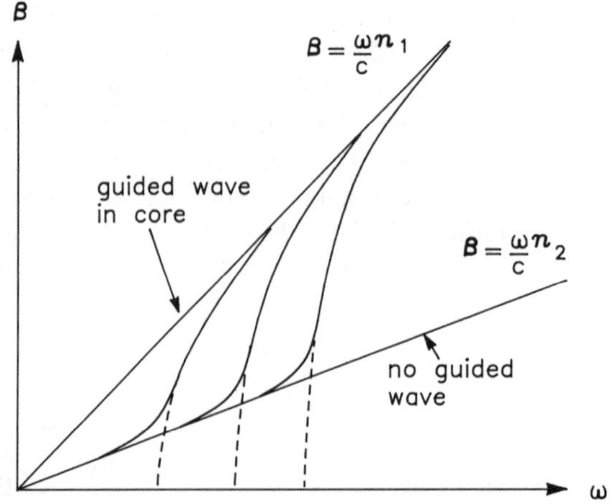

Figure 13.3. General form of $\beta\omega$ propagation characteristics for dielectric or optical waveguide

characteristics of β against ω are, however, different. It will be recalled that in metal waveguides, rays near to cut-off bounce rapidly between the walls so that the angle of incidence on the walls is small. Such rays do not satisfy the critical angle condition, equation (13.1), and will pass into the outer region. They will not be totally internally reflected and no guiding action takes place. Thus *cut-off* will occur when the rays are at the critical angle, $\theta = \theta_c$, and not at $\theta = 0°$. The result will be that the β against ω propagation characteristics will have the general form sketched in Figure 13.3. The top part of the picture is the same as for metal waveguides. Propagation is, however, cut-off at the $\beta = \omega n_2/c$ line since the wave is entirely in the cladding. There are no modes in the region between $\beta = 0$ and $\beta = \omega n_2/c$.

Both geometric optics ray theory and the electromagnetic wave theory are useful for describing different aspects of propagation in optical waveguides. The wave theory is necessary in order to find the fields and to obtain a complete picture of the propagation characteristics. The ray theory is useful when the wave theory is too complicated. The next section will develop the wave theory for the dielectric waveguide analogy of the parallel plate waveguide.

13.3 DIELECTRIC SLAB WAVEGUIDE

The simplest type of dielectric waveguide is the symmetric slab waveguide shown in Figure 13.4 with a rectangular slab or sheet of dielectric surrounded by another dielectric. This is not often used in practice because of the difficulty of manufacturing and supporting the configuration but it serves to illustrate the principles of dielectric waveguide propagation. In the slab waveguide analysis it will be assumed that the waveguide is infinite in the transverse y direction, as in the parallel plate waveguide analysis of Section 10.3. This is justified partly because the waveguides can be made large in wavelengths at optical frequencies and partly for the pragmatic reason that an exact analysis of a rectangular dielectric slab waveguide is extremely difficult.

The procedure for finding the propagation characteristics is the same as for metal waveguides. The appropriate electric and magnetic fields are found in the core and cladding regions and then the boundary conditions are used to eliminate the unknown amplitude coefficients. The analysis of dielectric waveguides requires two boundary conditions instead of the one boundary condition of zero tangential electric field which

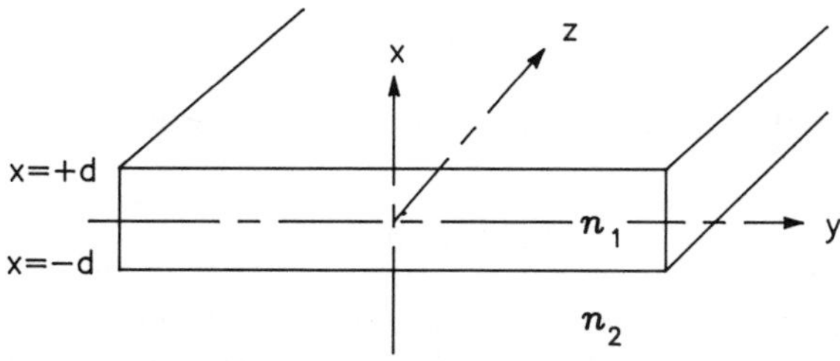

Figure 13.4. Geometry of symmetric dielectric slab waveguide

was used with metal conductors. The two boundary conditions are

(a) The tangential components of the electric field must be equal across a dielectric boundary.

(b) The tangential components of the magnetic field must be equal across a dielectric boundary.

It is possible to combine these two boundary conditions into one boundary condition by using transverse wave impedances and matching the impedances across the boundary. The dielectric slab waveguide supports TE modes and TM modes. The boundary condition for TM modes is

$$\frac{E_{z1}}{H_{y1}} = \frac{E_{z2}}{H_{y2}} \quad \text{at } x = \pm d \tag{13.2}$$

and for TE modes

$$\frac{H_{z1}}{E_{y1}} = \frac{H_{z2}}{E_{y2}} \quad \text{at } x = \pm d \tag{13.3}$$

where E_{z1}, E_{y1}, H_{z1} and H_{y1} are the fields in the core region and E_{z2}, E_{y2}, H_{z2} and H_{y2} are the fields in the cladding region. There is another complication in dielectric waveguides. Two propagation equations are needed to describe the TE modes and two propagation equations are needed to describe the TM modes. The reason is that the transverse fields can have either a peak on axis or a null on axis and these two possibilities cannot be represented by one equation. There must be one propagation equation where the transverse electric fields have odd symmetry about the $x = 0$ axis and the described by a sine function. The other propagation equation has transverse electric fields which have even symmetry about the $x = 0$ axis and the electric fields are described by a cosine function.

The principles are the same for TE and TM modes and both symmetries so the characteristics of the even TE modes will be derived. The fields in the inner region of Figure 13.4 ($-d < x < d$) are given by

$$E_{y1} = A\cos(Ux) \tag{13.4a}$$

$$H_{x1} = -\frac{\beta}{\omega\mu_0} A\cos(Ux) \tag{13.4b}$$

$$H_{z1} = -\frac{\mathrm{j}U}{\omega\mu_0} A\sin(Ux) \tag{13.4c}$$

where the exponential factor $\exp[\mathrm{j}(\omega t - \beta z)]$ has been assumed and

$$U^2 = \frac{\omega^2}{c^2} n_1^2 - \beta^2 \tag{13.5}$$

U is the wavenumber in the core. This takes the same form as the wavenumber of a metal waveguide, but must be given a different symbol because there is a second wavenumber for the cladding. In the cladding region where $x > \pm d$ the fields are not travelling waves and must be described by an exponentially decaying function:

$$E_{y2} = B\exp[-(|x| - d)] \tag{13.6a}$$

$$H_{x2} = -\frac{\beta}{\omega\mu_0}B\exp[-(|x| - d)] \tag{13.6b}$$

$$H_{z2} = -\frac{jW}{\omega\mu_0}B\left(\frac{-x}{|x|}\right)\exp[-(|x| - d)] \tag{13.6c}$$

where

$$W^2 = \beta^2 - \frac{\omega^2}{c^2}n_2^2 \tag{13.7}$$

W is the wavenumber for the cladding. Applying the boundary condition, equation (13.3), and simplifying to eliminate the coefficients A and B leads to an equation which describes the behaviour of ω and β for a given set of fixed parameters, n_1, n_2 and d. This is the *propagation equation* for even TE modes:

$$\tan(Ud) = \frac{W}{U} \tag{13.8}$$

The propagation equation for odd TE modes can similarly be obtained by starting with

$$E_{y1} = A\sin(Ux) \tag{13.9}$$

which leads to the propagation equation

$$\tan(Ud) = -\frac{U}{W} \tag{13.10}$$

The TM modes are given by similar equations. That for even TM modes is

$$\tan(Ud) = \frac{n_1^2 W}{n_2^2 U} \tag{13.11}$$

and that for odd TM modes is

$$\tan(Ud) = -\frac{n_2^2 U}{n_1^2 W} \tag{13.12}$$

The propagation characteristics plotted as βd against $\omega d/c$ are sketched for TE modes in Figure 13.5. The lower asymptote in Figure 13.5 is for the case when all the fields travelling in the outer region whilst the upper asymptote is for the case when all the fields are travelling in the slab. The TE_0 and TE_2 modes are the even solutions and the TE_1 and TE_3 are the odd solutions. The figure follows the form shown in Figure 13.3, but is not very helpful because the curves are too bunched up. This is bound to happen if $n_1 \approx n_2$, as is the case in optical waveguides. It is better to use a secondary parameter, V, which is defined by adding equations (13.5) and (13.7)

$$(U^2 + W^2)d^2 = V^2 \tag{13.13}$$

or

$$V^2 = (n_1^2 - n_2^2)\frac{\omega^2 d^2}{c^2} \tag{13.14}$$

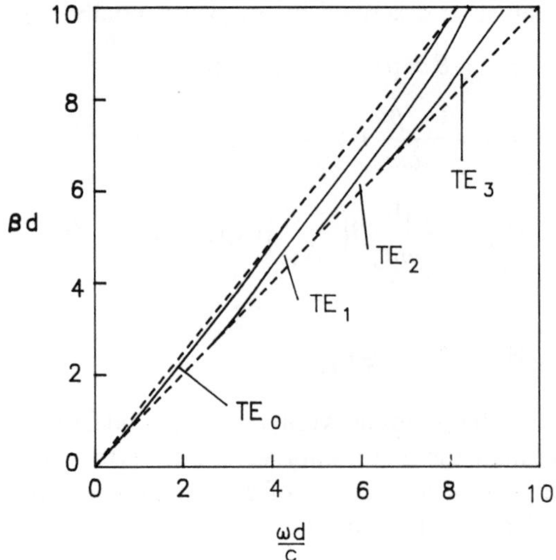

Figure 13.5. Propagation characteristics (βd against $\omega d/c$) for TE modes in dielectric slab waveguide

The distance d has been used to make the equation dimensionless. The *V number* is a useful parameter in optical waveguide design and is often quoted in descriptions of optical fibres,

$$V = \frac{\omega}{c} d(n_1^2 - n_2^2)^{1/2} \tag{13.15}$$

The square root term contains the description of the material properties and is called the *numerical aperture*, NA,

$$\mathrm{NA} = (n_1^2 - n_2^2)^{1/2} \tag{13.16}$$

The propagation characteristics coming from equations equations (13.8) and (13.10) to (13.12) can now be plotted as Ud against V, as shown in Figure 13.6. The TE curves are universal, but eh TM curves depend on the ratio n_2/n_1. Equations (13.11) and (13.12) show that if $n_2 \Rightarrow n_1$, the TM propagation equation tends to the TE propagation equation.

The method of obtaining these curves needs a bit of explanation. The propagation equations cannot be solved in closed form because the equations cannot be written in the form $U = f(V)$. They must be solved numerically on a computer using a iterative procedure. This can be demonstrated with the even TE mode propagation equation. A single equation combining equations (13.8) and (13.13) is

$$\tan(Ud) - \left[\left(\frac{V}{Ud}\right)^2 - 1\right]^{1/2} = 0 \tag{13.17}$$

The objective is to find the roots of this equation by inserting values for Ud and V such that the left-hand side is equal to zero. Normally one parameter, say V, is fixed and the other parameter, Ud, is varied. The value of the function on the left-hand side of

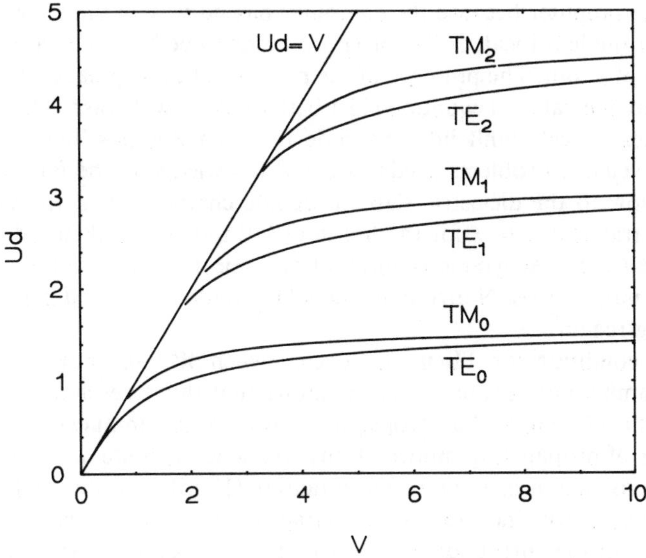

Figure 13.6. Propagation characteristics (Ud against V) for dielectric slab waveguide, $n_1/n_2 = 1.5$

equation (13.17) for $V = 6$ is plotted in Figure 13.7. There is a root around $Ud = 1.3$ and a second root around $Ud = 4.0$. Between these two roots the equation has a pole at $Ud = 1.6$. a computer root searching procedure needs to step in increments of Ud until a root is located and then iteratively find the value of Ud to a specified accuracy.

The iterative procedure is the same for all waveguide propagation equations, except those for parallel plate, circular and rectangular metal waveguides. It is not easy to

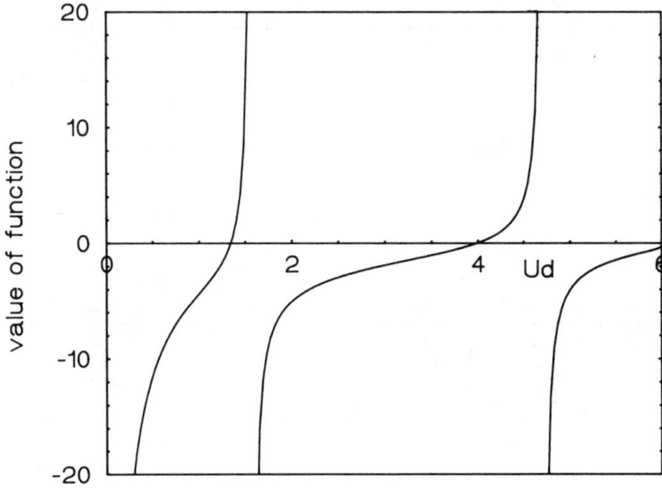

Figure 13.7. Computation of roots of propagation equation $f(Ud, V) = 0$

automate on a computer because the computer has no way of knowing where the root for a particular mode is located. The program needs to be helped by a reasonable guess at the location of a root. The presence of the pole does not help because if the computer tries to calculate the value of equation (13.17) at a pole, it will crash. The program needs to have a series of tests built into the code to avoid this possibility. The numerical solution of waveguide problems needs care and knowledge of the form of the modes.

Returning now to the dielectric slab waveguide characteristics, the variation of the transverse electric field with x for the first four TE modes are sketched in Figure 13.8. These show that the even symmetry modes have a peak on axis and the odd symmetry modes have a null on axis. Notice that the field is finite at $x = \pm d$ and decays quickly in the cladding region.

The cut-off condition for TE modes occurs when $W = 0$, or from equation (13.7), $\beta = \omega n_2/c$. Examination of equation (13.8) shows that this gives a valid solution for all values of ω. The TE_0 mode thus propagates down to zero frequency.

The number of propagating modes at any frequency is finite and is proportional to the V number, as shown in Figure 13.6. Equation (13.15) indicates that the number of propagating modes will increase as the refractive index of the slab increases or the refractive index of the surrounding dielectric decreases or the slab width increases. If $V < \pi/2$ there is only one propagating mode and if $V \gg 2\pi$, a large number of modes will exist. Rearranging equation (13.15) with this condition shows that in order to have a multi-mode waveguide, the slab half width should be chosen so that

$$d \gg \frac{\lambda}{NA} \tag{13.18}$$

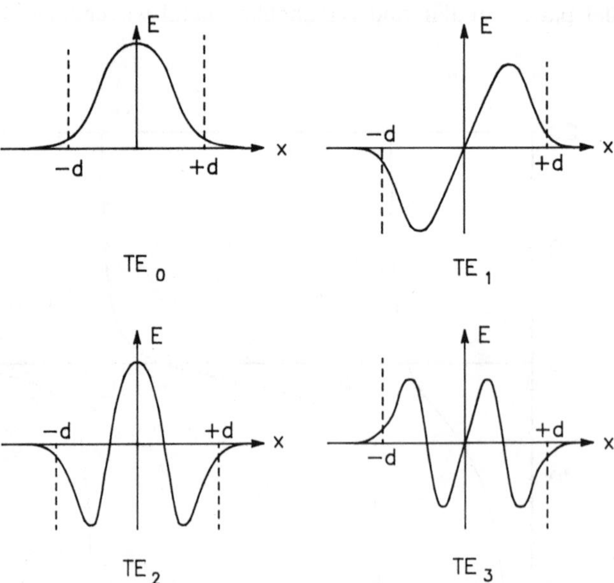

Figure 13.8. Amplitude of transverse electric fields against distance for TE modes in dielectric slab waveguide

Example 13.1 Dielectric slab waveguide

Find the thickness of a slab waveguide which operates at 1.5 μm and uses two glasses with refractive indices of 1.47 and 1.45. For both single TE mode propagation and for multi-mode propagation.

The numerical aperture is given by equation (13.16) as $\text{NA} = (1.47^2 - 1.45^2)^{1/2} = 0.242$. For single mode propagation, $V = \text{NA} \times 2\pi d/\lambda = \pi/2$ so $d = 1.55\ \mu\text{m}$. This is just larger than a wavelength. For multi-mode operation equation (13.18) gives $d \gg 6.2\ \mu\text{m}$. A value of $d = 10\mu\text{m}$ would be reasonable. Note that if the two refractive indices are closer together, then the thickness increases. This is a useful way of making the dimensions more manageable.

Practical forms of dielectric slab waveguides at optical frequencies are made by thin film deposition techniques which result in typical profiles shown in Figure 13.9. The dielectric waveguide is now asymmetric with respect to the centre line and the analysis becomes more complicated than that developed in this section. The principles, however, remain the same and only the detailed nature of the propagation characteristics are modified.

There are two forms of the slab waveguide which are used in integrated optics, *step-index* and graded-index waveguides. The first has the profiles shown in Figure 13.9 with a sharp change in relative permittivity at the boundary. The second type has a gradual change in relative permittivity across the boundary. This can be made by diffusing a material into a substrate so that the crystalline structure is modified near to the surface. The process is similar to that used to produce integrated circuits. The permittivity then changes in a controlled but graded form from the centre of the waveguide. The step-index waveguide is made on a semiconductor substrate and is more difficult to make but gives better characteristics for optical modulators and other optical processing components.

13.4 OPTICAL FIBRES

Optical fibre waveguides are increasingly used for the transmission of large numbers of telecommunications circuits over medium to long distances. They are also used for short distance instrumentation and measurement applications. In future telecommunication systems, optical fibres will take over from coaxial cable and microwave radio links as the main provider of circuits for all but the local consumer circuit. Submarine optical cables are being installed for transcontinental telecommunications and compete with satellite communications. Satellites are more versatile because orbit configurations and coverage areas can be changed during a satellite lifetime, but optical cables have higher signal capacity which makes them economically attractive.

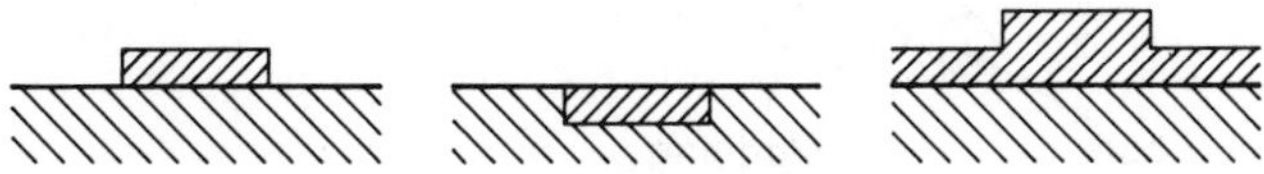

Figure 13.9. Cross-section of practical dielectric slab waveguides

All optical fibre waveguides are cylindrical dielectric waveguides. The cross-section of an optical fibre waveguide is shown in Figure 13.10. The principles of waveguide propagation in circular dielectric fibre are the same as for dielectric slab waveguide, but the analysis is more complicated due to the cylindrical geometry and the multiple layers of dielectric. It will be discussed in the next section. The optical waveguide shown in Figure 13.10 has three regions. The inner core is the main guiding region and is surrounded by the cladding. This is surrounded by a protective sleeve which has the function of strengthening the delicate glass fibre and protecting it from external mechanical forces.

One of the distinguishing features of optical waveguides is that the refractive index of the material used for the core and the cladding differ by less than a few percent. It is the difference in refractive index that is the important parameter, not the absolute values. For this reason a dimensionless parameter, Δ, is used to specify the difference in permittivity

$$\Delta = \frac{n_1^2 - n_2^2}{2n_1^2} \tag{13.19}$$

For $n_1 \approx n_2$, equation (13.19) becomes $\Delta \approx 1 - n_2/n_1$.

The dielectric used to make optical fibres is glass, although plastic can be used for low quality fibres. Glass is a generic term for a silica, SiO_2, doped with other materials. The number and proportions of the other materials determine the refractive index. The refractive index can vary between 1.4 and 1.8 but for waveguides, a range of between 1.4 and 1.6 is used. Typical glasses use silica doped with phosphorus, boron or germanium. It is possible to make the glass to have a refractive index to within 1% of a specified value.

Most optical fibre waveguides operate in the wavelength range of 0.8 to 1.7 μm. The reason for this choice is concerned with the losses which occur in glasses and will be explained in Section 13.6. Note that the frequency range is below the visible part of the spectrum so the waves in optical waveguides cannot be seen by the eye.

Optical fibres can be divided into three classes, *single-mode* fibres, *multi-mode* fibres and *graded-index* fibres. Typical refractive index profiles are shown in Figure 13.11. Single-mode and multi-mode fibres are *step-index* fibres. The single mode fibre has a narrow core which will only support one mode. This has the best electrical characteristics

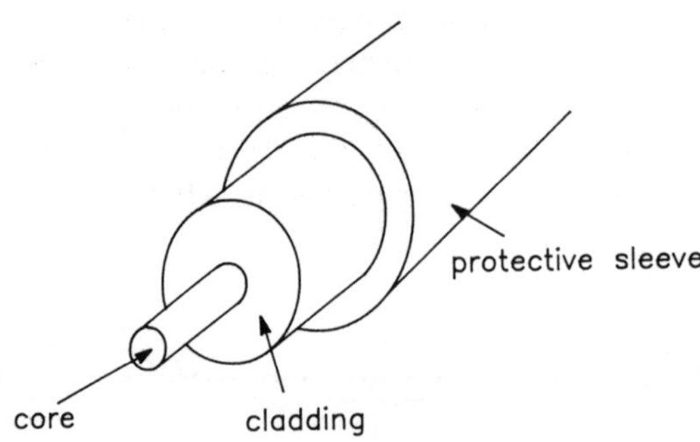

Figure 13.10. Optical fibre waveguide

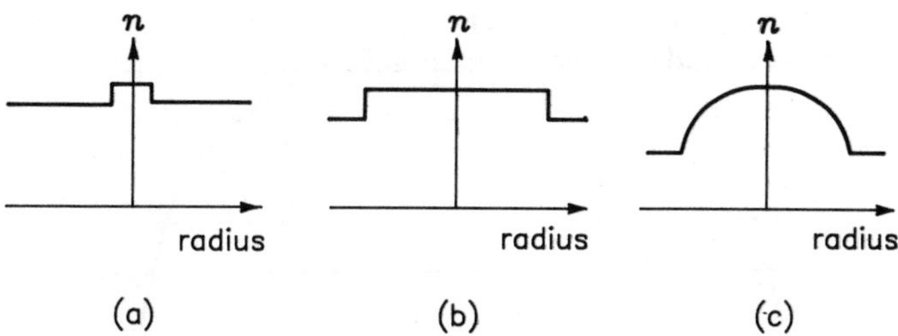

Figure 13.11. Refractive index profiles against radius for optical fibres. (a) Single-mode step index. (b) Multi-mode step index. (c) Graded index fibre

but the very narrow core makes the manufacture expensive. Multi-mode fibres can support a large number of modes because the waveguide diameter is large in wavelengths. A sketch of the ray picture is shown in Figure 13.12. The multi-mode fibre has a large number of ray paths because of the large number of modes. The individual modes are not excited, rather power is transported in a bundle of all possible modes. At the receiver, power is received in a bundle. This restricts the frequency bandwidth which can be transmitted due to an effect called dispersion which will be described in Section 13.8. Multi-mode optical fibres are, however, considerably easier to manufacture with lower tolerances than the single-mode fibre.

Graded index fibres are a compromise between single-mode fibres and multi-mode fibres. It has a gradual change in refractive index in the radial direction, with the peak value on the axis. There may be no sharp boundary from the core to the cladding. A typical profile has a refractive index profile of the form

$$n(r) = n_1\left[1 - 2\Delta\left(\frac{r}{a}\right)^k\right] \tag{13.20}$$

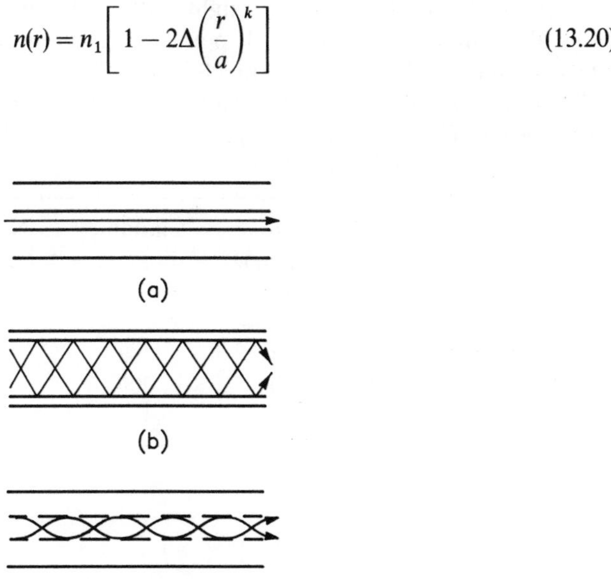

Figure 13.12. Rays in optical fibre. (a) Single-mode. (b) Multi-mode. (c) Graded index

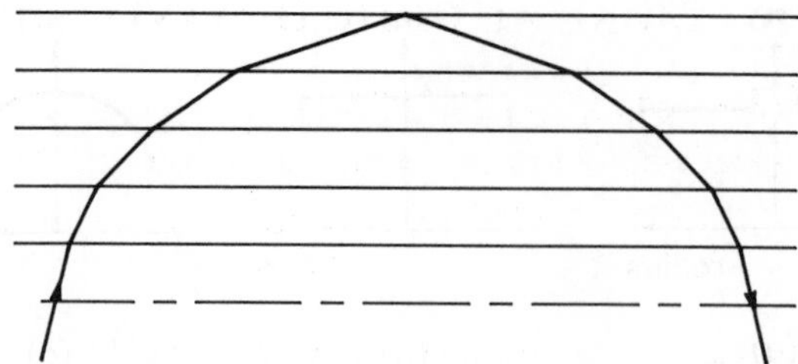

Figure 13.13. Graded index fibre represented as a series of discrete layers with constant refractive index, showing ray path

where a is the radius of the core–cladding boundary and k is an integer. If $k = 2$, then the profile is parabolic.

The rays in a graded index fibre follow a smooth sinusoidal type path, as shown in Figure 13.12. The reason can be understood if the continuously varying refractive index is represented as a series of discrete layers with constant refractive index in each layer, Figure 13.13. Using Snell's law to trace the path of a ray shows that it follows a path as sketched in the figure.

The ray pictures sketched in Figure 13.12 assume that the waveguide is planar and thin. The cylindrical geometry makes the real ray picture more complicated, because rays can travel in three dimensions along the waveguide. There are two classes of rays, *meridional* rays and *skew* rays, or helical rays. Meridional rays pass through the axis and are the ones shown in Figure 13.12. Skew rays are much more numerous than meridional rays. They do not pass through the axis and follow a helical path as they bounce off the core boundary. The ray picture as viewed in a cross-section of the fibre is shown in Figure 13.14. The straight paths between reflections are actually angled paths along the fibre so that each point of reflection around the boundary is further along the waveguide as the ray travels a helical path.

The three types of optical fibre waveguide differ in their performance. A summary is shown below. The values are typical of fibres within each type but there are wide variations in diameters and attenuation.

	Core diameter (μm)	Cladding diameter	Numerical aperture	Attenuation (dB/km)
Single mode	3	100	0.1	1
Multi-mode	100	200	0.3	20
Graded index	50	125	0.3	5

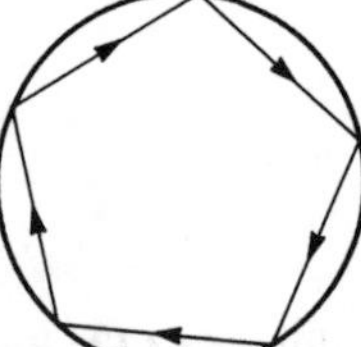

Figure 13.14. Cross-section of core of multi-mode or graded index fibre showing skew rays

13.5 MODES IN CYLINDRICAL OPTICAL WAVEGUIDE

The exact analysis of modes in a circular dielectric waveguide is complicated because the cylindrical waveguide is bound in two dimensions instead of the one dimension of the planar waveguide. Moreover cylindrical functions are inherently more complicated than planar functions. The propagation equation is composed of a combination of Bessel functions and the propagation characteristics must be solved numerically as described in Section 13.3. The mode set is also more complicated because both E_z and H_z have finite values. This means that pure TE and TM modes do not, in general, exist. It is still possible to divide the solutions into two orthogonal sets of modes. These are called *hybrid* modes and are designated HE_{pq} modes and EH_{pq} modes depending on whether the transverse magnetic or transverse electric fields are dominant. The subscript p refers to the number of variations in the radial direction. Modes with $p=0$ are circularly symmetric and are independent of the orientation of the waveguide. Other modes are described by either $\sin(p\phi)$ or $\cos(p\phi)$, where ϕ is the azimuthal angle. Modes with $p=1$ have two maxima of field around the circumference and the orientation of the waveguide is significant. When $p=0$, the HE and EH modes devolve back to TM_{0q} and TE_{0q} modes, respectively.

The transverse electric field patterns for the HE_{11}, HE_{21}, TE_{01} and TM_{01} are sketched in Figure 13.15. Comparison with the field patterns for modes in hollow circular metal waveguides, Figure 12.18, will show the similarity. The dominant mode is the HE_{11} mode. This has, however, a more linear field pattern than the dominant TE_{11} mode of circular waveguide. This is a desirable attribute and produces improved characteristics.

WEAKLY GUIDED SOLUTIONS

The exact solutions for cylindrical optical waveguide are complicated. It is possible to simplify the solutions somewhat by making use of the fact that all optical fibre waveguides

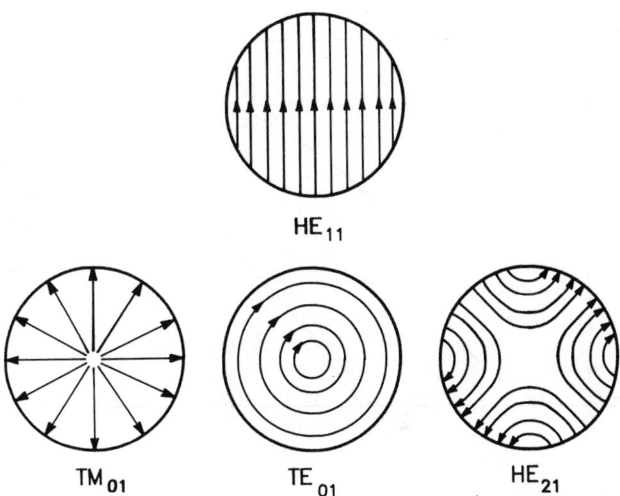

Figure 13.15. Transverse electric field patterns for modes in cylindrical optical waveguide

operate with a small refractive index difference between the core and the cladding. These are called *weakly guided* solutions and the theory will now be outlined.

Reference to Figure 13.3 indicates that if $n_1 \approx n_2$, the propagation curves of β against ω will lie between two asymptotic lines which are almost on top of each other. This means that

$$\beta \approx \frac{\omega}{c} n_1 \approx \frac{\omega}{c} n_2 \tag{13.21}$$

Then the wavenumbers U and W for the core and cladding regions, equations (13.5) and (13.7) will both be approximately zero. This then leads to $E_z \approx 0$ and $H_z \approx 0$. The fields are nearly all transverse to the direction of propagation, which is the condition for plane waves. Thus the weakly guided waves are essentially plane wave solutions with the electric and magnetic fields related by the wave impedance. There is an additional feature which also follows from the plane wave nature of the modes. The fields will be linearly polarised and will have an electric field which is in one direction only, as shown for the HE_{11} mode in Figure 13.5. This means that the fields in the cylindrical waveguide can be approximated by E_x and H_y, or by E_y and H_x, just as for the plane waves studied in Chapter 5.

The full mode set of HE, EH, TE and TM modes may now be approximated by a new set of modes called *linearly polarised*, LP_{pq}, modes. A group of the exact modes combine together to form one single LP mode. This can be seen from the exact propagation characteristics of the first few modes shown in Figure 13.16. Groups of the exact modes are almost on top of each other and are said to be degenerate. The correspondence between the exact and the linearly polarised modes is

The LP_{01} mode is the HE_{11} mode
The LP_{11} mode is the $TE_{01} + TM_{01} + HE_{21}$ modes
The LP_{21} mode is the $EH_{11} + HE_{31}$ modes
The LP_{02} mode is the HE_{12} mode.

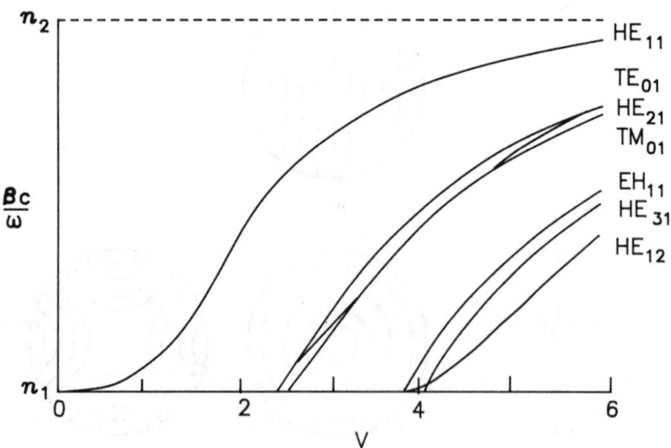

Figure 13.16. Propagation characteristics for modes in cylindrical optical waveguide obtained from exact analysis

The analysis of the LP modes under the weakly guidance approximation is done by the same procedure as that used for any other waveguide. The existence of only one electric or magnetic field component means, however, that the wave equation in cylindrical coordinates can be written using a scalar variable ψ where ψ is given by

$$\frac{d^2\psi}{dr^2} + \frac{1}{r}\frac{d\psi}{dr^2} + \frac{1}{r^2}\frac{d^2\psi}{d\phi^2} + U^2\psi = 0 \tag{13.22}$$

where $U^2 = (\omega^2/c^2)n_1^2 - \beta^2$ and n_1 is the refractive index of the core.

The general solution will have the form

$$\psi = \psi(r)\begin{matrix}\cos(p\phi)\\ \sin(p\phi)\end{matrix}\exp[j(\omega t - \beta z)] \tag{13.23}$$

substituting into equation (13.22) gives

$$\frac{d^2\psi(r)}{dr^2} + \frac{1}{r}\frac{d\psi(r)}{dr} + \left(U^2 - \frac{p^2}{r^2}\right)\psi(r) = 0 \tag{13.24}$$

The solutions to this equation are Bessel functions. In the core region they take the form

$$\psi_1 = AJ_p(Ur)\begin{matrix}\cos(p\phi)\\ \sin(p\phi)\end{matrix}\exp[j(\omega t - \beta z)] \tag{13.25}$$

where A is an amplitude coefficient and $J_p(Ur)$ is a Bessel function of the first kind and order p.

The fields in the cladding region need to satisfy the boundary condition that the fields are zero at infinite radius. The solutions take the form

$$\psi_2 = BK_p(Wr)\begin{matrix}\cos(p\phi)\\ \sin(p\phi)\end{matrix}\exp[j(\omega t - \beta z)] \tag{13.26}$$

where $W^2 = \beta^2 - (\omega^2/c^2)n_2^2$, B is an amplitude coefficient and $K_p(Wr)$ is a modified Bessel function of the second kind. These functions decays exponentially to zero at infinite radius.

The characteristic equation is obtained by matching the tangential electric and magnetic fields across the core–cladding boundary at $r = a$. After manipulation this leads to

$$U\frac{J_{p-1}(Ua)}{J_p(Ua)} = -W\frac{K_{p-1}(Wa)}{K_p(Wa)} \tag{13.27}$$

This propagation equation must be solved iteratively on a computer. It yields results which define the propagation behaviour of optical waveguide. By suitable choice of axes a set of normalised propagation characteristics can be obtained. These are shown in Figure 13.17 where the V number was defined earlier and $b = [(\beta c/\omega)^2 - n_1^2]/(NA)^2$.

The cut-off condition, $W = 0$, requires that $J_{p-1}(Ua) = 0$, from equation (13.27). The cut-off frequencies of LP modes are thus given by the roots of $J_p(Ua)$, however for $p = 0$ and $q = 1$, there is no cut-off frequency, so the LP_{01} mode propagates down to zero frequency. Since this is the dominant mode and has a nearly pure linear polarised electric field, it is the mode used in single mode fibres. The bandwidth of this mode is set by

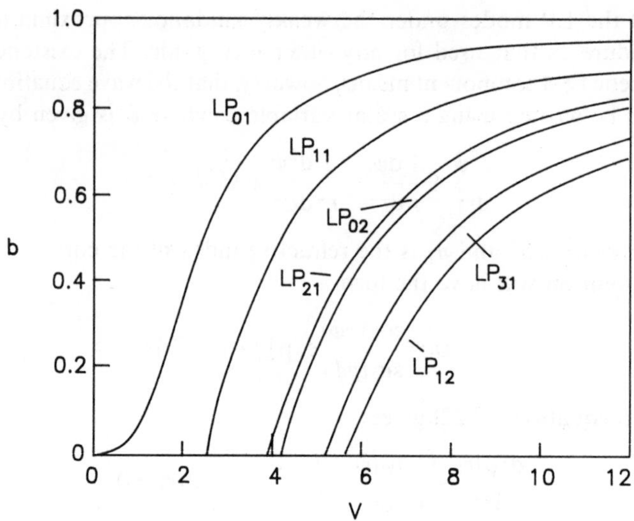

Figure 13.17. Normalised propagation characteristics for LP modes in cylindrical optical waveguide

the cut-off frequency of the next mode. This occurs when $J_0(Ua) = 0$, which is at $Ua = 2.405$. The cut-off condition is given by $Ua = V$ so the cut-off frequency of the LP_{11} mode is

$$V = \frac{2\pi a(n_1^2 - n_2^2)^{1/2}}{\lambda} = 2.405 \tag{13.28}$$

Example 13.2 Design of a single mode optical fibre waveguide

Design a single mode optical fibre for operation at $\lambda = 1.3\,\mu m$ with a core refractive index of 1.460.

The index difference Δ should be less than 0.01 for the weakly guided solutions to apply. Choosing $\Delta = 0.005$ and using equation (13.19) gives $n_2 = 1.453$. The waveguide then has a numerical aperture of $NA = 0.143$. The cut-off condition, equation (13.28) is $a/\lambda = 2.405/(2\pi NA) = 2.68$. Thus the diameter of the core must be less than $2a = 7.0\,\mu m$. It is better not to operate near the cut-off value, so a diameter of $2a = 5\,\mu m$ might be chosen. This gives a V number of 1.73.

13.6 TRANSMISSION LOSSES IN OPTICAL FIBRES

The transmission loss or attenuation is an important parameter in optical fibre waveguides for telecommunications. It determines the maximum length of waveguide before regeneration of signal is needed and hence the spacing between repeater stations. Some typical attenuations of glasses are given below

Material	Attenuation (dB/km)
Window glass	10 000
Optical glass for lenses	300
Pure water	100
Silica glass (1970)	20
(1975)	2
(1985)	0.5

When optical fibres were discovered in the mid 1960s the typical attenuation of glass was about 10 dB/m^{-1} or 10 000 dB/km. A major break through occurred in 1970 when researchers at Corning Glass Co. announced a laboratory glass with an attenuation of 20 dB/km. Since then, further strides have been made in glass material technology which has removed many of the impurities that cause losses so that it is possible to routinely produce glass fibres with attenuations less than 0.5 dB/km. This value can be put into perspective by noting that 6 km of fibre is needed to reduce the power level of the signal by half. This attenuation is not the limit and silica glasses have been made in the laboratory with attenuations below 0.2 dB/km.

The transmission losses in optical fibre waveguides come from three mechanism.

- Absorption in the glass.
- Scattering by imperfections in the glass.
- Radiation from the core due to bending or geometric non-uniformities.

The lowest losses have been found in silica-based glass, so the losses will be described for silica. The *absorption* is due to inherent electronic and molecular vibrations and to the presence of contaminants. The electronic absorption occurs at ultraviolet wavelengths and sets a basic limit on the minimum attenuation which can be achieved. It is shown in Figure 13.18. The molecular absorption occurs at infrared wavelengths and similarly sets a lower limit on attenuation. Impurity ions would also cause absorption, but these can be removed by careful manufacture. The minimum absorption loss is wavelength dependent with a minima in the infrared region of 1.1 to 1.8 μm, as shown in Figure 13.18.

Scattering losses for fused silica are caused by microscopic inhomogeneities which modify the refractive index. They are partly caused by the process of cooling glass from being a pure liquid when hot to its normal supercooled liquid state. The scattering losses can be reduced by cooling the glass over a longer interval of time. The minimum attenuation is set by random variations in the scatterers which are proportional to λ^{-4}. It gives the curve shown in Figure 13.18.

The result of the absorption and scattering losses is to define a low-loss window centred around 1.4 μm. Typical attenuations for a high grade single-mode fibre are shown

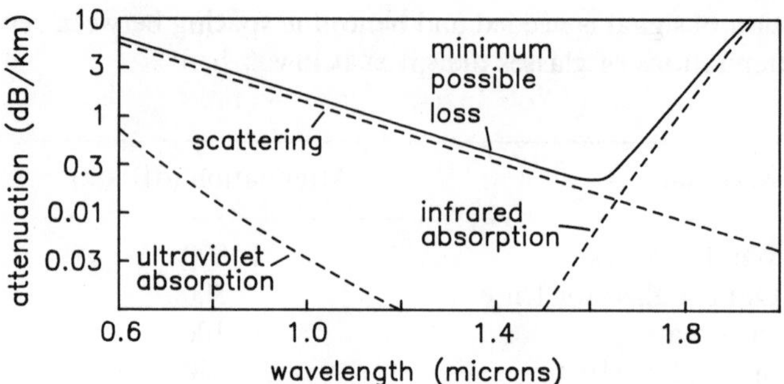

Figure 13.18. Attenuation factors in glass

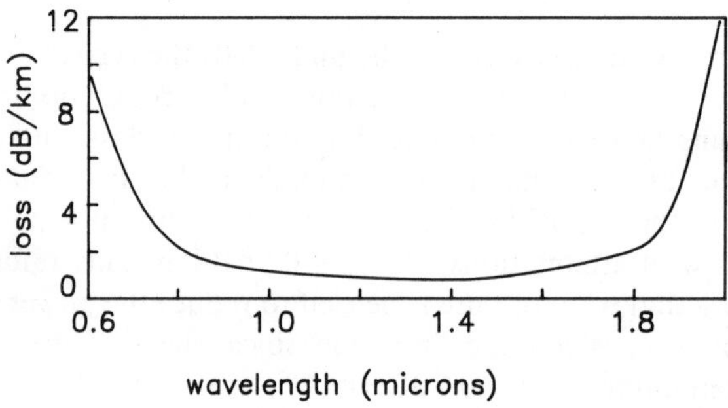

Figure 13.19. Attenuation against wavelength for typical optical glass

in Figure 13.19. The useful window is at wavelengths above the visible range so that the typical operating band for optical communications lies between 0.8 and 1.7 μm.

The third type of loss is the *radiation bending losses.* This is only present when the waveguide is bent in a curve. The electromagnetic analysis of bent waveguide is a difficult problem because the bending means that waves are no longer travelling in straight lines and power couples between modes and radiates away from the core. This can be seen by considering the ray picture shown in Figure 13.20. A ray which just satisfies the total

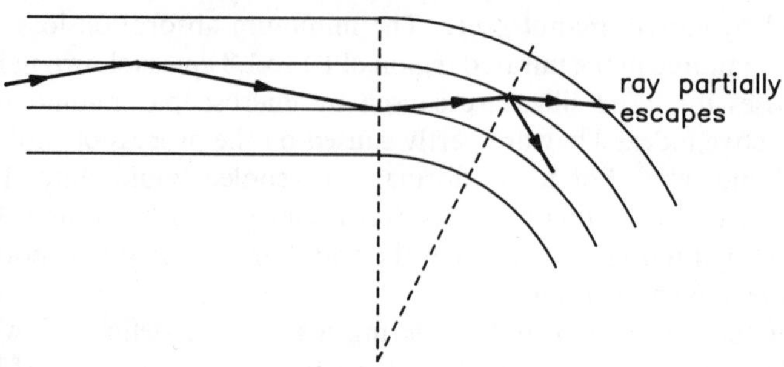

Figure 13.20. Mechanisms for radiation bending loss in optical fibre

internal condition along a straight waveguide will enter the bend and hit the boundary at a higher angle of incidence. The result is that it will be below the critical angle and will be transmitted into the cladding. In a gradual bend most of the power will be guided around the bend, but there will always be a small amount which radiates away from the core region into the cladding and protective regions. The amount of loss will depend on the radius of the bend and the additional attenuation can be estimated from the relation

$$\delta\alpha = 10\log_{10}\left(1 - \frac{1}{\Delta}\frac{a}{R}\right) \quad \text{(dB/km)} \tag{13.29}$$

where R is the radius of the bend and a the radius of the core. This shows that the bending loss can be reduced by increasing the bend radius, reducing the core diameter or increasing Δ. This means a larger difference in refractive index between core and cladding, which is not desirable from the propagation point of view.

13.7 COUPLING POWER INTO OPTICAL WAVEGUIDE

One of the problems of using optical waveguides is that of coupling power into the waveguide. This can introduce a significant amount of loss in an optical waveguide system unless the coupler is designed to be efficient. The coupling process will be briefly described for excitation at the end of an optical waveguide which could be either a slab waveguide or an optical fibre waveguide. The situation is sketched in Figure 13.21. There are three causes of coupling loss:

(a) The radiation from a source will be over a solid angular region, so that only some of the power will fall onto the end of the core.

(b) There will be a mismatch between the air and the dielectric waveguide, or between the source and the dielectric if they are butt jointed.

(c) The waveguide will only accept power which falls within a narrow cone.

The first of these causes comes from the fact that any optical source radiates its power over a cone of angles. There are two basic types of sources used with optical waveguides. The first source is a *light emitting diode*, or LED. This radiates between 2 and 10 mW

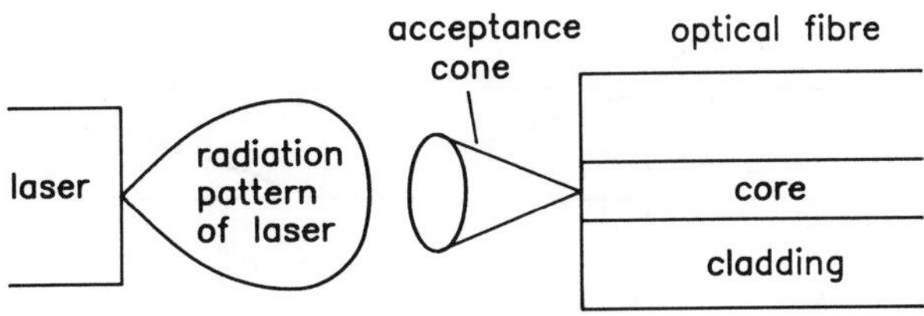

Figure 13.21. Schematic of process of coupling power into an optical waveguide

of power over a broad solid angle, so only a portion will fall onto the narrow core of an optical waveguide. The amount can be improved by using a lens to focus the energy. This will, however, require very precise positioning and one of the benefits of an LED is that it is simple and cheap. An LED radiates over a relatively broad spectrum of wavelengths. The other type of source is a *semiconductor laser*. This takes a variety of forms. It typically radiates the same power levels as an LED but has a very narrow line width of less than 1 nm. The laser is more expensive than the LED so it is usually worth joining it directly to the core through a coupling lens. This may not remove all the mismatch because the refractive indices of the laser, coupler, and core may be different.

The power loss due to different refractive indices can be evaluated by using the reflection formulae developed in Chapter 6. For normal incidence the amount of power lost when a plane wave is incident in air onto a dielectric is given by the *mismatch loss*,

$$\mathrm{ML} = 20\log_{10}\left(1 - \left|\frac{1-n_1}{1+n_1}\right|\right) \quad \mathrm{dB} \tag{13.30}$$

where n_1 is the refractive index of the core. If the incident medium is a dielectric then n_1 is the ratio of the core refractive index to the incident medium refractive index.

The waveguide has a restricted angle of acceptance because of the need to satisfy the total internal reflection condition at the core to cladding boundary. This can be illustrated using a ray diagram. Consider the situation shown in Figure 13.22 where a ray is incident on the core of an optical waveguide. The ray in incident in a medium with refractive index n_a at an angle ψ_1. The transmitted angle, ψ_t, into the core is given by Snell's law as

$$\sin\psi_i = \frac{n_1}{n_a}\sin\psi_t \tag{13.31}$$

The ray will hit the boundary between the core and the cladding at an angle θ_i given by $\cos\theta_i = \sin\psi_t$. This must be greater than the critical angle so that the ray will be totally internally reflected, or $\sin\theta_i \geqslant n_2/n_1$.

Substituting into equation (13.31) and rearranging

$$\sin\psi_i = \frac{n_1}{n_a}(1-\sin^2\theta_i)^{1/2} = \frac{n_1}{n_a}\left(1-\frac{n_2^2}{n_1^2}\right)^{1/2}$$

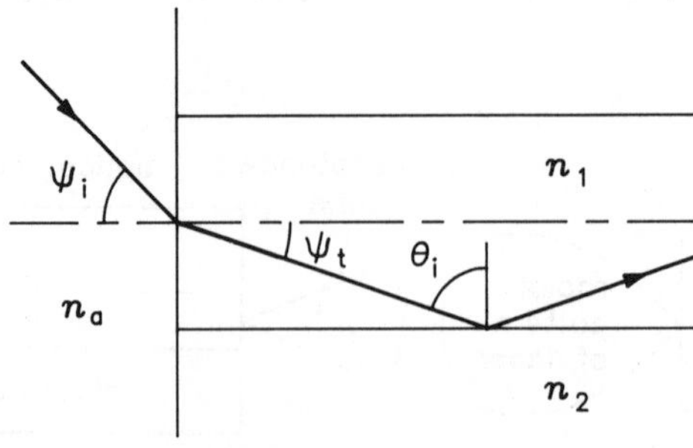

Figure 13.22. Ray geometry for coupling power into an optical fibre

or

$$\sin \psi_i = \frac{1}{n_a}(n_1^2 - n_2^2)^{1/2} = \frac{NA}{n_a} \qquad (13.32)$$

If the incident medium is air then $\sin \psi_i$ equals the numerical aperture. The numerical aperture can thus be thought of as a measure of the ability of the waveguide to accept power over a cone of angles, $\pm \psi_i$.

Example 13.3 Coupling into an optical waveguide

Find the acceptance cone for an optical waveguide with $n_1 = 1.6$ and $n_2 = 1.5$, if the incident medium is air. Also find the amount of power coupled to the waveguide if a 2 mW LED is used in which 30% of the power falls onto the end of the waveguide.

From equation (13.32), $NA = 0.56$ and $\psi_i = 33.8°$. The incident ray must therefore be within a cone of angle of $\pm 33.8°$ for the ray to enter the waveguide. The cone of angles decreases as the refractive indices of the core and cladding become closer to each other. If $n_1 = 1.5$ and $n_2 = 1.485$ then $NA = 0.21$ and $\psi_i = 12.2°$.

The power coupled into the waveguide is the LED power, minus the power not falling onto the waveguide, minus the power that is reflected by the mismatch. In dBs the LED power is +3 dBm. The loss due to power not falling onto the waveguide is $10\log(0.3) = -5.2$ dB. The mismatch loss is given by equation (13.30) with $n_1 = 1.6$, or -2.27 dB. The power coupled to the waveguide is therefore $+3 - 5.2 - 2.27 = -4.47$ dBm or 0.36 mW.

13.8 SIGNAL DISTORTION DUE TO DISPERSION

The frequency bandwidth available at optical frequencies is very large, particularly when viewed as information channels. Even with the restrictions caused by the attenuation characteristics of glass, described in Section 13.6, the frequency space remains huge. For instance, the wavelength band from 1.1 to 1.4 μm contains a bandwidth of over 58 THz. There is, however, a major limitation on how much of this bandwidth can be used in one optical waveguide. This is the distortion which occurs to analogue and digital signals by the *dispersion* in the optical waveguide. Dispersion is caused by any frequency dependent characteristics of the glass, the modes of propagation and the waveguide. The effect of dispersion on signals is best illustrated with digital pulses. The presence of dispersion means that different frequency components of the pulse take different amounts of times to travel along the waveguide. This causes pulses to broaden as they travel along the waveguide, as illustrated in Figure 13.23. Two pulses which are close together in time will overlap and become indistinguishable. There is thus a minimum time between pulses which will be a function of the amount of dispersion and the length of the waveguide.

Dispersion occurs in all types of waveguides, but in microwave waveguides, the bandwidth which is used is relatively narrow, so the effect can normally be ignored. In optical waveguides, the broad bandwidths mean that dispersion can often be a limiting factor in determining the maximum use of the waveguide for signal transmission. The measure of dispersion is the group velocity. It will be recalled from earlier chapters, that

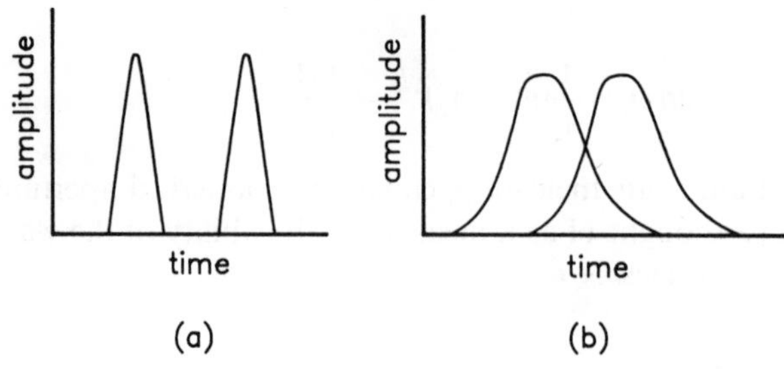

Figure 13.23. Effect of dispersion on two pulses. (a) At input to waveguide. (b) At output of waveguide

group velocity is the velocity of energy travel for a signal. If the group velocity changes with frequency over the bandwidth of a signal, then distortion will occur as described above. The group velocity is defined as

$$v_g = \frac{d\omega}{d\beta} \tag{13.33}$$

For a plane wave, the group velocity is equal to the phase velocity which is equal to the velocity of the wave in the medium ($v_g = c/n$). For a wave in a waveguide, the group velocity is usually a more complicated function of the actual velocity of the waves as they bounce along the waveguide.

There are three types of dispersion which can occur in optical waveguides

- Material dispersion.
- Waveguide dispersion.
- Multipath dispersion.

Each of these will now be briefly described.

Material dispersion arises because the refractive index of glass changes with frequency. The amount of change varies considerably from one silica compound to another and is a function of the amount of impurity. The refractive index is written as $n(\omega)$ to show the frequency dependence. The relationship between ω and β in a weakly guided optical waveguide where $n_1 \approx n_2$ is, from equation (13.21)

$$\omega = \frac{c\beta}{n(\omega)} \tag{13.34}$$

Differentiating this expression to find the group velocity gives

$$v_g = c\left(\frac{1}{n(\omega)} - \frac{\beta}{n(\omega)^2}\frac{dn(\omega)}{d\beta}\right) \tag{13.35}$$

This shows that v_g is a complicated function of $n(\omega)$ and small changes in the refractive index with frequency can have a considerable influence on the group velocity and hence on dispersion. Sometimes it is possible to choose silica compounds for the core and cladding which have material dispersion properties of equal magtnitude but opposite sign so that at some frequency a zero dispersion region exists.

Waveguide dispersion is caused by the variations of group velocity due to the non-linear propagation characteristics of guided modes. This occurs in all waveguides except those with a linear relationship between ω and β. TEM modes in metal waveguides, or the HE_{11} (LP_{01}) mode at low frequencies in optical fibre waveguide satisfy this exception. Examination of Figure 13.5, shows that in general β is not a linear function of ω. Thus equation (13.33) shows that the group velocity will change as frequency changes. In a single-mode optical fibre waveguide, the group velocity will show a marked non-linearity with frequency but it is usually possible to find a frequency region with approximately linear characteristics.

Multipath dispersion is sometimes called *modal dispersion* and is due to the different path lengths travelled by signals in different modes. Multipath dispersion only occurs in multi-mode optical waveguides which have either a step index profile or graded index profile. Thus all three types of dispersion occur in multi-mode waveguides but only material dispersion and waveguide dispersion occur in single-mode waveguide.

Multipath dispersion is a serious limitation on the use of multi-mode waveguide. This can be demonstrated by considering the ray picture in a multi-mode waveguide and considering the path length travelled by the rays. Figure 13.24 shows two extreme rays. An axial ray will travel the minimum distance whilst an edge ray at the critical angle will travel the maximum distance. The maximum frequency which can be transmitted along the waveguide can be found by evaluating the time difference between the axial ray and the edge ray.

The axial ray takes a time t_{min} to travel a distance L with velocity v,

$$t_{min} = \frac{L}{v} = \frac{L}{c} n_1 \tag{13.36}$$

Similarly the edge ray takes a time t_{max} to travel a distance l,

$$t_{max} = \frac{L}{v} = \frac{L}{c \cos\theta_t} n_1 \tag{13.37}$$

but $\cos\theta_t = \sin\theta_i = n_2/n_1$ so the total time delay per unit length is

$$\frac{\Delta t}{L} = t_{max} - t_{min} = \frac{n_1}{c}\left(\frac{n_1}{n_2} - 1\right) \tag{13.38}$$

The equation shows that the multipath distortion will be reduced by using an optical waveguide with a small difference between the refractive indices of the core and the cladding.

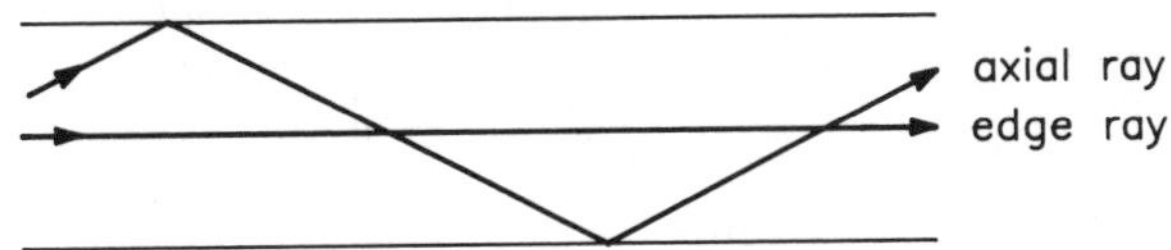

Figure 13.24. Axial and edge rays in a multi-mode optical waveguide

Example 13.4 Bandwidth due to dispersion in an optical fibre

Find the bandwidth over a 1 km length of multi-mode optical fibre waveguide with (a) $n_1 = 1.6$, $n_2 = 1.5$, and (b) $n_1 = 1.5$ and $n_2 = 1.485$.

Applying equation (13.35) directly for case (a) gives $\Delta t/L = 0.36\ \mu s\ km^{-1}$. This means that if a sharp pulse is sent along the waveguide then after 1 km it will occupy 0.36 μs. If a 2:1 safety margin is used then on a 1 km length of waveguide, the minimum time between pulses must be 0.72 μs or a maximum frequency of 1.4 MHz.

In case (b) with a smaller difference in refractive index between core and cladding, $\Delta t/L = 51$ ns which leads to an equivalent bandwidth of 10 MHz.

A multi-mode graded index fibre reduces the multipath dispersion because the edge rays travel a curved path (Figure 13.12) which reduces the difference between the axial path length and the edge ray path length. An improvement of up to ten times in the usable bandwidth is possible with graded index fibre waveguide.

13.9 OPTICAL WAVEGUIDE SYSTEMS

A complete optical waveguide communication system takes the form shown in Figure 13.25. The transmitter consists of a light source and a modulator of the base band digital or analogue signals onto the optical wave. The light source will be either an LED for a low cost, low capacity, system or a semiconductor laser for a high capacity system. The modulation is often done by modulating the signal onto the current feeding

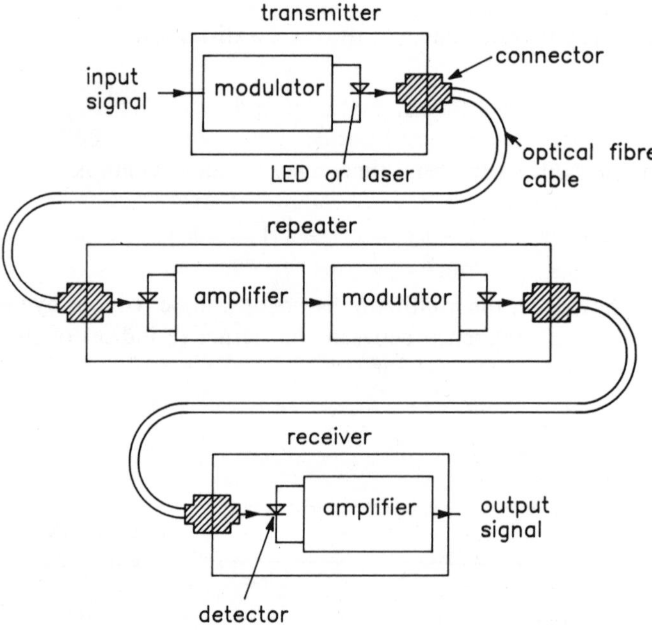

Figure 13.25. Optical waveguide communication system

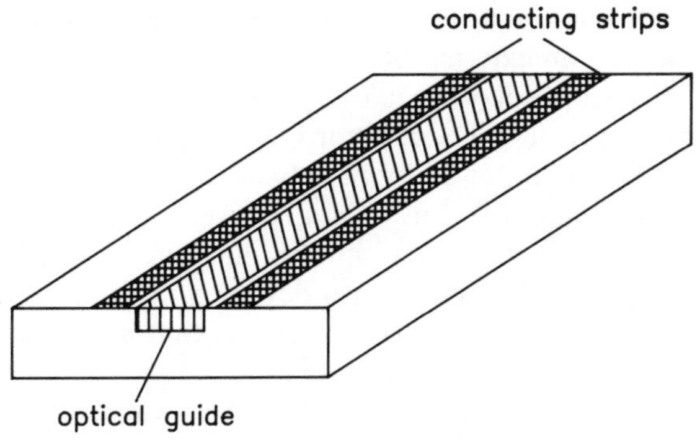

Figure 13.26. Optical modulation with an optical slab waveguide

the LED or laser. This has the advantage of simplicity. The alternative is to use an optical modulator. This can be done in many ways using integrated optics. One method uses a slab waveguide with electrodes on either side to change the electric field in the waveguide in response to the base band signal, Figure 13.26.

The next component in the system is a coupler or connector from source to the waveguide. The electromagnetic principles of coupling into an optical waveguide were discussed in Section 13.7. This is a special case of the more general need to connect lengths of optical waveguide together. Connection of separate lengths of cable can be done either by *splicing* two lengths together in a semi-permanent manner, or with a *connector*. Optical connectors must have a very high level of mechanical precision because of the small dimensions at optical wavelengths. In order to keep the attenuation to less than one dB, the axes must be laterally located within about 5% of the diameter of the core. The axes must also be aligned to within about 2°. In addition the ends of both fibres must be perfectly flat. This is done by polishing using fine abrasives.

An optical fibre has a cladding diameter of between 100 and 500 μm. The small dimensions are one of the advantages of optical fibres, but this brings with it mechanical problems of fragility, and ruggedness. Extra protection is needed to make an optical cable which will withstand handling. There are many cable configurations in use. Figure 13.27 shows one scheme. This has a central steel reinforcing core with a number of optical fibres around the core. Each of the optical fibres has a buffer sleeve around

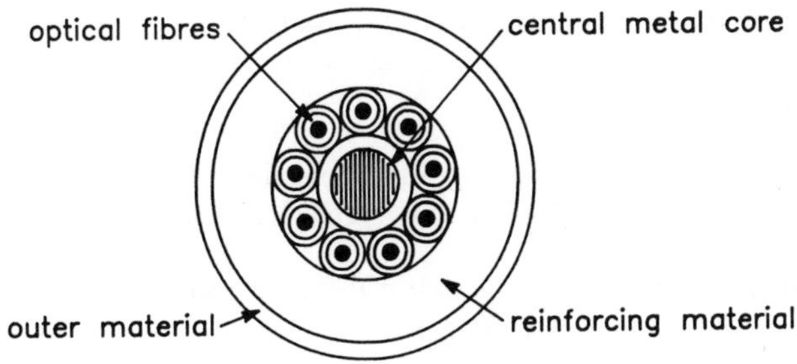

Figure 13.27. Cross-section of an optical cable with nine optical fibres

the cladding to provide additional support. The ring of fibres is surrounded by layers of packaging material and an external plastic envelope.

The receive end of an optical waveguide system, consist of a connector from the optical fibre to an optical detector. The detector will probably be either a PIN photodiode or an avalanche photodiode, followed by base band electronic processing. If repeaters are needed in the system they will consist of a receiver and transmitter back to back. In the future, optical amplifiers which can be integrated directly into the optical waveguide will probably be used. These are new devices which show considerable promise and are an example of the continuing developments that are taking place in optical waveguides, integrated optics and optoelectronics.

The amount of signal which is detected will be the source power less a number of losses. These losses can be summarised as follows

- Coupling or connector loss into the waveguide, as described in Section 13.7.
- Transmission loss along the optical fibre, as described in Section 13.6.
- Coupling or connector loss out of the waveguide.
- Noise in the detector.

The noise in the detector sets a lower limit to the sensitivity of the system. There are two sources of noise at optical frequencies, thermal noise and quantum, or shot noise. Thermal noise is the same as that discussed for microwave systems and is caused by the effective load resistor connected to the diode detector. Thermal noise usually predominates in photodiodes. Quantum noise arises from the intrinsic fluctuations in the optical generation of charge carriers. It is a fundamental factor which cannot be eliminated. In optical amplifiers it is the predominant source of noise and can cause equivalent noise temperatures of over 10 000 K. For this reason, it is better to use direct detection with a photodiode rather than optically amplify the signals.

PROBLEMS

1. Derive the propagation equations for TM modes in dielectric slab waveguide, (equations (13.11) and (13.12)).

2. Locate the first roots of the propagation equation for the odd TE modes in slab waveguide by plotting the equation against Ud for $V=6$.

3. An optical slab waveguide has $n_1 = 1.55$, $n_2 = 1.54$ and thickness of $d = 8\,\mu$m. Find (a) the V number at $\lambda = 1\,\mu$m, (b) the numerical aperture, (c) the single mode bandwidth of the TE_0 mode.

4. Find the number of modes which can propagate at $\lambda = 1.5\,\mu$m in an optical slab waveguide with $n_1 = 1.65$, $n_2 = 1.55$, $d = 4\,\mu$m.

5. Design a single TE mode optical slab waveguide to operate at 800 nm which uses glasses with refractive indices of 1.55 and 1.50.

6. Sketch the transverse distribution of electric field in both the core and cladding regions of a dielectric slab waveguide for the first four lowest order modes.

7. Millimetre waves may be guided by a dielectric slab waveguide. Consider a dielectric slab with $\varepsilon_1 = 2.5$ and $\varepsilon_2 = 1$. What should its thickness be in order that only the TE_0 mode may be excited at frequencies up to 300 GHz?

8. Demonstrate that when a mode propagates at the cut-off for the dielectric slab waveguide so that the rays are at the critical angle for total internal reflection, there is no field decaying outside the slab and propagation is at the velocity of light.

9. Plot the refractive index profile against radius for a graded index fibre which has a parabolic profile which has $n_1 = 1.555$ and (a) $n_2 = 1.55$, (b) $n_2 = 1.50$.

10. The graded index fibre of (a) in the last problem is simulated by four steps of equal radial distance which have the mean refractive index of the profile at the centre of each step. Use Snell's law to sketch the ray path for rays which are just trapped by the outermost layer.

11. A step index optical fibre has $n_1 = 1.55$, $n_2 = 1.52$ and $a = 5\,\mu\text{m}$. Find the cut-off wavelength under the weakly guided approximation for the LP_{01}, LP_{11} and LP_{21} modes. (The roots of the appropriate Bessel functions are given in Section 12.5.)

12. Choose an appropriate operating wavelength for a single mode optical fibre with $n_1 = 1.55$ and $n_2 = 1.52$. The operating wavelength should be 20% above the cut-off wavelength of the next propagating mode.

13. Find the core diameter required for a single mode of operation in a fibre guide at an infrared wavelength of 1.1 μm if the core index is 1.54 and the cladding index is 1.535.

14. Use the graph of attenuation in optical fibre in Figure 13.18 to estimate the lowest proportion of power which can be lost over a 20 km optical fibre at wavelengths of (a) 800 nm and (b) 1.4 μm.

15. An optical fibre with an attenuation of 0.3 dB/m is 100 m long and contains two right angle bends of radius 10 mm and 5 mm. The fibre has $n_1 = 1.58$, $n_2 = 1.57$ and a core diameter of 30 μm. Find the total attenuation.

16. Find the maximum bending radius for an optical fibre so that the radiation bending loss is less than 0.5 dB with a fibre which has $a = 4\,\mu\text{m}$, $n_1 = 1.455$, $n_2 = 1.45$.

17. An optical fibre guide has a core of refractive index 1.53 and a cladding of refractive index 1.51. Find the maximum angle θ at which rays will enter the fibre and be trapped.

18. Find the maximum acceptance angle from a medium of $n = 1.3$ for an optical fibre when the core has a refractive index of 1.6 and the cladding has a refractive index of 1.5.

19. A step index fibre of diameter 50 μm has $n_1 = 1.53$ and $n_2 = 1.45$. Find the number of reflections along a 1 m length of the fibre for a ray which enters (a) at the maximum acceptance angle, (b) at half the maximum acceptance angle.

20. A 5 mW laser is used to couple power into an optical fibre waveguide with an attenuation of 0.5 dB/km^{-1} and a core refractive index of 1.50. After 10 km, the power is detected by a photodiode attached to the fibre. The laser is in air and 60% of its power falls onto the end of the fibre. Find the power received by the photodiode.

21. Find the phase velocity and the group velocity for a fibre waveguide with a dispersion relationship $\beta = A\omega^{1/2}$ where A is a constant.

22. A glass used in an optical waveguide has a refractive index given by $n(\lambda) = 1.8 - 0.3\,\lambda$, where λ is in μm. Calculate the time delay due to material dispersion over a length of 1 km at (a) $\lambda = 1.1\ \mu$m, (b) $\lambda = 1.5\ \mu$m.

23. Calculate the time delay between an axial ray and one that enters a 500 m long optical fibre with $n_1 = 1.49$ at an angle of 15°.

24. Find the multipath delay, in ns/km, for a multimode fibre with $n_1 = 1.5$ and $n_2 = 1.485$.

25. Calculate the maximum rate at which data can be sent along an optical fibre with a cladding index of 1.46 and a core index of 1.51.

14 Antennas

14.1 INTRODUCTION

Antennas form the link between the guided wave and the free space parts of a radio or microwave system. The guided parts are cables or waveguides to and from the transmitter and receiver. The free space parts of a system are any of the systems described in Chapters 7 and 8. This chapter gives an introduction to the principles of antennas.

Antenna theory and design is a big subject in its own right and there are many books which describe the theory in detail. Some of these are listed in the references and bibliography at the end of the book. Modern antenna design makes considerable use of advanced electromagnetic wave theory to help to solve problems. The size of microwave antennas means that few approximations can be made in the theory and this inevitably results in advanced theoretical techniques. considerable recourse is made to numerical and computational methods to solve the theory and this needs powerful computers. After theoretical designs have been developed, they must be built. This adds another layer of knowledge because the antenna is often the most physically exposed part of a microwave system, so mechanical and environmental data is important. Finally the antenna must be tested which involves carefully designed antenna test ranges. Thus antenna design is a mix of many disciplines

The principles to be described below concentrate on microwave antennas, but free space optical systems also use radiators. These are not usually called antennas, but the same principles apply. Fortunately the short optical wavelength by comparison with the size of objects means that geometric optics approximations can be made which considerably simplifies the design problems and removes much of the complexity of microwave antenna design.

14.2 PURPOSE OF ANTENNAS AND TYPES

The purpose of a transmitting antenna is to efficiently transform the currents in a circuit or waveguide into radiated radio or microwave energy. The purpose of a receiving antenna is to efficiently accept the radiated energy and convert it to guided form for detection and processing by a receiver. The design and construction of an antenna usually involves compromises between the desired electromagnetic performance and the mechanical size, mass and environmental characteristics.

Antennas for radio and microwave systems fall into two broad categories depending on the degree to which the radiation is confined. Microwave radio relay and satellite communications use *pencil beam* antennas, where the radiation is confined to one narrow beam of energy, Figure 14.1. Mobile communications and broadcasting are more likely to require antennas with *omni-directional* patterns in the horizontal plane and toroidal patterns in the vertical plane.

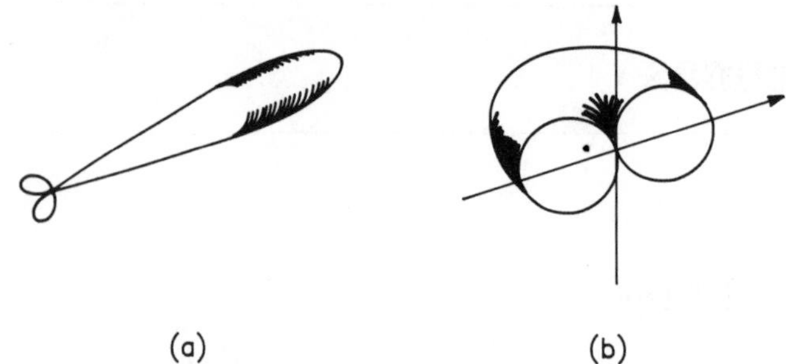

Figure 14.1. Radiation from antennas. (a) Pencil beam. (b) Toroidal beam

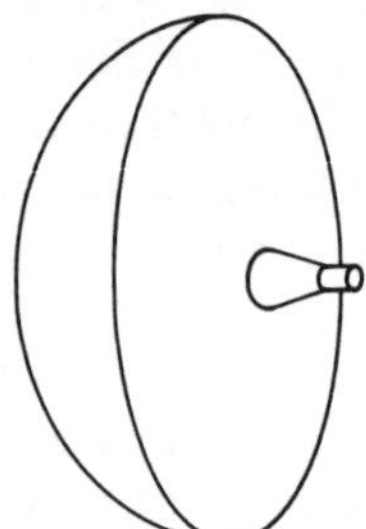

Figure 14.2. Reflector antenna and feed horn

Pencil beam antennas either consist of an aperture antenna or an array antenna. At microwave frequencies, the most common type of pencil beam antenna is a medium to large size *reflector* antenna. This consists of a reflector, or mirror, which collimates the signals from a *feed horn* at the focus of the reflector, Figure 14.2. At higher frequencies, lens can be used in place of the reflector. Reflectors, lenses and feed horns are *aperture antennas* because the basic radiating element is an aperture. The design problem is to first determine the aperture fields which will yield the specified radiation characteristics and secondly to design the reflectors, lenses and horns to produce the aperture fields. Aperture antennas can be designed to meet very stringent specifications.

Omni-directional antennas consist of elements which are small in wavelengths, such as *dipoles and monopoles.* The radiation characteristics are influenced by the presence of surrounding objects. Non-electromagnetic factors such as the size are often as important in the design as the electromagnetic radiation properties. For this reason the design of omni-directional antennas is partly an empirical process in which expertise and previous experience play an equal part with theoretical knowledge.

In between the large aperture antennas and the small element antennas lies *array antennas* which consist of two or more elements. The radiation from an array antenna is determined principally by the physical spacing and electrical signals driving the elements rather than the radiation characteristics of the elements themselves.

14.3 BASIC PROPERTIES OF ANTENNAS

There are some basic properties of antennas which apply to all types of antennas and are universally used to describe antennas. These are listed below and will be briefly described.

- Reciprocity
- Field regions surrounding antennas
- Radiation patterns
- Gain and directivity
- Efficiency of an antenna
- Polarisation properties
- Impedance properties
- Frequency characteristics

RECIPROCITY

The principle of *reciprocity* is one of the most important properties of an antenna. It states that the properties of an antenna when acting as a transmitter are identical to the properties of the same antenna when acting as a receiver. For this to apply, the medium in between the two antennas must be linear, passive and isotropic, which is always the case for free space microwave systems. The principle of reciprocity is actually derived by using circuit theory and assuming that the free space region can be represented as a transmission line. This concept was used in Chapter 11. Antenna reciprocity means that the description of the action of an antenna can be done in whichever mode is most convenient. Usually this is when the antenna is acting as a radiator of signals. An antenna which is designed for reception of plane waves will still be described as if it was a radiator of waves.

FIELD REGIONS SURROUNDING ANTENNAS

The fields which are radiated by an antenna change their characteristics as distances increases from the antenna, Figure 14.3. Immediately next to the antenna, in the *reactive* region, the fields are due to the conduction current term in Maxwell's equations (see Chapter 3). The reactive components predominate and the electric and magnetic fields decay with the square or the cube of distance from the antenna. This is the region that causes mutual coupling between components.

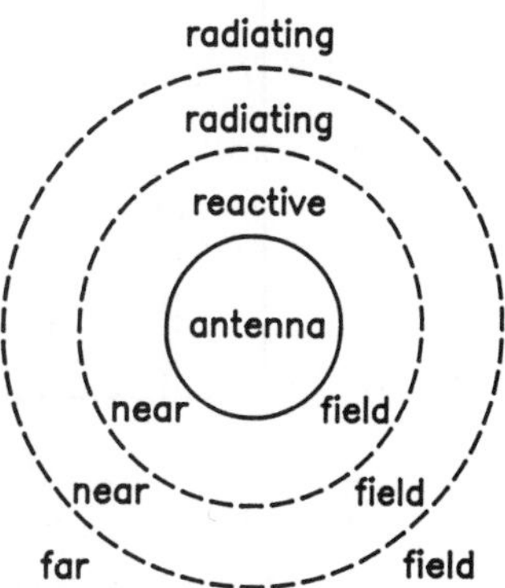

Figure 14.3. Field regions surrounding an antenna

Beyond the reactive region, the fields are radiating and are divided into two subregions. The first is called the *radiating near-field* region or the *Fresnel* region. In this region the waves coming from the antenna are spherical waves and the source of the spherical waves can be physically observed. This means that amplitude and phase of the wavefronts depends on the distance from different parts of the antenna.

Not all antennas have a near-field region, but all antennas have the main radiating region—the *radiating far-field* region, or *Fraunhofer* region. In this region the waves are essentially plane waves and there are only transverse electric and magnetic field components. The amplitude of the electric and magnetic fields decay linearly with distance. The angular radiation characteristics of the antenna are independent of distance from separate parts of the antenna. To an observer in the far-field region, the antenna appears as a point source. The far-field region is the region of main concern for microwave systems because transmitters and receivers are usually separated by long distances. The boundary between the near-field and far-field regions is not precise and was discussed in Section 5.1.

RADIATION PATTERNS

The directional selectivity of an antenna is represented by the *radiation pattern*. It is a plot of the relative strength of the radiated field as a function of the observation angle. A pattern taken along the principal direction of the electric field is called an *E-plane* cut, the orthogonal plane is called an *H-plane* cut. The most common plot is the rectangular decibel plot, Figure 14.4, which can have scales of relative power and angle chosen to suit the antenna being characterised. Other types of plots are the polar plot, Figure 14.5. This is used for small antennas and has the advantage that it gives a *birds eye* view of the radiation pattern. It is, however, inflexible and not suitable for most

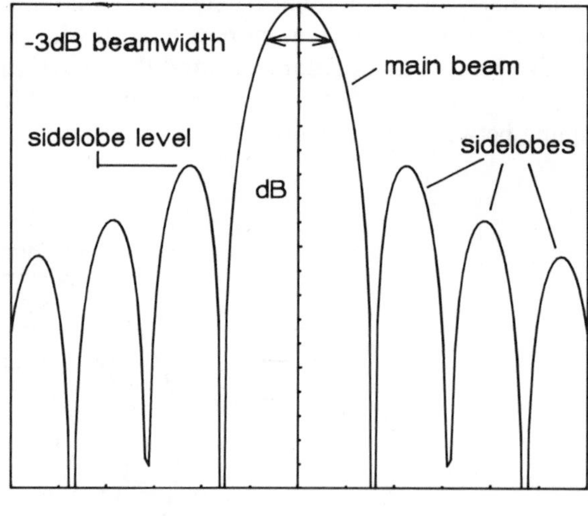

Figure 14.4. Radiation pattern—rectangular decibel plot

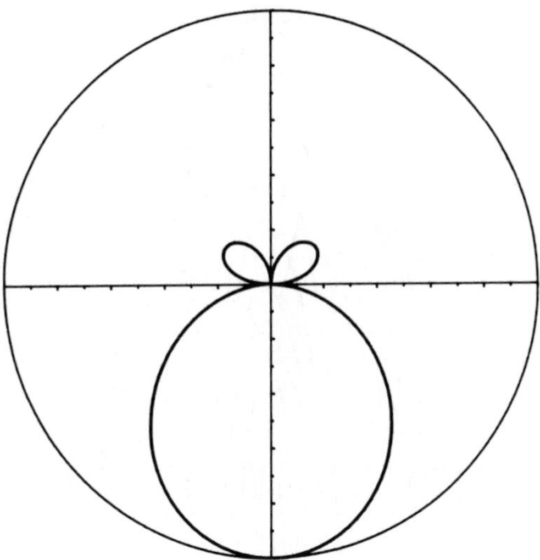

Figure 14.5. Radiation pattern—polar field plot.

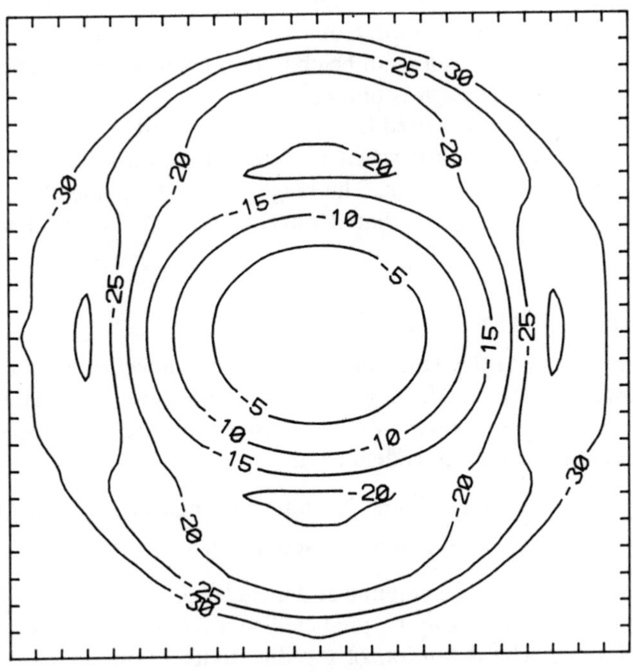

Figure 14.6. Radiation pattern—contour plot

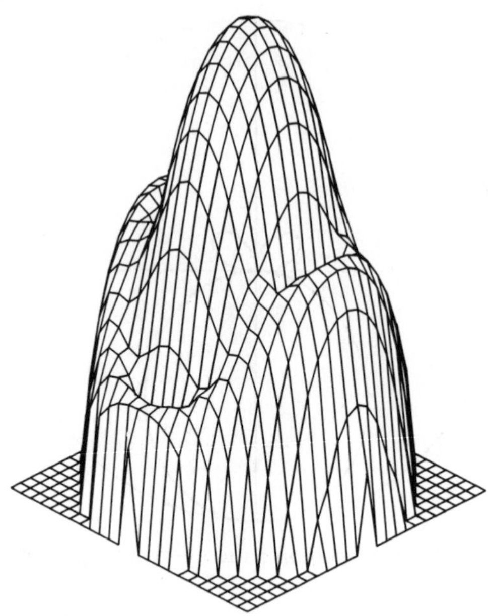

Figure 14.7. Radiation pattern—three-dimensional isometric plot

high gain microwave antennas as the beam is too narrow to be seen as other than a thin line. There are also two-dimensional contour plots, Figure 14.6, and three-dimensional isometric plot, Figure 14.7. Contour plots are useful for cases where the radiated beam has a non-circular shape. Isometric plots give a good visual representation of a radiation pattern but the scales can be chosen to display the pattern in many ways, so their use for engineering design is limited.

A radiation pattern is characterised by the *main beam* and *sidelobes*, Figure 14.4. The angular width of the main beam is specified by the *beamwidth* between the -3dB, or half power, points on the main beam. The height of the first sidelobe is an important measure of how much energy is radiated outside the main beam. This is called the *sidelobe level.*

GAIN AND DIRECTIVITY

The *gain* of an antenna was described in Chapter 7 in connection with the calculations of how much power was received from a transmitter. The gain is a measure of the ability of an antenna to concentrate power into a narrow angular region of space. The power gain in a specified direction (θ, ϕ) is defined as

$$G(\theta, \phi) = \frac{4\pi \text{ power radiated per unit solid angle in direction } \theta, \phi}{\text{Total power accepted from source}} \tag{14.1}$$

This is an inherent property of an antenna and includes dissipative losses in the antenna. The dissipative losses cannot easily be predicted so a related parameter, the *directivity*, is used in calculations. The definition of the directivity is similar to that of the gain except that the denominator is replaced by the total power radiated. The gain is a parameter which is measured whereas the directivity is calculated. For most microwave antennas the losses in the antenna are small so it is usually a good approximation to equate directivity and gain.

If the direction of the gain is not specified, the peak value is assumed. The value of the gain is normally quoted in dBs. The definitions given above are in effect specifying the gain relative to a loss-less isotropic source. This is sometimes stated explicitly by using the symbol dBi.

The gain, G, an antenna can be related to the effective aperture area, A, of an antenna and this was done in Section 7.3 by

$$G = \frac{4\pi A}{\lambda^2} \tag{14.2}$$

The effective area is a fictional area which, if used in receive mode, collects enough power to produce a gain G. Equation (14.2) is a useful equation because it enables the peak gain of any antenna to be calculated. For instance if an antenna has a rectangular area of 50 mm by 100 mm, then the peak gain at 10 GHz ($\lambda = 300$ mm) is $G = 10\log_{10}[4\pi \times 50 \times 100/(30)^2] = 18.4$ dB.

EFFICIENCY OF AN ANTENNA

The gain calculated in this example would not be realised in practice for an aperture of the specified area because the distribution of electric and magnetics fields across the aperture area would not be uniform. The physical area would have to be greater than the effective area. The ratio of the two areas is given by the efficiency, η,

$$\eta = \frac{\text{effective area of antenna}}{\text{physical area of antenna}} \times 100\% \tag{14.3}$$

The efficiency is a useful indication of how well a microwave antenna makes uses of its physical area. Normal aperture antennas have efficiencies in the range 50 to 80%.

POLARISATION PROPERTIES

The polarisation characteristics of plane waves were described in Section 5.11. These will be partially determined by the polarisation properties of an antenna. Communication and radar antennas can radiate waves which are either linearly polarised or circularly polarised. In many modern systems, two orthogonal polarisations are transmitted simultaneously in order to double channel capacity. This could be a horizontally polarised wave and a vertically polarised wave or two circular polarised waves with opposite directions of rotation of the electric vector. Antennas are required to have very pure polarisation properties so that very little energy leaks into the other polarisation. The leakage is measured by the amount of *cross-polarisation*, which is the difference between the E plane radiation pattern and the H-plane radiation pattern.

IMPEDANCE PROPERTIES

As far as circuit designers are concerned the antenna looks like a load impedance which is at the end of a waveguide or cable. Antennas have complex impedance properties so that maximum power transfer will occur when the antenna forms a conjugate match to the transmission line. The impedance of an antenna is made up of a number of parts. It is divided into a *self impedance* and a *mutual impedance*. The mutual impedance accounts for the influence of nearby objects and for any mutual coupling to other antennas in the reactive field region. The self impedance consists of the *radiation resistance*, the *loss resistance* and the *self reactance*. The self reactance is the imaginary part of the impedance of the structure of an antenna. The loss resistance is the ohmic losses in the antenna structure. The radiation resistance measures the power absorbed by the antenna from the incoming plane waves. It is one of the most significant parameters for small antennas because the radiation resistance can be very small and this makes it difficult to match to a normal transmission line.

FREQUENCY CHARACTERISTICS

An antenna is both a *spatially selective filter* and *a frequency selective filter*. The spatially selective characteristics are measured by the radiation pattern. The frequency selective characteristics are measured by the *bandwidth* of the antenna. This measures the frequency range over which the antenna operates. The upper and lower frequencies can be specified in terms of a number of possible parameters—gain, polarization, beamwidth and impedance. It is usually specified as a percentage of the centre frequency. The bandwidth of practical antennas varies from less than 1% to over 100%.

14.4 RADIATION FROM APERTURES

Apertures form the largest category of microwave antennas and they also represent almost all optical antennas. The radiation from apertures illustrates most of the significant properties of pencil beam antennas. The radiation characteristics can be determined by simple mathematical relationships, although the derivation is beyond the scope of this book. If the electric field across an aperture, Figure 14.8, is $E_a(x, y)$ and the aperture is large in wavelengths, then the radiated far-fields, $E_p(\theta, \phi)$, are given by

$$E_p(\theta, \phi) = \int_{-\infty}^{\infty} \int_{-\infty}^{\infty} E_a(x, y) \exp[jk(x \sin\theta \cos\phi + y \sin\theta \sin\phi)] \, dx \, dy \qquad (14.4)$$

Equation (14.4) is in the form of a Fourier integral. For high or medium gain antennas the pencil beam radiation is largely focused to a small range of angles around $\theta = 0$. In

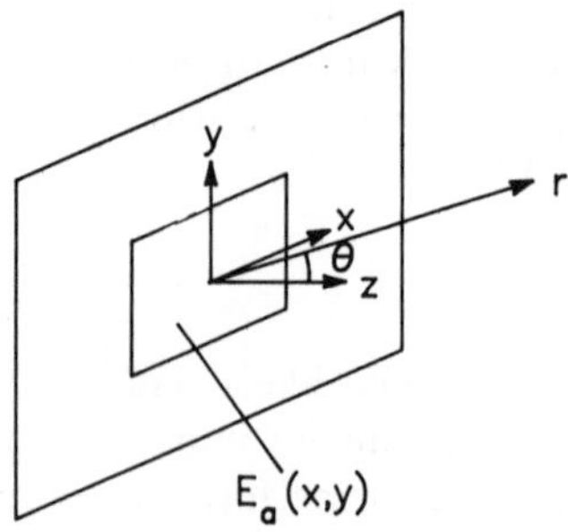

Figure 14.8. An aperture antenna

this case equation (14.4) shows that the distant radiated fields, and the aperture fields are the Fourier transformation of each other. Fourier transforms have been widely studied in engineering and science and their properties can be used to understand the radiation characteristics of aperture antennas. Simple aperture distributions have analytic Fourier transforms, whilst more complex distributions must be solved numerically on a computer.

The simplest aperture is a one-dimensional line source distribution, $E_a(x)$ of length a, Figure 14.9. This serves to illustrate many of the features of aperture antennas. If the amplitude and phase of the electric field across the aperture is constant so that $E_a(x) = 1$, the radiated field in the $\phi = 0$ plane is given from equation (14.4) as

$$E_p(\theta) = \int_{-a/2}^{a/2} \exp(jkx \sin\theta)\, dx \tag{14.5}$$

Evaluating the integral gives

$$E_p(\theta, \phi) = \frac{\sin(\pi u)}{\pi u} \tag{14.6}$$

where $u = (a/\lambda) \sin \theta$.

The $\sin(\pi u)/\pi u$ type distribution occurs widely in antenna theory. It is plotted in Figure 14.10. The following deductions can then be made about the line aperture.

(a) The radiation pattern can be plotted in normalised form so that one pattern is correct for all sizes of apertures. This is generally true of all aperture antennas.

(b) The width of the main beam, at the -3 dB level is $0.88\,\lambda/a$. Thus the beamwidth is inversely proportional to the aperture width. Again this is generally true for all apertures although the proportionality factor will vary.

(c) The first sidelobe level is at -13.2 dB below the peak of the main beam. This is quite high and is a disadvantage of a uniform aperture distribution. The level can be reduced considerably by a tapered aperture distribution where the field is greatest at the centre of the aperture and tapers to a lower level at the edge of the aperture. If $E_a(x) = \cos(\pi x/2a)$ then the first sidelobe level is at -23 dB. The energy which was in the sidelobes moves to the main beam with the result that the beamwidth broadens to $1.2\,\lambda/a$. In practice almost all antennas have natural

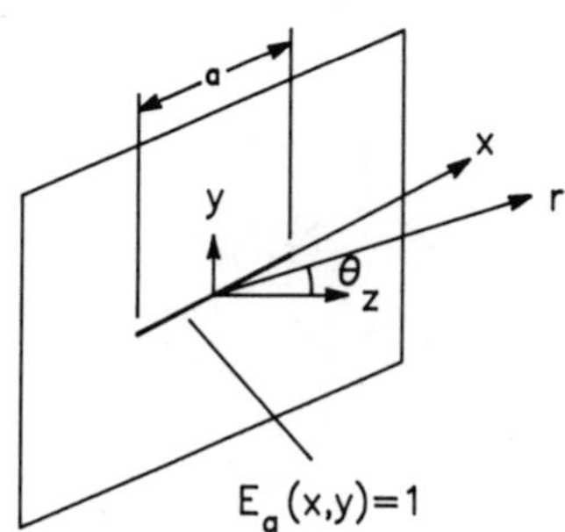

Figure 14.9. Line source aperture with uniform illumination

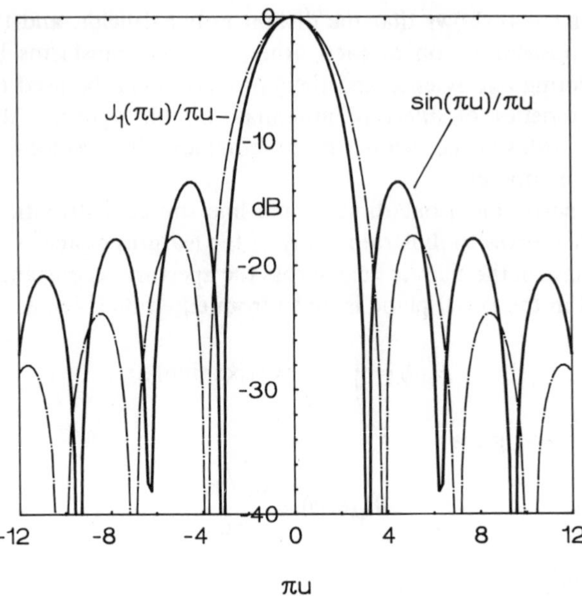

Figure 14.10. Radiation pattern of a uniformly illuminated line source aperture, $\sin(\pi u)/\pi u$, and a circular aperture, $J_1(\pi u)/\pi u$.

tapers across the aperture which result from bounary conditions and waveguide modes.

(d) The nulls in the radiation pattern are periodic at $(a/\lambda)\sin\theta = \pm 1, \pm 2 \ldots.$

(e) The phase pattern, which is not plotted in Figure 14.10, is constant along each lobe but changes sign by π or 180° at every null.

The above deductions were obtained for a simple line source but they immediately apply to rectangular apertures which can be formed from two line source distributions in orthogonal planes.

Circular apertures, Figure 14.11, form the largest single class of aperture antennas

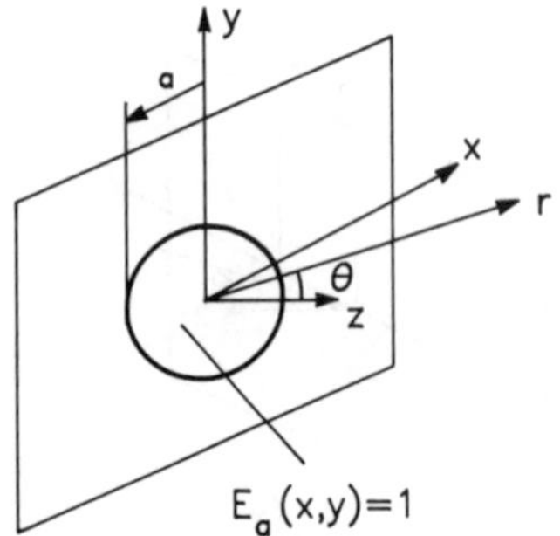

Figure 14.11. Circular aperture with uniform illumination

because it is easy to make reflectors, lenses and horns with circular cross-sections. The ideal circular aperture has a radiation pattern which is independent of the azimuthal angle, ϕ. The simplest case is a uniformly illuminated electric field, $E_a(r) = 1$, which gives a radiated field

$$E_p(\theta, \phi) = \frac{2J_1(\pi u)}{\pi u} \tag{14.7}$$

where $J_1(\pi u)$ is a Bessel function of zero order, $u = (2a/\lambda)\sin\theta$ and a is the radius of the aperture. Equation (14.7) can be compared to equation (14.6) and is also plotted in Figure 14.10. The first sidelobe level is at -17.6 dB so shows a worthwhile improvement over the line source. The beamwidth is slightly wider at $1.02\lambda/a$.

Practical aperture antennas consist of *reflector antennas, lenses* and *horns*. Reflector antennas are widely used in satellite communications, microwave radio relay, radar and radio astronomy.

The axi-symmetric parabolic reflector with a feed at the focus of the paraboloid is the simplest type of reflector antenna. The geometry is shown in Figure 14.12. The paraboloid has the property that energy from the *feed horn* at the focus F goes to the point P on the surface where it is reflected parallel to the axis to arrive at a point A on the imaginary aperture plane. The paraboloidal mirror was studied in Chapter 9. The equation describing the surface is

$$r^2 = 4F(F - z) \tag{14.8}$$

where F is the focal length. At the edge of the reflector the relationship between the focal length and the diameter D is given by,

$$\frac{F}{D} = \tfrac{1}{4}\cot\left(\frac{\theta_0}{2}\right) \tag{14.9}$$

The depth of the paraboloid is usually specified by its F/D ratio. Common sizes for microwave reflector antennas are between $F/D = 0.25$, which gives $\theta_0 = 90°$, to $F/D = 0.5$ which gives $\theta_0 = 53°$. Ray optics indicates that the path length from F to P to A in Figure 14.12 is equal to twice the focal length and this is true for all positions of A. Hence the phase across the aperture is constant. The amplitude across the aperture plane will peak at the centre and taper towards the edge for two reasons. Firstly because the feed will have a tapered radiation pattern and secondly because the action of a

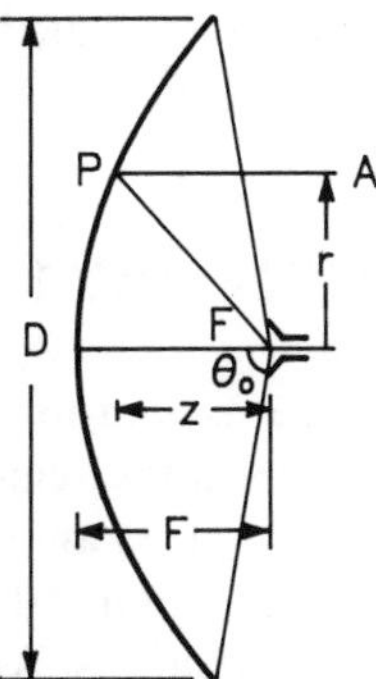

Figure 14.12. Parabolid reflector antenna

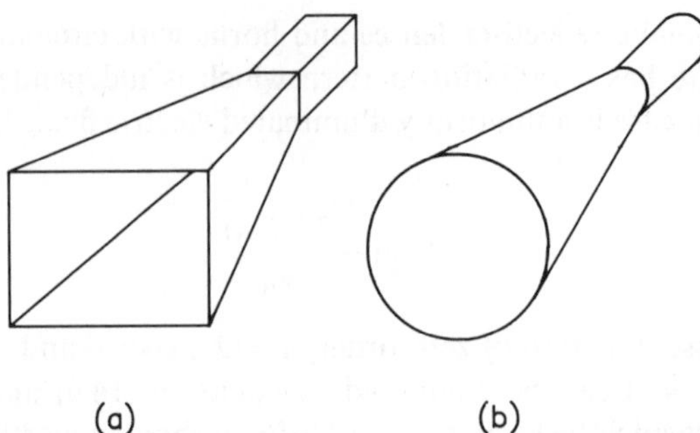

Figure 14.13. Horns. (a) Pyramidal. (b) Conical

parabola in transforming a spherical wave from the feed into a plane wave across the aperture introduces a path loss which is a function of angle θ. The aperture electric field is then

$$E_a(\theta, \phi) = E_F(\theta, \phi) \cos^2\left(\frac{\theta}{2}\right) \tag{14.10}$$

where $E_F(\theta, \phi)$ is the radiation pattern of the feed at the focus.

The peak gain of a reflector antennas can be calculated from equation (14.2) as $G = (\pi D/\lambda)^2$. The efficiency is determined by the uniformity of the aperture field, the amount of power which radiates outside the reflector angle θ_0 and any loss due to blockage by the feed horn at the focus. Often high efficiency is not the main criteria and low sidelobes may be specified as more important.

Lenses are circular aperture antennas which refract the power from a feed horn at the lens focus. Otherwise they are equivalent to reflector antennas. They are not often used at microwave frequencies because plastic lenses would be too heavy by comparison with a parabolic reflector. As frequency increases, they start to be used and at sub-millimetre waves, infra red and optical frequencies they displace reflectors as the prime choice. The geometric optics properties of lenses were studied in Chapter 9, but they can also be treated as a circular aperture and the above theory applies.

There are a wide variety of shapes of horn antennas, of which the rectangular horn and conical horn are the most common, Figure 14.13. They are small aperture antennas whose radiation characteristics and gain can be determined using the above aperture theory. Some horns are designed to have good polarisation properties, by partially filling the horn with a dielectric or by corrugating the walls of the horn.

14.5 RADIATION FROM SMALL ANTENNAS

Small antennas are needed for mobile communications operating at frequencies from HF to the low microwave region. Most of these are derivatives of the simple dipole, Figure 14.14. The dipole is an electric current element which radiates from the currents flowing along a small metal rod. The radiation pattern is always very broad with energy radiating in most directions. An important design parameter is the impedance of the

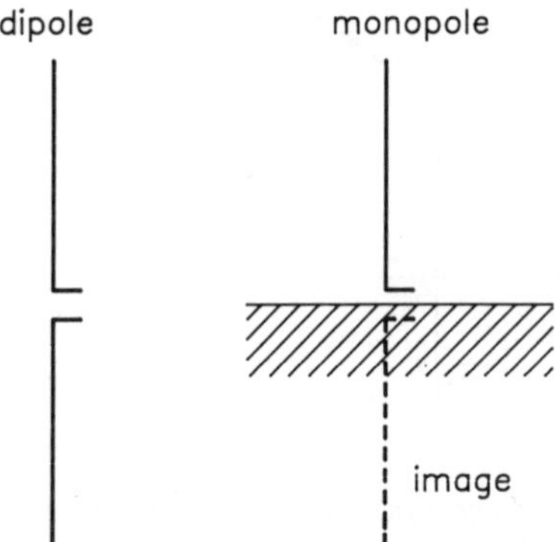

Figure 14.14. Dipole and monopole

dipole which can vary considerably depending on the exact size and shape of the rod. This means that the impedance matching between the antenna and the transmitting or receiving circuit becomes a major design constraint.

The radiation fields from a dipole are obtained by integrating the radiation from an infinitesimally small current element over the length of the dipole. This depends on knowing the current distribution which is a function not only of the length but also of the shape and thickness of the rod. The principles can, however, be illustrated with a very small dipole which has a constant current, I_0, along its length, l, Figure 14.15. The radiation characteristics have to be found using an intermediate vector potential, $\boldsymbol{A}$,

$$\boldsymbol{A} = \hat{z} \int_{-l/2}^{l/2} I_0 \frac{\mathrm{e}^{-\mathrm{j}kr}}{4\pi r} \mathrm{d}z \tag{14.11}$$

where $k = 2\pi/\lambda$. This equation is sometimes called the radiation integral. The vector potential $\boldsymbol{A}$ is related to the magnetic fields by $\boldsymbol{H} = \nabla \times \boldsymbol{A}$. After evaluation using spherical coordinates and vector algebra, the radiated fields are given by

$$\begin{aligned} H_\phi &= \frac{I_0 l \sin\theta}{4\pi k}\left(\frac{\mathrm{j}k}{r} + \frac{1}{r^2}\right)\exp[\mathrm{j}(\omega t - \beta r)] \\ E_\mathrm{r} &= \frac{60 I_0 l \cos\theta}{k}\left(\frac{k}{r^2} - \frac{\mathrm{j}}{r^3}\right)\exp[\mathrm{j}(\omega t - \beta r)] \\ E_\theta &= \frac{\mathrm{j}30 I_0 l \sin\theta}{k}\left(\frac{k^2}{r} - \frac{\mathrm{j}k}{r^2} - \frac{1}{r^3}\right)\exp[\mathrm{j}(\omega t - \beta r)] \\ E_\phi &= H_\theta = H_\mathrm{r} = 0 \end{aligned} \tag{14.12}$$

These fields apply at all distances out from the small dipole. The terms containing r illustrate how the fields vary with distance. In the near-field region the $1/r^2$ and $1/r^3$ terms dominate. If the power density is evaluated using Poynting's vector ($P = \mathrm{Re}[\boldsymbol{E} \times \boldsymbol{H}]/2$) it will be found to be zero. This indicates that the waves in the near

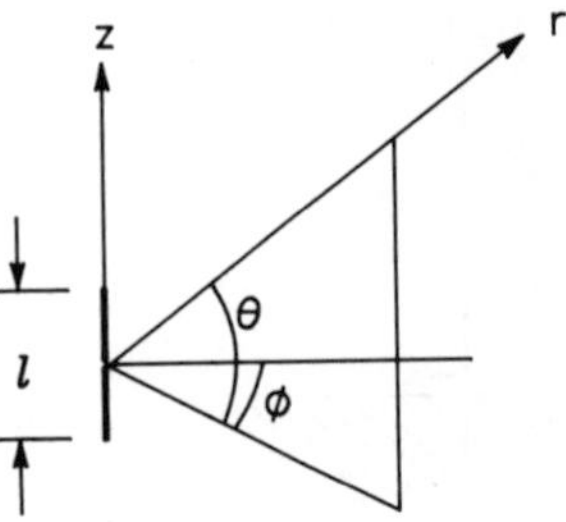

Figure 14.15. Very small dipole

field region are standing waves and not travelling waves. The electric and magnetic fields are reactive, not radiating.

Further away from the antenna, the $1/r^2$ and $1/r^3$ terms become very small and can be ignored. This is the far-field region. In the far-field region $E_r = 0$ and the two field components are transverse to the r direction, which is the direction of travel. Thus the fields become plane waves, as expected. In this region the electric and magnetic fields are given, from equation (14.12) as

$$E_\theta = \frac{j30 I_0 lk \sin\theta}{r} \exp[j(\omega t - \beta r)]$$

$$H_\phi = \frac{E_\theta}{120\pi} \qquad (14.13)$$

The polar radiation pattern of the electric field is plotted against angle in Figure 14.16. This has a broad radiation pattern with nulls along the axis of the dipole and maxima in the orthogonal plane.

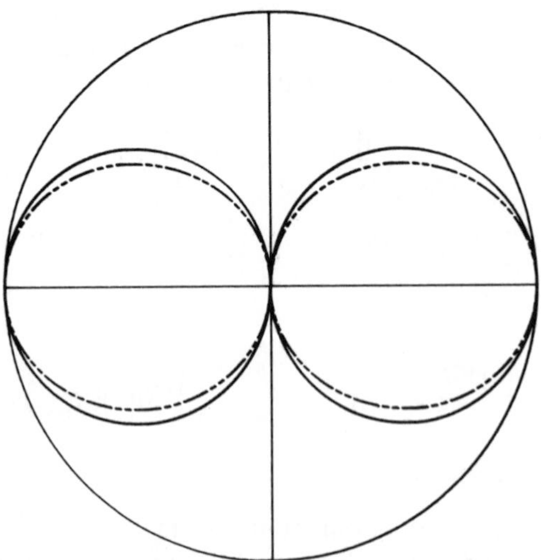

Figure 14.16. Far field radiation patterns of very small dipole (________) and half wavelength dipole _____ - - _____)

The impedance of the dipole is determined by the radiation resistance, R_r. For the very short dipole, this is

$$R_r = 80\pi^2\left(\frac{l}{\lambda}\right)^2 \tag{14.14}$$

If $l = 0.02\lambda$, then equation (14.14) gives $R_r = 0.32\,\Omega$. This is very small and illustrates the matching problem that exists at low frequencies where antennas may be short in wavelengths.

The radiation resistance increase as the length increases, but the patterns are no longer easy to evaluate. A dipole which is half a wavelength long has improved properties by comparison with a short dipole. It has a radiated electric far field given by

$$E(\theta) = \mathrm{j}60I_0\frac{\cos\left[(\pi/2)\cos\theta\right]}{\sin\theta}\frac{\mathrm{e}^{-\mathrm{j}kr}}{r} \tag{14.15}$$

The half wave dipole has a radiation pattern, Figure 14.16, which is a similar to a very short dipole, but a bit narrower. It has a half power beamwidth of 78°. The radiation resistance must be evaluated numerically. For an infinitely thin dipole it has a value of $73 + \mathrm{j}\ 42.5\,\Omega$. For finite thickness the imaginary part can become zero in which case the dipole is easily matched to a coaxial cable of impedance $75\,\Omega$. The half wave dipole has a gain of 2.15 dB and is very widely used as a simple antenna, particularly at VHF and UHF.

A monopole is a dipole divided in half at its centre feed point and fed against a ground plane, Figure 14.14. The ground plane acts as a mirror and consequently the image of the monopole appears below the ground. Since the fields extend over a hemisphere the power radiated is half that of the equivalent dipole. The radiation resistance is also half that of the equivalent dipole with the same current. The gain of a monopole is twice that of a dipole. The radiation pattern above the ground plane is the same as that of the dipole.

14.6 RADIATION FROM ARRAYS

Array antennas consist of a number of discrete elements which are usually small in size. Typical elements are horns, dipoles, and microstrip patches. The discrete sources radiate individually but the pattern of the array is largely determined by the relative amplitude and phase of the excitation currents on each element and the geometric spacing apart of the elements. The total radiation pattern is given by

$$\text{Total pattern} = [\text{Pattern factor}] \times [\text{Array factor}] \tag{14.16}$$

where the *pattern factor* is the radiation pattern of one element of the array, and the *array factor* is the radiation pattern of the array assuming each element is formed of isotropic radiators. A significant feature of the pattern multiplication in equation (14.16) is that if there is a null, or zero field, in either the pattern factor or the array factor, it will appear in the total pattern. This means that the total pattern is determined by whichever of the two factors changes most rapidly with angle. Normally this is the array factor.

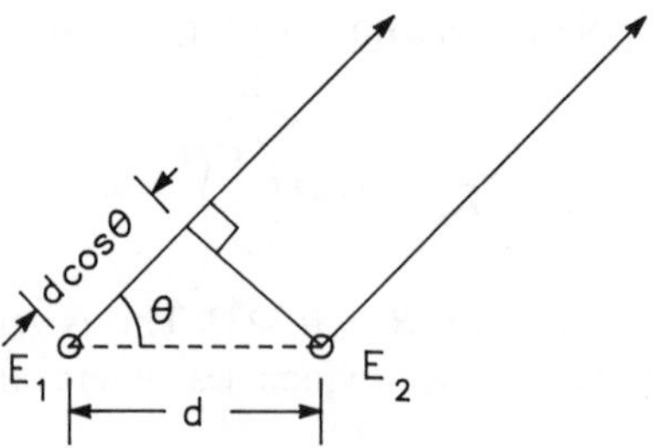

Figure 14.17. Array of two isotropic elements

The principles of arrays can be understood by considering a simple array of two isotropic elements which are fed by currents which produce electric fields of amplitude E_1 and E_2, Figure 14.17. If the observation point is far enough away such that the lines from the two elements can be considered to be parallel, then the difference in path length between the two rays will be $kd\cos\theta$, where $k = 2\pi/\lambda$ and d is the spacing apart of the elements. θ is the angle measured from the plane of the array. The array factor is then

$$E = E_1 + E_2 \exp[\mathrm{j}(\delta + kd\cos\theta)] \tag{14.17}$$

δ is the phase difference between the current driving the two elements. When $E_1 = E_2$ and $\delta = 0$, equation (14.17) simplifies to

$$E = 2\cos\left(\frac{kd}{2}\cos\theta\right) \tag{14.18}$$

The pattern for very small spacings will be almost omnidirectional and as the spacing is increased the pattern develops nulls and peaks. This is shown in Figure 14.18 which shows the polar patterns for spacings of $\lambda/4$, $\lambda/2$, $3\lambda/4$ and λ. For a spacing of $d = \lambda/4$, the radiation pattern is still nearly omni-directional. At a spacing of half a wavelength, a null appears along the array axis, Figure 14.18(b), and a peak broadside to the array. This is called a broadside array. Above $d = \lambda/2$, an extra pair of lobes appears. At $d = \lambda$ all four lobes have equal peak amplitudes, Figure 14.18(d). These extra lobes are generally undesirable because the aim of an antenna is to send power in only one or two directions. Consequently the ideal array spacing is half a wavelength.

If a phase difference, δ, of 180° exists between the two elements then the pattern shown in Figure 14.19 results. This is compared with the pattern for no phase difference. Now the main beam is along the direction of the array and the array is called an end-fire array. This illustrates one of the prime advantages of an array. By changing the electrical phase it is possible to make the peak beam direction occur in any angular direction. The angle at which the main beam occurs is determined by the phase difference between elements. Changing the relative amplitudes, phases and spacings can produce a wide variety of patterns so that it is possible to synthesise almost any specified radiation pattern. Array theory is largely concerned with synthesising an array factor to form a specified pattern.

The array factor for an array of more than two elements follows the same principle as those just described. The principle of pattern multiplication in equation (14.16) can

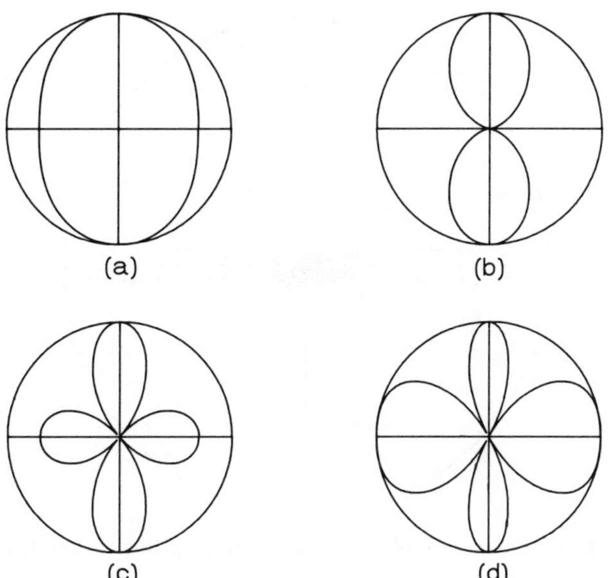

Figure 14.18. Radiation patterns of two element array fed with equal amplitude and phases. (a) $d = \lambda/4$. (*b*) $d = \lambda/2$. (c) $d = 3\lambda/4$. (d) $d = \lambda$.

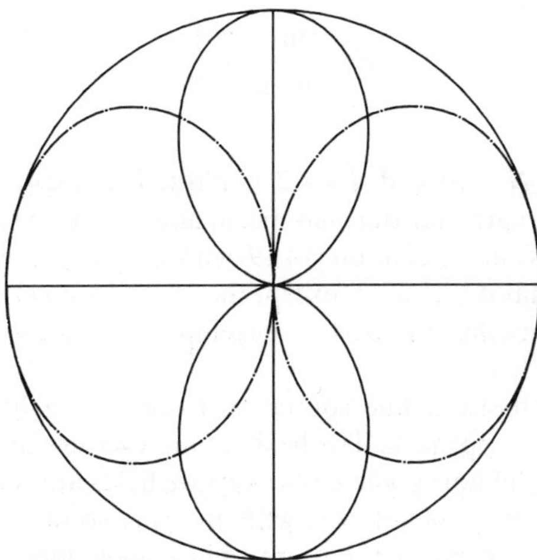

Figure 14.19. Radiation pattern of two-element array with $d = \lambda/2$. Broadside array, $\delta = 0°$ (————). Endfire array, $\delta = 180°$ (——-—)

be used to break large arrays into smaller parts. For instance the array factor for a four element array spaced $\lambda/2$ apart is

[array factor of a 2 element array spaced $\lambda/2$ apart] × [pattern factor] ×
[array factor of a 2 element array spaced λ apart]

A single array factor for an N element linear array can be determined by extending

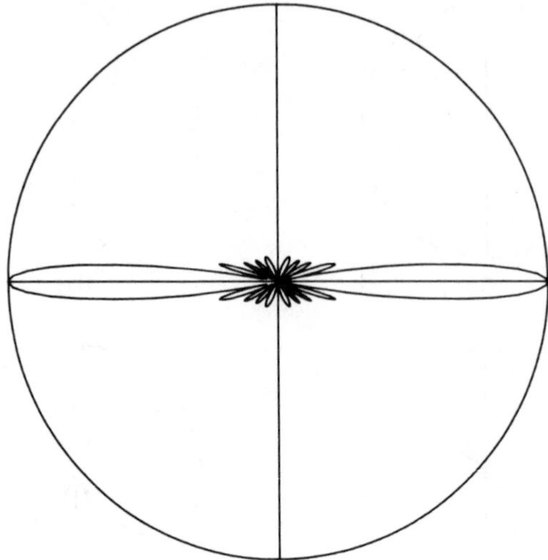

Figure 14.20. Radiation pattern of a ten-element array with elements spaced $\lambda/2$ apart

the principle of equation (14.17). The normalised pattern for an N element array with equal amplitudes is given by

$$E = \frac{\sin(N\Psi/2)}{N\sin(\Psi/2)} \tag{14.19}$$

where $\Psi = \delta + kd\cos\theta$.

The array factor for $N = 10$ and $d = \lambda/2$ is plotted in Figure 14.20. The larger the number of elements, the narrower the main beam and the more sidelobes exist in a fixed angular region. Comparison of equation (14.19) with the pattern of a line source aperture, equation (14.6) shows that it is similar, except that the sine function in the denominator of equation (14.19) means that the pattern envelope does not decay continuously as θ is increased.

It is possible to synthesise a line source aperture with a linear array. There is a significant benefit to this approach. The fields of an aperture antennas are determined by the reflectors, lenses and horns where the aperture fields are constrained by boundary conditions. This constraint does not exist with an array so that a much larger range of radiation patterns can be produced. Patterns which have most of the radiated power in the main beam and very little power in the sidelobes can be designed.

PROBLEMS

1. An antenna has a far field radiation pattern given by $E = \cos(0.75\pi\sin\theta)$ for $0 \leqslant \theta \leqslant 90°$. Plot the radiation pattern (a) as a rectangular plot of power against angle θ, (b) as a rectangular plot of dBs against angle, (c) as a polar plot of power against angle.

2. For the pattern plotted in Problem 1, find the position of the first null, the -3 dB beamwidth and the sidelobe level.

3. Calculate the peak gain of an antenna with a circular diameter of 0.8 m at 5 GHz, 25 GHz and 100 GHz.

4. Find the size of a 2×1 rectangular shaped antenna with an efficiency of 55% which will give a gain of 35 dB at 12 GHz.

5. Show that the -3 dB beamwidth of a uniformly illuminated line source is given by $0.88\lambda/a$, where a is the width of the source.

6. Find the length of a uniformly illuminated line source required to give a beamwidth of 10°. How many sidelobes are present between $\theta = 0°$ and $\theta = 90°$ for this line source?

7. Confirm that the first sidelobe level of a $\sin(\pi u)/\pi u$ pattern is at -13.2 dB.

8. Plot the phase pattern of a 4λ uniformly illuminated aperture for $0° \leqslant \theta \leqslant 90°$.

9. A square uniformly illuminated aperture can be treated as two orthogonal line source distributions. Show that for this case the peak gain in dB's is given by $45 - 20\log_{10}(\Delta\theta)$, where $\Delta\theta$ is the -3 dB beamwidth in degrees.

10. A line source with a triangular distribution has a radiation pattern given by $E_p(\theta) = |\sin(\pi u/2)/(\pi u/2)|^2$, where $u = (a/\lambda)\sin\theta$. Plot the pattern against u to find (a) the position of the first null, (b) the -3 dB beamwidth and (c) the first sidelobe level.

11. Prove that the ratio of the focal length to diameter of a parabolic reflector antenna is given by $4F/D = \cot(\theta_0/2)$ where θ_0 is the subtended angle at the feed.

12. Find the distance for the focus to any point on the surface of a parabola in terms of the angle between the axis and the ray direction. Hence confirm equation (14.10) for the aperture distribution of a parabolic reflector antenna.

13. Find the difference in dBs between the field at the centre of a parabolic reflector antenna and at the edge of the aperture for $F/D = 1.0$, 0.5, 0.25.

14. The fields of a very small dipole are given by equation (14.12). In the near field region only the dominant term of each field component is significant. Show that under these circumstances, there is no real power flow.

15. The radiation resistance of a very small dipole is given by $R_r = 2P/I_0$, where

$$P = \frac{1}{2}\int_0^{2\pi}\int_0^{\pi} E_\theta H_\phi r^2 \sin\theta \, d\theta \, d\phi$$

and E_θ and H_ϕ are given by equation (14.13). Show that the radiation resistance is given by $R_r = 80\pi^2(l/\lambda)^2$.

16. Find the half power beamwidth of a very short dipole and a half wave dipole.

17. Two half wave dipoles are spaced half a wavelength apart so that the ends of the dipoles nearly touch. Assuming no mutual interaction and equal amplitudes and phases

for the currents fed to each dipole, use pattern multiplication to sketch the polar radiation pattern.

18. Repeat Problem 17 with the two dipoles parallel to each other so that they are on opposite sides of a square.

19. Plot the polar radiation pattern of two equal amplitude isotropic elements spaced $\lambda/2$ apart and with a phase difference of 90° between elements.

20. Show that the direction of the main beam of a two element array occurs at an angle θ_0 given by $\delta + kd\cos\theta_0 = 0$, where δ is the phase difference between elements, $k = 2\pi/\lambda$ and d is the separation distance. Confirm this result from the plot drawn in the last problem.

21. Show that the pattern of an array of two isotropic elements spaced quarter of a wavelength apart and fed with equal amplitudes but with a phase difference of 90°, is asymmetric with the peak direction along the axis of the array.

22. Use the results of problems 19 and 21 to sketch the pattern which would be produced by four isotropic elements spaced $\lambda/4$ apart and with a progressive phase delay of 90° between elements.

23. Derive the array factor for an N element linear array (equation (14.19)) by extending the principle used to obtain the normalised pattern of a two-element array to N uniformily spaced elements.

Bibliography

1.1 FUNDAMENTALS AND GENERAL TEXTS

Carter, R. G., *Electromagnetism for Electronic Engineers*, Chapman and Hall, London, 1986.
Cheng, D. K., *Field and Wave Electromagnetic*, Addison Wesley, Reading, MA, 1989.
Christopoulos, C., *An Introduction to Applied Electromagnetism*, Wiley, Chichester, 1990.
Duffin, W. J., *Electricity and Magnetism*, McGraw-Hill, London, 1990.
Hayt, W. H., *Engineering Electromagnetics*, McGraw-Hill, New York, 1989.
International Telephone and Telegraph Co. Inc., *Reference Data for Radio Engineers*, 7th edn., Howard Sams, Indianapolis, 1985.
Jordan, E. C. and Balmain, K. G., *Electromagnetic Waves and Radiating Systems*, Prentice-Hall, New Jersey, 1968.
Kip, A. F., *Fundamentals of Electricity and Magnetism*, McGraw-Hill, New York, 1969.
Krauss, J. D., *Electromagnetics*, McGraw-Hill, New York, 1991.
Lorrain, P., Corson, D. P. and Lorrain F., *Electromagnetic Fields and Waves*, Freeman, New York, 1988.
Neff, H. P., *Introductory Electromagnetics*, Wiley, New York, 1989.
Ramo, S., Whinnery, J. R. and van Duzer, T., *Fields and Waves in Communication Electronics*, Wiley, New York, 1984.
Sander, K. F. and Reed, G. A. L., *Transmission and Propagation of Electromagnetic Waves*, Cambridge University Press, 1978.
Shen, L. C. and Kong, J. A., *Applied Electromagnetism*, Brooks Cole, Monterey, CA, 1983.

HISTORICAL

Baker, W. J., *A History of the Marconi Company*, Methuen, London, 1974.
Clarke, A. C., Extraterrestrial relays, *Wireless World*, **51**, Oct 1945, 305–308.
Clerk Maxwell, J., *A Treatise on Electricity and Magnetism*, Clarendon Press, Oxford, 1873 (Dover, New York, 1954).
Hertz, H., *Electric Waves*, Macmillan, London, 1983 (Dover, New York, 1962).
IEEE, *Historical perspectives on microwave technology*, Special issue of *IEEE Trans. on Microwave Theory and Techniques*, **MTT-32**, Sept. 1984.
Kao, K. C. and Hockham, G. A., *Dielectric fiber surface waveguide for optical frequencies*, *Proc IEEE*, 1966, **117**, 1151–1158.
Marconi, G., Wireless Telegraphy, *J. I. E. E.*, **28**, 1899, 273.

MICROWAVE SYSTEMS

Baden Fuller A. J., *Microwaves*, Pergamon, London, 1984.
Eaves, J. L. and Reedy, E. K., *Principles of Modern Radar*, Van Nostrand Rheinhold, New York, 1987.
Lynn, P., *Modern Radar Systems*, Macmillan, London, 1987.
Edwards, T. C., *Foundations for Microstrip Circuit Design*, Wiley, Chichester, 1987.
Flock, W. L., *Electromagnetics and the Environment*, Prentice-Hall, New Jersey, 1979.
Hall, M. P. M., *Effects of the troposphere on radio communications*, Peter Peregrinus (IEE), London, 1979.

INIRC, Guidelines on limits of exposure to radiofrequency electromagnetic fields in the frequency range from 100 kHz to 300 GHz., *Health Phys.* 1988, **54**, 115–123.
Krauss, J. D., *Radio Astronomy*, McGraw-Hill, New York, 1966.
Maral, G. and Bousquet, M., *Satellite Communication Systems*, Wiley, Chichester, 1986.
Metaxas, A. C. and Meredith R. J., *Industrial Microwave Heating*, Peter Peregrinus (IEE), London, 1983.
Pozar, D. M., *Microwave Engineering*, Addison-Wesley, Reading, MA, 1990.
Pratt, T. and Bostain, C. W., *Satellite Communications*, Wiley, New York, 1986.
Pritchard, W. L. and Sciulli, J. A., *Satellite Communication Systems Engineering*, Prentice-Hall, New Jersey, 1986.
Skolnik, M. I., *Introduction to Radar Systems*, McGraw-Hill, New York, 1981.
Sonnenberg C. J., *Radar and Electronic Navigation*, Butterworth, London, 1988.
Ulaby, F. T., Moore, R. K. and Fung, A. K., *Microwave Remote Sensing: Active and Passive*, Addison-Wesley, Reading, MA, 1981.

MICROWAVE WAVEGUIDES AND COMPONENTS

Carter, R. G., *Electromagnetic Waves—Microwave Components and Devices*, Chapman and Hall, London, 1990.
Collin, R. E., *Field Theory of Guided Waves*, McGraw-Hill, New York, 1963 (IEEE Press, 1990).
Collin, R. E., *Foundations for Microwave Engineering*, McGraw-Hill, New York, 1966.
Combes, P. F., Graffeuil, J. and Sautereau, J-F., *Microwave Components, Devices and Active Circuits*, Wiley, Chichester, 1987.
Gupta, K. C., Garg, R. and Bahl, I. J., *Microstrip Lines and Slot-lines*, Artech House, 1979.
Rizzi, P. A., *Microwave Engineering—Passive Circuits*, Prentice-Hall, New Jersey, 1988.

OPTICS AND OPTICAL SYSTEMS

Born, M. and Wolf, E., *Principles of Optics*, Pergamon, London, 1975.
Driscoll, W. G. and Vaughan, W., (Eds), *Handbook of Optics*, McGraw-Hill, New York, 1978.
Fincham, W. H. A. and Freeman, M. H., *Optics*, Butterworth, London, 1980.
Hecht, E., *Optics*, Addison-Wesley, Reading, MA, 1974.
Horne, D. F., *Optical Instruments and Their Applications*, Adam Hilger, Bristol, 1980.
Jenkins, F. A. and White, H. E., *Fundamentals of Optics*, McGraw-Hill, New York, 1981.
Jones, K. A., *Introduction to Optical Electronics*, Wiley, New York, 1988.
Pedrotti, F. L. and Pedrotti, L. S., *Introduction to Optics*, Prentice Hall, New Jersey, 1987.
Smith, F. G. and Thompson, J. H., *Optics*, Wiley, Chichester, 1988.
Wilson, J. and Hawkes, J. F. B., *Optoelectronics—An Introduction*, Prentice-Hall, London, 1989.
Yu, F. T. S. and Khoo, I., *Fundamentals of Optical Engineering*, Wiley, New York, 1990.

OPTICAL COMMUNICATIONS AND WAVEGUIDES

Barnoski, M. K., (Ed), *Fundamentals of Optical Fiber Communications*, Academic Press, New York, 1974.
Gowar, J., *Optical Communication Systems*, Prentice Hall, New York, 1984.
Halley, P., *Fibre Optic Systems*, Wiley, Chichester, 1987.
Kao, C. K., *Optical Fibre*, Peter Peregrinus (IEE), London, 1988.
Keiser, G., *Optical fiber communication*, McGraw-Hill, New York, 1983.
Lee, D. L., *Electromagnetic Principles of Integrated Optics*, Wiley, New York, 1986.
Midwinter, J. E., *Optical fibers for transmission*, Wiley, London, 1979.
Sandbank, C. P., *Optical fibre communication systems*, Wiley, London, 1980.
Senior, J., *Optical Fiber Communications*, Prentice Hall, new York, 1985.
Snyder, A. W. and Love, J. D., *Optical waveguide theory*, Chapman and Hall, London, 1983.

ANTENNAS

Balanis, C. A., *Antenna Theory, analysis and design*, Harper and Row, New York, 1982.
Collin, R. E., *Antennas and Radiowave Propagation*, McGraw-Hill, New York, 1985.
Griffiths, J., *Radiowave Propagation and Antennas—An Introduction*, Prentice Hall, New York, 1987.
Rudge, A. W., Milne, K., Olver A. D. and Knight P. (Eds), *The Handbook of Antenna Design*, Peter Peregrinus (IEE), London, 1982.
Stutzman, W. L. and Thiele G. A., *Antenna Theory and Design*, Wiley, New York, 1981.

Appendix 1 Essential symbols and units

Symbol	Quantity	SI units	Abbreviation
$\boldsymbol{B}$	Magnetic flux density	weber/metre^2	Wb/m^2
C	Capacitance	farad	F
c	Velocity of light	metres/second	m/s
$\boldsymbol{D}$	Electric flux density	coulomb/metre^2	C/m^2
d	Distance	metre	m
dB	Decibel ($= 10\log_{10}$(power ratio))	dimensionless	
$\boldsymbol{E}$	Electric field intensity	volt/metre	V/m
$\boldsymbol{F}$	Force	newton	N
f	Frequency	hertz	Hz
G	Conductance	siemen	S
$\boldsymbol{H}$	Magnetic field intensity	ampere/metre	A/m
I	Current	ampere	A
$\boldsymbol{J}$	Current density	ampere/metre^2	A/m^2
j	Imaginary number ($\mathrm{j}^2 = -1$)	dimensionless	
K	Wavenumber ($K^2 = \omega^2\varepsilon\mu - \beta^2$)	1/metre	1/m
k	Complex propagation constant	1/metre	1/m
L	Inductance	henry	H
l	Length	metre	m
n	Refractive index of a medium	dimensionless	
$\hat{\boldsymbol{n}}$	Unit vector normal to surface	dimensionless	
P	Power	watt	W
Q	Charge	coulomb	C
q	Charge	coulomb	C
R	Resistance	ohm	Ω
R_s	Surface resistance	ohm/square	Ω/sq
R	Range or radial distance	metre	m
r	Radius	metre	m
S	Poynting vector	watt/metre^2	W/m^2
s	Surface area	metre^2	m^2
T	Time period of sinusoidal wave	second	s
t	Time	second	s
V	Voltage	volt	V
VSWR	Voltage standing wave ratio	dimensionless	
v	Volume	metre^3	m^3
v	Velocity	metre/second	m/s
W	Work	joule	J
X	Reactance	ohm	Ω
x	Rectangular coordinate	metre	m
$\hat{\boldsymbol{x}}$	Unit vector in x direction	dimensionless	
Y	Admittance	siemen	S
y	Rectangular coordinate	metre	m
$\hat{\boldsymbol{y}}$	Unit vector in y direction	dimensionless	
Z	Impedance	ohm	Ω

Symbol	Quantity	SI units	Abbreviation
Z	Wave impedance	ohm	Ω
Z_0	Wave impedance of free space	376.7	Ω
z	Rectangular coordinate	metre	m
$\hat{z}$	Unit vector in z direction	dimensionless	
α	Attenuation constant	1/metre	1/m
β	Phase constant	1/metre	1/m
γ	Propagation constant	1/metre	1/m
$\tan\delta$	Loss tangent	dimensionless	
ε	Permittivity ($=\varepsilon_0\varepsilon_r$)	farad/metre	F/m
ε_0	Absolute permittivity of free space	8.854×10^{-12}	F/m
ε_r	Relative permittivity	dimensionless	
η	Intrinsic impedance of a medium	ohm	Ω
θ	Angle of rays at a boundary	radian or degree	rad or °
θ	Spherical coordinate	radian or degree	rad or °
λ	Wavelength	metre	m
μ	Permeability ($=\mu_0\mu_r$)	henry/metre	H/m
μ_0	Absolute permittivity of free space	$4\pi \times 10^{-7}$	H/m
μ_r	Relative permittivity	dimensionless	
ρ	Reflection coefficient	dimensionless	
ρ	Volume charge density	coulomb/metre^3	C/m^3
σ	Conductivity	siemen/metre	S/m
ϕ	Cylindrical or spherical coordinate	radian or degree	rad or °
ψ	Magnetic flux	weber	Wb
Ω	Ohm		
ω	Radian frequency ($=2\pi f$)	radian/second	rad/s

Appendix 2 Physical and mathematical constants

Velocity of light	$c = 2.997\,924\,58 \times 10^{8}$ m/s
Permittivity of free space	$\varepsilon_0 = 8.854 \times 10^{-12}$ F/m
Permeability of free space	$\mu_0 = 4\pi \times 10^{-7}$ H/m
Impedance of free space	$Z_0 = 376.7\,\Omega$
Boltzmann's constant	$k = 1.3805 \times 10^{-23}$ J/K
Planck's constant	$h = 6.626 \times 10^{-34}$ J s
Electron charge	$e = 1.603 \times 10^{-19}$ C
Electron mass	$m = 9.109 \times 10^{-31}$ kg
Pi	$\pi = 3.141\,592\,65$
Base of natural logarithms	$e = 2.718\,281\,83$
Radian	$1\ \text{rad} = 57.295\,7795^\circ$

Appendix 3 Material constants

The values listed below are representative of the named materials. The values for commercial products might differ slightly from those listed.

Table A3.1 Relative permittivity of materials at microwaves and lower frequencies.

Material	Relative permittivity
PTFE	2.1
Paraffin	2.1
Polyethylene	2.25
Polypropylene	2.5
Polystyrene	2.7
Wood	2 to 4
Paper	3
Rubber	3
Snow (dry)	3.3
Nylon	3.5
Glass	4
Ice (pure)	4
Quartz	5
Sodium chloride	5.7
Alumina	10
Silicon	12
Germanium	16
Diamond	16
Ethyl alcohol	25
Water (pure)	80
Titanium dioxide	100
Barium titanate	1200

Table A3.2 Refractive index of media at optical wavelengths.

Medium	Refractive index
Hydrogen	1.000 132
Air	1.000 293
Carbon dioxide	1.000 45
Water (pure)	1.333
Ethyl alcohol	1.361
Glass	1.4 to 1.8
Fused silica	1.46
Vitreous quartz	1.45
Crown	1.52
Crystal quartz	1.55
Dense flint	1.75
Sodium chloride	1.50
Diamond	2.42

Table A3.3 Relative permeability of materials.

Material	Relative permeability
Copper, aluminium plastics, wood	1.0
Nickel	50
Cast iron	60
Cobalt	60
Steel	300
Ferrite (typical)	1000
Iron (transformer)	3000
Iron (pure)	4000
Mumetal	20000

Table A3.4 Conductivity of materials.

Material	Conductivity (S/m)
Silver	6.17×10^7
Copper	5.80×10^7
Gold	4.10×10^7
Aluminium	3.82×10^7
Brass	1.50×10^7
Nickel	1.15×10^7
Bronze	1.10×10^7
Carbon steel	6×10^6
Lead	4.6×10^6
Germanium (pure)	2.2×10^6
Stainless steel	1.1×10^6
Mercury	1×10^6
Cast iron	1×10^6
Graphite	7×10^4
Carbon	3×10^4
Silicon (pure)	1.2×10^3
Ferrite (typical)	100
Water (sea)	4
Water (pure)	10^{-4}
Granite	10^{-6}
Bakelite	10^{-9}
Polystyrene	10^{-16}
Quartz	10^{-17}

Appendix 4 Useful mathematical relationships

TRIGONOMETRIC RELATIONS

In the following a and b are angles:

$$\sin(a \pm b) = \sin a \cos b \pm \cos a \sin b$$
$$\cos(a \pm b) = \cos a \cos b \mp \sin a \sin b$$
$$\sin(a + b) + \sin(a - b) = 2 \sin a \cos b$$
$$\cos(a + b) + \cos(a - b) = 2 \cos a \cos b$$
$$\sin(a + b) - \sin(a - b) = 2 \cos a \sin b$$
$$\cos(a + b) - \cos(a - b) = -2 \sin a \sin b$$
$$\sin 2a = 2 \sin a \cos a$$
$$\cos 2a = \cos^2 a - \sin^2 a = 2\cos^2 a - 1 = 1 - 2\sin^2 a$$
$$\sin^2 a + \cos^2 a = 1$$
$$\tan(a \pm b) = \frac{\tan a \pm \tan b}{1 \mp \tan a \tan b}$$

EXPONENTIAL AND HYPERBOLIC RELATIONS

$$\cos x = \tfrac{1}{2}(e^{+jx} + e^{-jx})$$
$$\sin x = \tfrac{1}{2j}(e^{+jx} - e^{-jx})$$
$$e^{\pm jx} = \cos x \pm j \sin x$$
$$\cos jx = \cosh x$$
$$\sin jx = j \sinh x$$
$$e^{\pm x} = \cosh x \pm \sinh x$$

SERIES EXPANSIONS

Trigonometric:

$$\sin x = x - \frac{x^3}{3!} + \frac{x^5}{5!} - \frac{x^7}{7!} + \cdots$$

$$\cos x = 1 - \frac{x^2}{2!} + \frac{x^4}{4!} - \frac{x^6}{6!} + \cdots$$

Exponential:

$$e^x = 1 + x + \frac{x^2}{2!} + \frac{x^3}{3!} + \frac{x^4}{4!} + \cdots$$

Binomial:

$$(1+x)^n = 1 + nx + \frac{n(n-1)}{2!}x^2 + \frac{n(n-1)(n-2)}{3!}x^3 + \cdots$$

APPROXIMATIONS FOR SMALL QUANTITIES

In the following approximate relations, $x \ll 1$:

$$(1 \pm x)^n \approx 1 \pm nx$$
$$(1+x)^{1/2} \approx 1 + \tfrac{1}{2}x$$
$$e^x \approx 1 + x$$
$$\ln(1+x) \approx x$$
$$J_1(x) \approx \frac{x}{2}$$

VECTOR RELATIONS

In the following relations, E, H, e, h and V are scalars, and $\boldsymbol{E}$ and $\boldsymbol{H}$ are vectors, where $\boldsymbol{E} = E_x\hat{\boldsymbol{x}} + E_y\hat{\boldsymbol{y}} + E_z\hat{\boldsymbol{z}}$.
Scalar product of two vectors:

$$\boldsymbol{E}\cdot\boldsymbol{H} = EH\cos\theta$$
$$\boldsymbol{E}\cdot\boldsymbol{H} = E_xH_x + E_yH_y + E_zH_z$$

Scalar product between unit vectors in rectangular coordinates:

$$\hat{\boldsymbol{x}}\cdot\hat{\boldsymbol{y}} = \hat{\boldsymbol{x}}\cdot\hat{\boldsymbol{z}} = \hat{\boldsymbol{y}}\cdot\hat{\boldsymbol{z}} = 0$$
$$\hat{\boldsymbol{x}}\cdot\hat{\boldsymbol{x}} = \hat{\boldsymbol{y}}\cdot\hat{\boldsymbol{y}} = \hat{\boldsymbol{z}}\cdot\hat{\boldsymbol{z}} = 1$$

Vector product of two vectors:

$$\boldsymbol{E}\times\boldsymbol{H} = EH\sin\theta$$

$$\boldsymbol{E}\times\boldsymbol{H} = \begin{vmatrix} \hat{\boldsymbol{x}} & \hat{\boldsymbol{y}} & \hat{\boldsymbol{z}} \\ E_x & E_y & E_z \\ H_x & H_y & H_z \end{vmatrix}$$

$$\boldsymbol{E}\times\boldsymbol{H} = (E_yH_z - E_zH_y)\hat{\boldsymbol{x}} + (E_zH_x - E_xH_z)\hat{\boldsymbol{y}} + (E_xH_y - E_yH_x)\hat{\boldsymbol{z}}$$
$$\boldsymbol{E}\times\boldsymbol{H} = -\boldsymbol{H}\times\boldsymbol{E}$$

Vector product between unit vectors in rectangular coordinates:

$$\hat{\boldsymbol{x}}\times\hat{\boldsymbol{y}} = \hat{\boldsymbol{z}} \quad \hat{\boldsymbol{y}}\times\hat{\boldsymbol{z}} = \hat{\boldsymbol{x}} \quad \hat{\boldsymbol{z}}\times\hat{\boldsymbol{x}} = \hat{\boldsymbol{y}}$$
$$\hat{\boldsymbol{x}}\times\hat{\boldsymbol{x}} = \hat{\boldsymbol{y}}\times\hat{\boldsymbol{y}} = \hat{\boldsymbol{z}}\times\hat{\boldsymbol{z}} = 0$$

Vector identities:

$$\nabla(e+h) = \nabla e + \nabla h$$
$$\nabla(eh) = e\nabla h + h\nabla e$$
$$\nabla\cdot(\boldsymbol{E}+\boldsymbol{H}) = \nabla\cdot\boldsymbol{E} + \nabla\cdot\boldsymbol{H}$$
$$\nabla\cdot(e\boldsymbol{H}) = \boldsymbol{H}\cdot(\nabla e) + e(\nabla\cdot\boldsymbol{H})$$

$$\nabla \cdot \nabla V = \nabla^2 V$$

$$\nabla \cdot (\boldsymbol{E} \times \boldsymbol{H}) = \boldsymbol{H} \cdot (\nabla \times \boldsymbol{E}) - \boldsymbol{E} \cdot (\nabla \times \boldsymbol{H})$$

$$\nabla \cdot (\nabla \times \boldsymbol{E}) = 0$$

$$\nabla \times (\boldsymbol{E} + \boldsymbol{H}) = \nabla \times \boldsymbol{E} + \nabla \times \boldsymbol{H}$$

$$\nabla \times (e\boldsymbol{H}) = (\nabla e) \times \boldsymbol{H} + e(\nabla \times \boldsymbol{H})$$

$$\nabla \times \nabla V = 0$$

$$\nabla \times (\boldsymbol{E} \times \boldsymbol{H}) = \boldsymbol{E}(\nabla \cdot \boldsymbol{H}) - \boldsymbol{H}(\nabla \cdot \boldsymbol{E}) + (\boldsymbol{H} \cdot \nabla)\boldsymbol{E} - (\boldsymbol{E} \cdot \nabla)\boldsymbol{H}$$

$$\nabla \times \nabla \times \boldsymbol{E} = \nabla(\nabla \cdot \boldsymbol{E}) - \nabla^2 \boldsymbol{E}$$

Gradient (grad) of a scalar, ∇V. The result is a vector:
Rectangular coordinates

$$\nabla V = \frac{\partial V}{\partial x}\hat{\boldsymbol{x}} + \frac{\partial V}{\partial y}\hat{\boldsymbol{y}} + \frac{\partial V}{\partial z}\hat{\boldsymbol{z}}$$

Cylindrical coordinates

$$\nabla V = \frac{\partial V}{\partial r}\hat{\boldsymbol{r}} + \frac{1}{r}\frac{\partial V}{\partial \phi}\hat{\boldsymbol{\phi}} + \frac{\partial V}{\partial z}\hat{\boldsymbol{z}}$$

Spherical coordinates

$$\nabla V = \frac{\partial V}{\partial r}\hat{\boldsymbol{r}} + \frac{1}{r}\frac{\partial V}{\partial \theta}\hat{\boldsymbol{\theta}} + \frac{1}{r\sin\theta}\frac{\partial V}{\partial \phi}\hat{\boldsymbol{\phi}}$$

Divergence (div) of a vector, $\nabla \cdot \mathbf{E}$. The result is a scalar:

Rectangular coordinates

$$\nabla \cdot \boldsymbol{E} = \frac{\partial E_x}{\partial x} + \frac{\partial E_y}{\partial y} + \frac{\partial E_z}{\partial z}$$

Cylindrical coordinates

$$\nabla \cdot \boldsymbol{E} = \frac{1}{r}\frac{\partial (rE_r)}{\partial r} + \frac{1}{r}\frac{\partial E_\phi}{\partial \phi} + \frac{\partial E_z}{\partial z}$$

Spherical coordinates

$$\nabla \cdot \boldsymbol{E} = \frac{1}{r^2}\frac{\partial (r^2 E_r)}{\partial r} + \frac{1}{r\sin\theta}\frac{\partial (E_\theta \sin\theta)}{\partial \theta} + \frac{1}{r\sin\theta}\frac{\partial E_\phi}{\partial \phi}$$

Curl of a vector, $\nabla \times \boldsymbol{E}$. The result is a vector:
Rectangular coordinates

$$\nabla \times \boldsymbol{E} = \left(\frac{\partial E_z}{\partial y} - \frac{\partial E_y}{\partial z}\right)\hat{\boldsymbol{x}} + \left(\frac{\partial E_x}{\partial z} - \frac{\partial E_z}{\partial x}\right)\hat{\boldsymbol{y}} + \left(\frac{\partial E_y}{\partial x} - \frac{\partial E_x}{\partial y}\right)\hat{\boldsymbol{z}}$$

Cylindrical coordinates

$$\nabla \times \boldsymbol{E} = \left(\frac{1}{r}\frac{\partial E_z}{\partial \phi} - \frac{\partial E_\phi}{\partial z}\right)\hat{\boldsymbol{r}} + \left(\frac{\partial E_r}{\partial z} - \frac{\partial E_z}{\partial r}\right)\hat{\boldsymbol{\phi}} + \frac{1}{r}\left(\frac{\partial (rE_\phi)}{\partial r} - \frac{\partial E_r}{\partial \phi}\right)\hat{\boldsymbol{z}}$$

Spherical coordinates

$$\nabla \times \boldsymbol{E} = \frac{1}{r\sin\theta}\left(\frac{\partial(E_\phi \sin\theta)}{\partial\theta} - \frac{\partial E_\theta}{\partial\phi}\right)\hat{r} + \frac{1}{r}\left(\frac{1}{\sin\theta}\frac{\partial E_r}{\partial\phi} - \frac{\partial(rE_\phi)}{\partial r}\right)\hat{\theta}$$

$$+\frac{1}{r}\left(\frac{\partial(rE_\theta)}{\partial r} - \frac{\partial E_r}{\partial\theta}\right)\hat{\phi}$$

Laplacian of a scalar, $\nabla^2 V = \nabla\cdot\nabla V$:
Rectangular coordinates

$$\nabla^2 V = \frac{\partial^2 V}{\partial x^2} + \frac{\partial^2 V}{\partial y^2} + \frac{\partial^2 V}{\partial z^2}$$

Cylindrical coordinates

$$\nabla^2 V = \frac{1}{r}\frac{\partial}{\partial r}\left(\frac{r\partial V}{\partial r}\right) + \frac{1}{r^2}\frac{\partial^2 V}{\partial\phi^2} + \frac{\partial^2 V}{\partial z^2}$$

Spherical coordinates

$$\nabla^2 V = \frac{1}{r^2}\frac{\partial}{\partial r}\left(\frac{r^2\partial V}{\partial r}\right) + \frac{1}{r^2\sin\theta}\frac{\partial}{\partial\theta}\left(\sin\theta\frac{\partial V}{\partial\theta}\right) + \frac{1}{r^2\sin^2\theta}\frac{\partial^2 V}{\partial\phi^2}$$

Divergence theorem:

$$\int_s \boldsymbol{E}\cdot d\boldsymbol{s} = \int_v \nabla\cdot\boldsymbol{E}\,\mathrm{d}v$$

Stokes theorem:

$$\int_l \boldsymbol{E}\cdot d\boldsymbol{l} = \int_s (\nabla\times\boldsymbol{E})\cdot d\boldsymbol{s}$$

Index